Social Insects

ECOLOGY AND BEHAVIOURAL BIOLOGY

M. V. Brian

LONDON   NEW YORK

Chapman and Hall

First published 1983
by Chapman and Hall Ltd
11 New Fetter Lane, London EC4P 4EE
Published in the USA by
Chapman and Hall
733 Third Avenue, New York NY10017

Printed in Great Britain at the University Press, Cambridge

ISBN 0 412 22920 X (cased)
ISBN 0 412 22930 7 (Science Paperback)

British Library Cataloguing in Publication Data

Brian, M.V.
  Social insects.
  1. Insect societies   2. Ecology
  I. Title
  595.705'24   QL496
  ISBN 0–412–22920–X
  ISBN 0–412–22930–7 Pbk

Library of Congress Cataloging in Publication Data

Brian, M. V. (Michael Vaughan), 1919–
  Social insects.

  (Science paperback)
  Bibliography: p.
  Includes index.
  1. Insect societies. I. Title. II. Series: Science paperbacks.
  QL496.B83   1983   595.7'0524   83–5299
  ISBN 0–412–22920–X
  ISBN 0–412–22930–7 (pbk.)

Contents

Preface

Here is a guide to the ecology of social insects. It is intended for general ecologists and entomologists as well as for undergraduates and those about to start research on social insects; even the experienced investigator may find the comparison between different groups of social insects illuminating.

Most technical terms are translated into common language as far as can be done without loss of accuracy but scientific names are unavoidable. Readers will become familiar with the name even though they cannot visualize the animal and could reflect that only a very few of the total species have been studied so far! References too are essential and with these it should be possible to travel more deeply into the vast research literature, still increasing monthly. When I have cited an author in another author's paper, this implies that I have not read the original and the second author must take responsibility for accuracy!

Many hands and heads have helped to make this book. I thank all my colleagues past and present for their enduring though critical support, and I thank with special pleasure: E. J. M. Evesham who fashioned the diagrams; J. Free, D. J. Stradling and J. P. E. C. Darlington who supplied photographs; D. Y. Brian and R. A. Weller who were meticulous on the linguistic side; and G. Frith and R. M. Jones who collated the references.

List of plates

CHAPTER 1

Introduction

'Ecology' is used in a broad sense to include population genetics, growth, development and differentiation of populations, behaviour and communication, relations within and between species and community structure. 'Social insects' is used in a narrow sense to include only eusocial forms, i.e. those with caste differentiation in one or both sexes. This means that only termites and the wasp group in which I include ants and bees are considered; social beetles, bugs, butterflies and spiders are reluctantly excluded. Individually, insects, though beautifully designed, are always small; they have never achieved the size that molluscs and vertebrates, especially dinosaurs and mammals, have achieved. There are good physiological reasons for this. Instead, they have co-operated, and formed organized groups of individuals that have evolved into societies so well integrated that they are effectively new individuals, superorganisms.

Their success is not in doubt. Social insects are, today, ubiquitous on dry land and attain high ecological activities. Termites (Isoptera), as social cockroaches, have evolved immense and highly differentiated societies from quite a simple insect plan. They are supreme converters in soil from the tropics to the desert; they stir and mix and, with the aid of bacteria, protozoa and fungi, recycle cellulosic fibres. Ants, bees and wasps (Hymenoptera) have evolved from higher, more subtle and sensitively designed insects that tolerate and use light and now range from the Equator to the polar regions. They too began as scavengers and hunters but have gone on to establish many mutualistic associations with plants and other insects as protectors, predators, scavengers, honey-dew producers, pollinators and seed dispersers. Their societies may be minute or immense, mobile or stationary with high powers of communication and integration. Nevertheless, non-social insects continue to live in the ecological spaces between societies or in association with social forms and the question continually returns: why has sociality arisen and evolved to its full so far in only the termites and the wasp group? Why not in pseudoscorpions, spiders, bugs, beetles and moths, all of which show considerable parental care but not prolonged domination and manipulation over their offspring? There still seems to be room for new forms.

Termites originate in a cockroach-like stock of exopterygote insects. All have filiform antennae and biting mandibles; and their thorax carries two large elaborately veined wings that give them a weak clumsy flight used for dispersal. The thorax is continuous with the tubular abdomen. They pair for life and produce larvae that are miniature adults in all but their flight and reproductive systems. They grow whilst moulting and finally inflate their wings and genitalia from external rudiments. Juveniles work in the nest; in the most advanced families they harden and darken and are fit to go out in daylight but this is achieved at the cost of permanent sterility. All are diploid with an XY chromosome sex determination and both sexes are social and polymorphic, though in higher termites genetic sex differences influence the type of caste that develops.

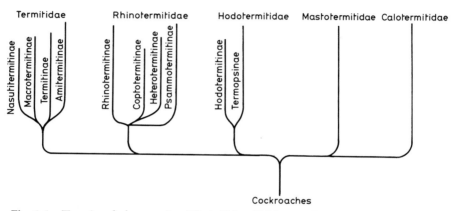

Fig. 1.1   Termite phylogeny simplified. (After Krishna and Weesner, 1969, 1970.)

Termites are subdivided into 5 families all of which comprise wholly social species (Fig. 1.1). The lowest, most cockroach-like one, the Mastotermitidae had a world-wide distribution in the Eocene but is today restricted to Australia as the species *Mastotermes darwiniensis*. Its miniature features include five-jointed tarsi, anal lobes on the hind wing and an eggcase. Another primitive but still widespread family is the Calotermitidae ('dry wood' termites) which excavate in logs as they lie exposed on the soil surface. The Hodotermitidae, which has fossils that go back to the mid-Cretaceous in Labrador, retains some primitive characters. There are two sub-families, the Termopsinae that live in rotting wet wood ('damp-wood' termites) and the Hodotermitinae ('harvester termites') that cut and store pieces of grass in their nests (Fig. 1.2). The Rhinotermitidae are named from the fact that they have evolved small soldiers with snouts for the ejection of repellants. The highest family, the Termitidae, includes a great many termites with pigmented and sclerotized neuter castes. The Amitermitinae is the simplest sub-family (Fig. 1.3); the Termitinae which originate in the Ethiopian region have some remarkable soldiers with snap closing mandibles; the Macroter-

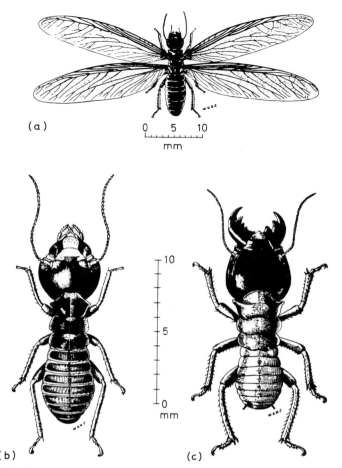

(a)

0 5 10
mm

10

5

0
mm

(b)   (c)

Fig. 1.2 The harvester termite *Hodotermes mossambicus*. (a) Winged sexual, (b) worker, (c) soldier. (From Harris, 1961.)

mitinae culture a fungus on combs and the Nasutitermitinae have soldiers that shoot jets of toxic and sticky material; Krishna and Weesner (1970) and Hermann (1979–82) provide more details on termite biology. Today, termites are the main class of converter animals in all but the cooler belts of the world. They have descended deep into soil and ascended to the tops of trees. They scout out and concentrate on any piece of cellulosic litter that becomes available and carry it into sealed air-conditioned mounds for decomposition and energy release. Their effect in ecosystems has been summarized in a diagram by Wood and Sands (1978) reproduced here (Fig. 1.4).

Whilst the termites started to evolve social systems in or before the Cretaceous epoch, other insects evolved more flexible and sensitive

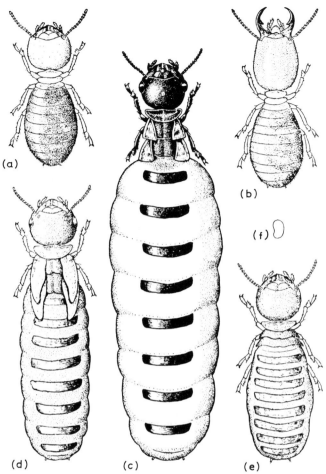

Fig. 1.3 (a) worker, (b) soldier and three types of reproductive female of *Amitermes hastatus* (Amitermitinae). The last comprises: (c) a deälate normal sexual, (d) a sexually mature nymph, (e) a sexually mature larva. There is also (f) an egg to scale. The worker is 5 mm long. (From Skaife, 1954.)

individual forms with stronger more manoeuvrable flight: beetles, flies and the wasp-like Hymenoptera. The last of these orders is still largely composed of solitary species that place their eggs with great skill in or on host plants or invertebrates and others that excavate or build cells that are provisioned with small prey or pollen, given an egg and left to survive alone. The female evolves an astonishing degree of skill in placing her eggs in a protected natural or a specially constructed nutritive environment (solitary wasps and bees). In contrast the male specializes in finding and copulating with females; since the

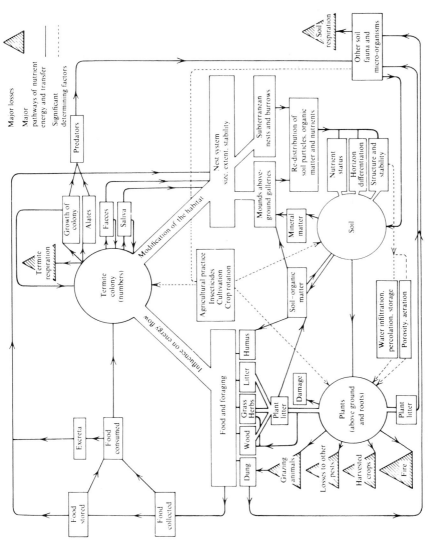

Fig. 1.4 A scheme of the role of termites in ecosystems that summarizes their actions through nest construction and food transformation. (From Wood and Sands, 1978.)

female can store sperm for her whole life, after copulating he dies. Males are haploid and can develop parthenogenetically from unfertilized eggs whereas females are diploid and normally arise from fertilized eggs. To fertilize or not, to open the sperm valve or leave it shut, brings sex under environmental and individual female control in both parasitoid and social groups of Hymenoptera.

The skill of the females depends on the highly articulated, responsively motored body, and a concentrated central nervous system fed by highly refined sense organs. Antennae are elbowed with convergent tips that can measure space; mouth parts (in bees) evolve into sucking tubes; the 'neck' is flexible; fore and hind wings are hooked together and act in unison; legs carry grooming combs or pollen baskets (in bees); a waist is formed by constricting the anterior segments of the abdomen so that its tip can be brought round between the legs as far as the mouth. The segmental rings telescope and the narcotizing ovipositor of parasitic Hymenoptera is used to inject poison for defence. Eggs are simply extruded from the genital opening and taken off by the jaws. This female skill, added to their ability to control the sex of offspring, has enabled females to form unisexual societies.

Three distinct groups of social Hymenoptera are recognized. First the Sphecoidea, a series of digger-wasps that hunt for special prey, sting it and take it into underground cells; these culminate in hairy pollen-eating bees (Apoidea). Second, the Vespoidea that have a stronger thorax, and transport prey to cells which are often above ground, and in the most evolved cases (Vespidae) are made of carton. Third, the Formicoidea (one family, the Formicidae) which start from ectoparasitic wasps of the Tiphioid group that live by capturing insect larvae in soil and paralysing and laying eggs on them, and evolve into the diverse ants. Of these three, only the ants are all social, for the bees and wasps have many solitary species.

The modern bees (Apoidea) have evolved social life in three families. The Halictidae (Fig. 1.5) show a great range of organization right up to the eusocial state; for the most part they dig out nests underground and make cells provisioned with balls of pollen. The Anthophoridae, also mainly solitary, have one tribe (Ceratini) that is social and unusual for bees in making open-plan nests in plant cavities. The Apidae are all social and include the important pollinating Bombinae (bumble-bees) and Apinae which has a tropical tribe without stings (Meliponini) and a worldwide tribe that stings and has an enormous capacity to make and store honey (Apini). The Apinae all have long tubular tongues and hairs to brush-up pollen, and are amongst the most highly evolved of all social insects. One species, *Apis mellifera*, is the only social insect to have evolved a mutualistic relation with *Homo sapiens* and as a result now shares almost the same wide geographical distribution (Fig. 1.6).

The modern wasps (Vespoidea, Vespidae) apart from the subsocial oriental

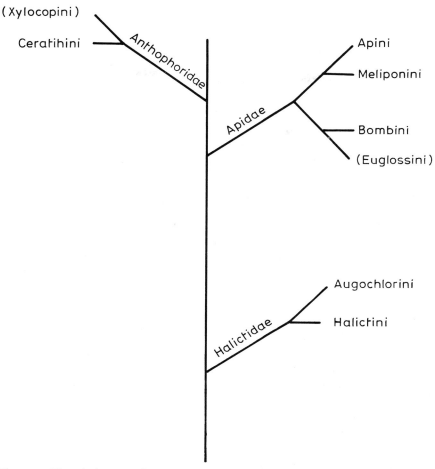

Fig. 1.5 The phylogeny of social bees. (After Michener, 1974.) Only tribes with social members are shown; related ones that are not social at all are in brackets.

tropical group that has a very thin petiole (Stenogastrinae) are grouped into two sub-families. The Polistinae are worldwide but reach their peak of diversity and social organization in the tropics with the tribe Polybiini where they are famous for their sociable queens (polygyny) and for the variety of their nest design. The Vespinae are the common temperate wasps with a big size difference between queens and workers which make thickly-lagged paper nests.

Ants started as soil-living predators but have extended on to the soil surface and up trees where they feed on many plant products. As have termites (also originally soil insects), they have evolved the capacity to break off their wings once dispersal and copulation are over, and their neuter castes are primarily

Fig. 1.6 Cave drawing from Spain showing one of the first human beings to take honeycombs from a bee nest in a cliff hole *ca.* 7000 years B.P. (From Budel and Herold, 1960.)

wingless. In many even the reproductive females are wingless and are sought out by flying males. Ant females are highly polymorphic in advanced species; apart from a great size difference between queens and workers made possible by their derivation from different sorts of eggs, the workers can come in many sizes whose function ranges from fine work in the nest to heavy work and defence that require strength. In contrast to termites it is unusual for ant 'soldiers' to be restricted to defence and unusual for them to appear amongst the first broods of a young queen. Ants have converged towards termites in many ways apart from wing dehiscence; a striking one is their mutualistic relationship with fungi which break down plant leaves. Indeed their change from specialized hunters to mixed vegetable feeders has brought them near to termites. In the future tidy and simple world ecosystem dominated by human beings, they are likely to have an assured role as agents of biological control; the weaver ant, *Oecophylla* has been used to protect citrus fruit in China for millenia. The single family (Formicidae) is divided into several sub-families. Two of these are considered to be close to the primitive stock: Myrmeciinae,

Ponerinae and four to be advanced: Dorylinae, Dolichoderinae, Formicinae and Myrmicinae. The phylogeny of these sub-families is very uncertain at present (Brown, personal communication 1982). Figures of the simplified phylogenies of other groups are given and more detail can be found in Hermann (1979–82).

The design of this book follows a sequence starting outside the society: first the collection of food, then its conversion into social insect material in special nests, and after that population growth. This leads on to reproduction and dispersal, and the formation of limited societies (colonies); after a pause to take in genetics and the origin of sociality, the book finally goes into the subject of speciation, species co-existence and the interlocking mosaic of colonies that spreads out over natural resources from nest centres. In ancient and sustained communities these evolve a complex of co-operative and competitive inter-relationships.

CHAPTER 2

Food

Food is needed to maintain the living tissues and enable them to grow, to build and repair structural parts, and to energize the whole system. Termites eat dead plant material in many of its guises: leaf litter, wood, dung, seeds, herbage. With the co-operation of symbiotic bacteria and protozoa they digest the cellulose, a fibrous polysaccharide, and release energy. Ants and social wasps by contrast start their evolution as predators; they digest the oils and carbohydrates to release energy and later use plant sugars to supplement these. This association with plants blossoms out until higher ants become vegetarians and eat seeds and green leaves. Bees, as a branch of the sphecoid wasps that became pollen eaters before they became social, show increasing skill as pollen collectors. Plants of course, since they capture sunlight and synthesize organic substances are at the base of the ecological pyramid and the great energy density that they provide enables some insects to sustain high resident populations that can dominate their environment from permanent fixed positions.

Foods must include vitamins. In the water-soluble group riboflavin (B2), cyanocobalamine (B12), L-ascorbic acid (C), and myo-inositol are generally needed by insects though this varies from species to species and some can obtain them through gut micro-organisms (Rockstein, 1978). The fat-soluble vitamins include some polyunsaturated fatty acids along with carotene (A) and α-tocopherol (E). These are well distributed in fresh plants and animals but may be scarce in the diet of termites; probably their gut symbionts and their habit of eating their dead help to conserve the supply.

2.1 Termites as decomposers

Primitively, termites may well have been polyphagous and eaten a variety of organic material. In many ways the most primitive species alive today, *Mastotermes darwiniensis* in Australia is polyphagous and eats dung and plant litter as well as wood. So does the African desert termite, *Psammotermes hybostoma*, which does not stop at fossil soils and rubbish of all sorts. To help digestion all termites have micro-organisms in hind gut pouches. In the more primitive families (Calotermitidae, Hodotermitidae and Rhino-

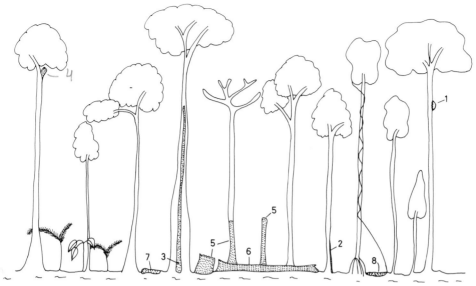

Fig. 2.1 The places where termites attack wood in Pasoh Forest, West Malaysia. (After Abe, 1979.) 1, nests around trunks, e.g. *Microcerotermes*; 2, surfaces near tree bases; 3, centres of trunks; 4, dead branches of living trees; 5, tree stumps or standing dead trees; 6, fallen trunks; 7, fallen branches and leaf litter; 8, humus in soil.

termitidae) obligate flagellate protozoa digest intracellularly, but in the Termitidae bacteria digest cellulose. The hind gut is lined with cuticle and is moulted periodically but a food exchange system ensures re-inoculation. This is an important advantage that comes from communal living and food sharing.

Today, termites play a major part in decomposing dead trees in the tropics (Fig. 2.1). They exercise choice over the type of wood they eat; in Australia one *Pinus* is not attacked by *Nasutitermes exitiosus* because the pinenes it contains are the alarm pheromones of this termite and their presence is disturbing. This may seem incredible yet many other species can eat this pine, and *N. exitiosus* attacks an even more aromatic wood *Eucalyptus marginata* which is avoided by most other termites. Soft tissues seem to be preferred to hard ones (Wood, 1978). Stage of decomposition is also important (Fig. 2.2) and fungal excretions sensed by the termite can either repel or attract depending upon the species (Sands, 1969); but fungi can also destroy repellent chemicals like turpentine. The size of logs affects their attractiveness too, thus *N. exitiosus* eat the logs and stumps of *Eucalyptus* in the dry forest of South Australia but two other species eat its twigs. Most wood eaters leave living wood alone and many make a marked distinction between dry wood (e.g. Calotermitidae) and wet wood (e.g., Termopsinae family Hodotermitidae).

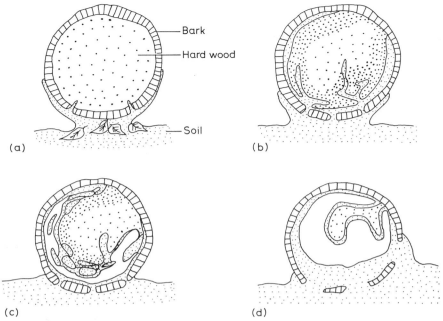

Fig. 2.2 Stages in the decomposition of a log. (a) Termites enter through beetle holes, take in soil and plaster it under log; (b) galleries are made and more soil brought in; (c) galleries are extended into chambers and the log sinks into the soil; (d) the log disintegrates, termites begin to go away and the log crumbles into the soil surface. (After Abe, 1980.)

Wood is largely a mixture of cellulose and lignin in various stages of decomposition. Cellulose is a polymer of glucose; lignin is a polymer of phenyl-propane units derived from cinnamyl alcohols. The ratio of lignin/cellulose varies from species to species but ranges from 0.4 to 0.8. *Calotermes flavicollis* of the primitive family Calotermitidae can digest 70–90% of the cellulose whereas *Nasutitermes ephratae* of the Termitidae, may use over 90% of it. Lignin digestion is much less efficient: no more than a third for *C. flavicollis* and around half for *N. ephratae* (LaFage and Nutting, 1978; Lee and Wood, 1971; Wood, 1978).

Plate 1 *Opposite*
Above: Fungus combs cultivated in an earth nest of the leaf-cutting ant, *Acromyrmex octospinosus*. The fungus comb is a sponge-like mass that receives new substrate on top and supplies fungal food bodies below. (Photograph courtesy of D.J. Stradling.)
Below: The fungus comb of the termite *Macrotermes michaelseni*. The comb occupies a chamber of the nest, where it rests on small conical pillars (see also Plate 10). (Photograph courtesy of J.P.E.C. Darlington.)

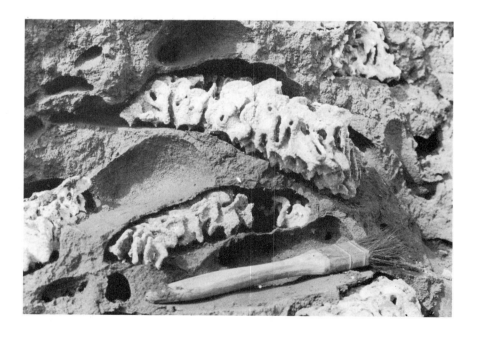

One subfamily of Termitidae, the Macrotermitinae, make a comb in which a basidiomycete fungus *Termitomyces* grows. There seems to be no doubt that this comb is built of partially digested faeces and is a structure that no solitary insect could create. Carbon powder (soot) mixed in sawdust is passed through the gut and put on top of the comb; it then takes 5–8 weeks to pass down to the bottom where it is eaten by the workers (Josens, 1971a). The comb contains very little of the clay gut contents of workers and does not look like their usual faeces. Symbiosis with a basidiomycete fungus that invades and breaks down lignin as well as cellulose adds to the use made of the fibrous litter collected by the workers. The fungus concentrates this into a protein-rich conidiophore whose cell wall of chitin is readily digested by the termite; just as readily, in fact, as its redundant nest mates are when necessary. Although the biochemical details are not yet fully understood there is no doubt that the termites eat and benefit from the fungus (LaFage and Nutting, 1978; Rohrmann and Rossman, 1980) (see Plate 1).

Macrotermes michaelseni in Kenya, collects herbs instead of dead wood especially in the middle of the dry season when the sexuals are growing. With four colonies ha^{-1} each spreading over 2500 m^2 the total number of collections in each m^2 is around 15 a year (Lepage, 1981b). *M. bellicosus* in primary savannah woodland in Nigeria at Mokwa (9° N) takes 177 kg ha^{-1} of wood and 64 kg ha^{-1} of leaf litter each year; this is about 6.4% of the total annual litter production. Termites together take 24% (Collins, 1981a, b) but the Macrotermitinae in general have a higher weight-specific consumption rate than other groups of termites, no doubt because their domestic fungus, unlike the wild micro-organisms, is active throughout the dry season.

Some Hodototermitidae cut dry grass at night in the open or by day under soil covers. Two subfamilies of the Termitidae are also able to do this: the Termitinae and Nasutitermitinae. *Hodotermes mossambicus* in South Africa selects grasses rather than shrubs and dry grass rather than fresh green grass.

Trinervitermes geminatus prefers fine-leaved grasses to coarse ones and rejects the repellent lemon grass; in Nigeria it uses a lot of *Andropogon* (Ohiagu and Wood, 1976; Wood, 1978). *T. trinervius* eats nine species of grass in African savannah leaving nearby legumes alone; it cuts dead grass though it might still be green but unlike some species has no store (Bodot, 1967). Perhaps the most specialist feeders are those which collect lichens from the tops of trees in the Orient (*Hospitalitermes*) or that eat the carton of other termite nests. In spite of being outwardly vegetarian, all termites eat and recycle their surplus or damaged nest mates.

Another feeding habit which has evolved several times in termites is eating organic matter (humus) in soil; three subfamilies of Termitidae do this: the Termitinae, Apicotermitinae and Nasutitermitinae. In soil-feeding species of the subfamily Termitinae the bacteria include spirochaetes and indigenous actinomycetes. The filaments of the latter are actually supported by reflexed

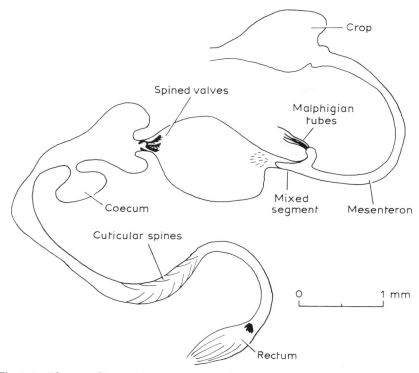

Fig. 2.3 The gut of *Procubitermes*, to show the enormous size and complexity of the hind gut, that is the region behind the Malphigian tubes. This includes the cuticular spines. (After Bignell *et al.*, 1980.)

cuticular spines in pouches of the hind gut (Fig. 2.3) and in the mid gut they inhabit the fluid space outside the peritrophic membrane, normally used by pathogens (Bignell *et al.*, 1981). Actinomycete bacteria in soil can secrete phenol oxidases that break down woody components of humus and it is likely that the gut inhabitants may be doing the same thing. These symbionts enable high conversion efficiencies to be obtained: some 54–93% compared with 27% for cockroaches, 30–40% for diplopods, 33% for isopods, or 14% for mites (Wood, 1978).

2.2 Wasps and ants as predators

Solitary wasps, whether Sphecoid or Vespoid, lay eggs on a prey in a burrow or cell after paralysing it with their sting. Sphecoid wasps show considerable prey specialization: grasshoppers, spiders, caterpillars and beetle grubs are used but each wasp is a specialist at paralysing them and transporting them home. The most primitive wasps, likely ancestors of all others, are grouped in the superfamily Scoliodea. To this group, which includes the family Tiphiidae, belongs the genus *Methocha* whose larvae feed on tiger-beetle larvae (*Cicindela*). These the adult stings in the neck and pulls down a burrow, then she stings again and lays one egg before filling up the burrow (Wilson and Farish, 1973).

This specialist hunter trait persists amongst primitive ants. Thus *Amblyopone pallipes* (Ponerinae) stings centipedes and collects its brood around them rather than take them into a nest (Traniello, 1978). Other Ponerinae, a mainly tropical subfamily with 52 genera, also show specialization on a particular invertebrate group (Brown, 1973). At Lamto, Ivory Coast, West Africa, stratification accompanies food specialization. In trees, *Platythyrea* concentrate on the caterpillars of noctuid moths (71% of their insect diet); on the soil surface *Leptogenys* eats woodlice (Isopoda) and *Megaponera* eats termites. In the soil two species of *Amblyopone* and one of *Apomyrma* eat centipedes (Geophilidae), two species of *Pletroctena* eat millepedes (Julidae), *Hypoponera* eats springtails (Collembola), *Centromyrmex* eats termites, *Paltothyreus* eats worms (Oligochaeta) and *Discothyreus* eats various sorts of Arthropod eggs. In addition, at least nine ant genera eat termites (Lévieux, 1977), and a whole tribe of ponerines, the Cerapachyini, specialize in eating other ants. Very few slugs and snails are eaten. *Megaponera foetens* at Mokwa, Nigeria, in moist savannah eats termites of the Macrotermitinae only; *Macrotermes bellicosus* (72%) is the most important with *Odontotermes* (22%) next. 'Predation on the first species results in an annual turnover slightly exceeding the standing crop populations' (Longhurst *et al.*, 1978). Thus these relatively primitive ants predate a wide range of other arthropods and worms including the social termites and ants. There is also a clear tendency as social life evolves for predators to go for smaller prey that can be overpowered without the use of their sting and take more. This is possible with food-sharing groups.

The subfamily Myrmicinae with about 130 genera over the world (Brown, 1973) illustrates this tendency further for it contains a tribe, the Dacetini, that specializes in eating springtails which are not large insects and many are needed to make one new ant. Workers stalk their prey and, when within a few millimetres of it fold their antennae back behind their jaws, which open very

---

Plate 2 *Opposite*
Above: The mouthparts of a last stage larva of *Myrmica* ($\times$ 70). This shows, from above clockwise: labrum, pointed mandibles, maxillary palps, and labium. The mouth lies between the labrum and the mandibles, and the salivary duct opens on the labium.
Below: The mouthparts of a last stage larva of *Myrmica* ($\times$ 1750) showing the broken tip of one mandible, and rows of epicuticular 'teeth' that may assist in the retention and ingestion of food. (SEM photographs courtesy of C. Hawkins, University of Southampton.)

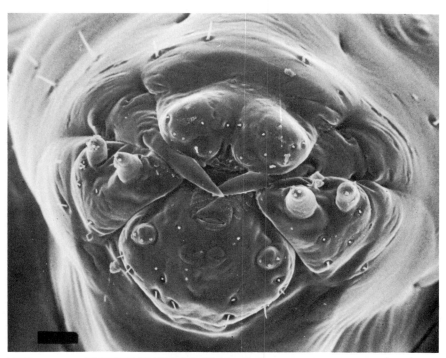

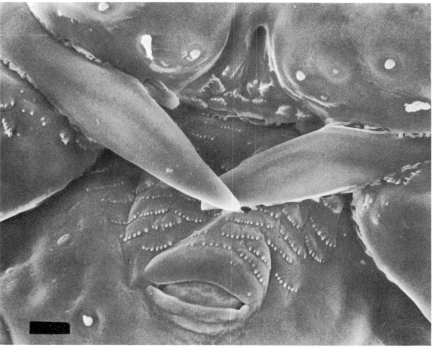

wide; then as the prey moves, they snap and transfix it with their sharp mandibular teeth. Only surface-living springtails are taken (Wilson, 1953). Other myrmicines specialize in eating termites; thus the genus *Carebara* with extremely small workers, only a few millimetres long, actually lives inside their nests. Species of this genus produce gynes up to 2000 times worker size (Wheeler, 1936), which after copulation, fly to new termitaria with workers clinging to their legs.

A great subfamily of hunting ants derived from basic Ponerinae is the Dorylinae. This is a much less diverse subfamily than the two discussed, with only three genera in the Africa/Orient region and six genera in Central America (Brown, 1973; Wilson *et al.*, 1967). Both highly specialized predators and very generalized ones occur, the latter at the apex of social organization. The genus *Aenictus* contains many species and ranges from Africa to New Guinea, most forage in the soil and are blind but some, e.g. *A. gracilis* and *A. laeviceps*, forage on the surface. They have been found to eat at least nine species of ant as well as a variety of other invertebrates, and have been classed as general ant predators (Schneirla, 1971; Wilson, 1964). More advanced forms of this subfamily hunt termites underground, e.g. *Anomma kohli* and the climax of their evolution comes with the huge colonies of many millions of polymorphic workers in search of prey during the period when the larvae are growing, e.g. *Anomma*, the driver ant in Africa and *Eciton*, the army ant in America. They flush out a great many types of invertebrate including other ants and wasps and can even attack and eat torpid vertebrates. Wilson (1971) has suggested that they have evicted the ant tribe Cerapachyini (Ponerinae) from all but the most inaccessible parts of the world such as Madagascar, Fiji and even Australia. Thus social insects, not just termites, form a major food source for ants.

2.3 Sugars as fuel save prey

Many wasps hunt for prey only when they have larvae since animal carcases are not easy to store. *Vespula* species hunt by flying close to plant foliage and flowers and they take a great variety of small soft-bodied insects using only their jaws; open flowers are also visited for their sugars (Proctor and Yeo, 1973; Spradbery, 1973a).

The huge bulldog-ants (*Myrmecia*) of Australia hunt and, using their sting

---

Plate 3 *Opposite*
Above: The mandibles of a *Myrmica* worker. This is a young individual, and the cuticular teeth are sharp and are gripping a sliver of food, (× 70).
Below: The mandibles of an old worker of *Myrmica* with worn teeth; one has been broken off completely, (× 70). (SEM photographs courtesy of C. Hawkins, University of Southampton.)

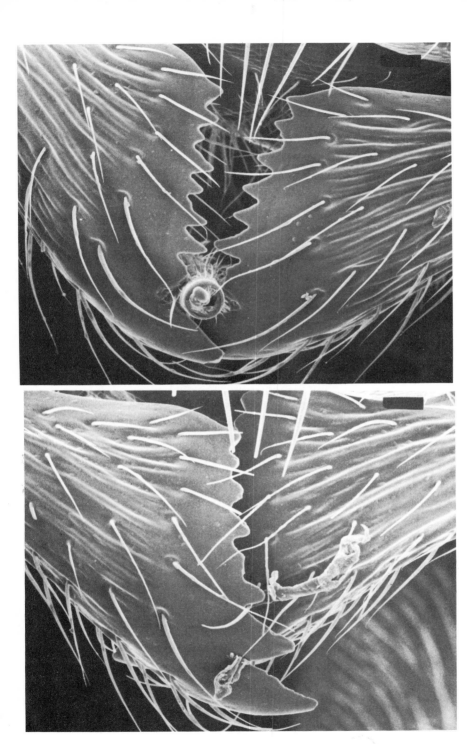

kill insects as big as bees; these are broken up and put near the large larvae which reach out and feed on them. After the larvae have pupated adults feed themselves by sucking up sugary exudates from plants. In the subfamily Myrmicinae the sting is reduced in primitive forms and modified in advanced forms. Even in the primitive genera it is not strong enough to paralyse effectively and the jaws are used to grip, bite, crush, tear or pierce holes in the cuticle of the prey animal; their juices are then sucked out. Fruit juices and nectar are collected and the sugars used for energy; this saves protein, much of which is scavenged rather than hunted as prey. *Myrmica* kept in the laboratory do quite well on a diet of insects and water but do much better (metamorphose more larvae, lay more eggs) if also given sucrose solution. They then eat fewer insects and pass a clear urine instead of one milky with uric acid granules. Evidently the sugars stimulate worker activity and are used to fuel worker movement whilst the protein is used for egg formation and larval growth (Brian, 1973a).

Honey-dew is another common source of sugars; it is the excretion of various plant-sucking bugs: mainly Aphidae, Coccidae and Aleurodidae. Three species of *Camponotus* (Formicinae) that live in trees in West Africa all use aleurodids extensively but use coccids and aphids as well and bugs from three other families less frequently (Lévieux, 1977). These ants collect vegetable gums and prey on larval insects of seven orders and adult insects of nine orders. Another group of plant bugs, the Membracidae, that feed on golden rod in eastern North America attract ants of several species. The biggest, a *Formica* is hostile to leaf-eating beetles and if present most of the year can improve the growth, and seed production of the golden rod (Messina, 1981).

Aphids shoot a liquid excretion from their anus. This usually lands on leaves which become sticky and develop moulds, unless it is collected by ants, wasps or bees. It contains not only the plant sap less what the aphids have taken out for themselves but additional excretions from the aphid (Way, 1963). An interesting example has recently been obtained using *Myzus persicae* cultured on pot-grown seedlings of radish (*Raphanus sativus*). The plants contain glucose and fructose yet the honey-dew has trehalose, melezitose and sucrose in addition, and whereas the plant has seven organic and five phenolic acids, the honey-dew has eight and eighteen, respectively. Other compounds such as the plant hormones auxin and gibberellin as well as cytokinins and growth inhibitors pass into the honey-dew from the plant but ammonia is added by the aphid and this raises the pH (Hussain *et al.*, 1974). From 9 to 23 amino-acids and amides are present in honey-dew comprising 2–3% dry matter; as expected, the composition varies with the species both of the bug and of the plant; seasonally too (Auclair, 1963). Many other compounds either in the plant or synthesized in the aphid are present in small quantities.

In temperate regions wasps and bees with short tongues use honey-dew regularly. By flying to the tops of forest trees in areas in which ants are scarce they can obtain a plentiful supply without competition; this is of value to them at times of year when suitable open flowers are either not readily available or are not secreting actively because of drought. Some very good honey is made from the excretions of plant bugs.

The casual collection of honey-dew from leaves has evolved into a direct donor–receiver relationship. Today there are some aphids that will save their liquid excretion until an ant touches them and then give it up without mess as a droplet that stands on hairs round their anus. They also restrain their normal defensive reactions such as ejecting sticky material from cornicles. Since ants repel predators and parasites and often build shelters against wind and rain, such co-operative bugs have a better chance of surviving and reproducing than others. Even when their relationships are still facultative, the bug populations are affected substantially by ant attendance, for Way (1963) who reviews this vast subject remarks that bugs may receive direct and indirect benefits other than protection: they take in more food and excrete more honey-dew, they aggregate more and denser populations develop. This has a tranquilizing effect so that fewer wander about, and in some way, not fully understood, they avoid producing their winged sexual forms as early as they normally would at such densities. Perhaps their young progeny are being culled by ants?

Wood ants (*Formica rufa* and allied species) collect an enormous amount of honey-dew from aphids feeding on both deciduous and coniferous forest trees in north temperate regions. A large polygyne colony of *Formica lugubris* living in the Jura mountains (Switzerland/France) hunts for prey of about its own size; many of these are bugs. Cherix (1980) suggests that the ants take so many insects that insectivorous birds are unable to live there. Many prey are plant bugs but honey-dew is also collected from several species of aphid, especially from *Cinara pruinosa* which feeds on spruce (*Picea abies*) from June onwards. Later *Cinara* too are used as prey and Cherix (1980) describes seeing workers attack aphids that leave their group and make peculiar movements; are they parasitized or just senile? Clearly the balance between protection and predation is delicate.

There have been many studies of the feeding habits of wood ants because of their potential use in the biological control of defoliating caterpillars; naturalists point out the green islands of leafy trees near wood ant nests. *Formica rufa* near its northern limit in the British Isles preys in spring on an aphid, *Drepanosiphum platanoides*, and its peak intake corresponds with the aphid peak of abundance. It never tends this species. Many flies, like *Bibio*, fall prey too and caterpillars, mainly of the winter moth (*Operophtera brummata*) are collected from oak trees (*Quercus*), especially in May (Skinner, 1980b). Honey-dew is collected from a different aphid, *Periphyllus testudinaceous*, on sycamore (*Pseudoplatanus* sp.) by more than 60% of foragers; more than half the energy intake is in this form. This ant is opportunistic within the general class of 'small insects on trees and on the ground' and will take whatever is abundant and easy to catch. The defoliating caterpillar of *Tortrix viridana*, a moth, is an interesting case: although Inozemtsev (1974) found it reduced by a third, Skinner found fewer taken than expected but he points out that they may be cut up and ingested first. If *F. rufa* is excluded from *Drepanosiphum platanoides* by sticky bands many more aphids build up (Skinner and Whittaker, 1981). The ants also reduce the caterpillar population especially in the lower branches and it is estimated that the area consumed by larvae is only 1.2%, in foraged, as against 8.5% in unforaged, trees. Thus *F. rufa* reduces caterpillar damage to leaves, and some populations of aphids but the main honey-dew producing aphid increases and the quantity of honey-dew produced is substantial (3.5×10^7 J/tree). For the tree, the cost/benefit ratio is very complicated (see also Wellenstein, 1980).

As an example of the complexity of ant–aphid defoliator–tree systems in tundra the work of Laine and Niemela (1980) in Finnish Lapland at 70° N on *Formica aquilonia* (the northern European wood ant) and birch (*Betula pubescens*) with its defoliating caterpillar *Oporima autumnata* is apt. The ant stabilizes its population in forests below 240 m altitude at around 5 nests ha⁻¹ using a birch-feeding aphid which excretes honey-dew. It is thus able to attack *Oporima* larvae as they appear and prevent them doing much damage, but its action is most intense near its nests, in fact the damage to leaves in a normal year fell from 67% to 43% at 40 m from the nest and in an 'outbreak' year it was unable to prevent serious defoliation except around its mound where a 'green island' was left. Thus the birch tree that harbours aphids as food and sheds its twigs for the ant to build nests is more likely to prosper than one that does not, but if it prospers too much it may shade the ant mound and cause it to go away – a very delicate and spatially shifting system in metastable equilibrium (Wielgolaski, 1975).

With these wood ants there may be a sharp distinction between aphids as prey and aphids for honey-dew, or a single species may serve both purposes at different seasons. So, too, is a dominant tropical formicine tree ant, *Oecophylla longinoda* fiercely predacious, but cultures several species of bugs. One of these, *Saissetia zanzibarensis* (Coccidae) it protects and treats gently and collects its honey-dew; but once the workers are satisfied with sugar fluids they become more aggressive and kill and eat the bugs for protein, feeding them to their larvae. Way (1954, 1963) was able to show a rough balance between the coccid and the ant populations. In Europe another formicine, *Lasius flavus*, lives almost entirely underground and has associations with many species of root-sucking aphid, that are obligatory. The aphids may never produce winged sexual forms and depend on the ant for dispersal as well as protection; even the aphid eggs are collected and stored and their young put on food plants by the ants. In a population studied by Pontin (1978) seven species of aphid live on underground roots of various grasses. Pontin found evidence of a stable predator/prey relationship in that some young aphids in their first instar are fed to ant larvae, and he calculated that a single colony of *Lasius flavus* occupying some 5–7 m² of grassland could survive and grow by virtue of this mutualistic relationship alone, though soil insects are certainly eaten too.

In a study of a six-species ant community in West Scotland, Muir (1959) found 14 species of aphid on 21 species of herb. Most of the aphids are obligate ant associates (myrecophiles) belonging to three polyphagous species. Thus *Forda formicaria* cultivated by five ant species feeds underground on twelve plants of which nine are grasses; *Tetraneura ulmi*, cultivated by five ant species feeds underground on nine plants, all grasses. The preferred and most common grasses are *Agrostis alba*, *A. tenuis* and *Festuca rubra*, on which all three myrmecophiles live. In spite of this the ants still use various adventitious aphids on a variety of herbs both above and below ground. Way (1963) gives many other examples of ant/bug associations.

The wood ant story suggests that trees may well gain from the presence of ants, indeed, they may even accommodate more honey-dew aphids than others. However, there is a more direct way which is under the control of the tree: to have nectaries on those parts of the shoot system that need protection, active only during the period crucial for defence. The North American black cherry (*Prunus serotina*) has nectaries on young leaves that secrete for about 3 weeks just after the buds open: the time when leaf-eating caterpillars of the genus *Malacosoma* hatch. *Formica obscuripes*, of the wood ant group, visits these nectaries and in so doing encounters a small caterpillar which it overpowers and takes to its nest; bigger ones escape. Tilman (1978) managed to show that the *Malacosoma* larvae can be so reduced in numbers in their early instars by this ant that there are not enough large ones later on to cause serious defoliation. Other examples can be found in Bentley (1977).

A perennial sunflower, *Helianthella quinquenensis* (Asteraceae) lives at heights over 2500 m in the Rocky Mountains and has nectaries on the margins of flower bracts which secrete a solution not only rich in sucrose but containing 19 amino acids at a concentration 16 times that in the floral nectary. These attract as many as 40 ants per head of various species mainly *Formica obscuripes* again. They come from nests in the soil and collect the extrafloral nectar leaving the floral nectar for pollinating bumble bees. Inouye and Taylor (1979) were able to show that above 3000 m the ants deter insects which try to lay eggs in the florets and thereby prevent seed-set (Fig. 2.4). As other plants of this family use systemic repellents against insect attack the question can be asked (but not yet answered): how do the two methods compare in ease of evolution, effectiveness and cost?

The tropical ginger plant (*Costus woodsonii*) secretes a nectar rich in

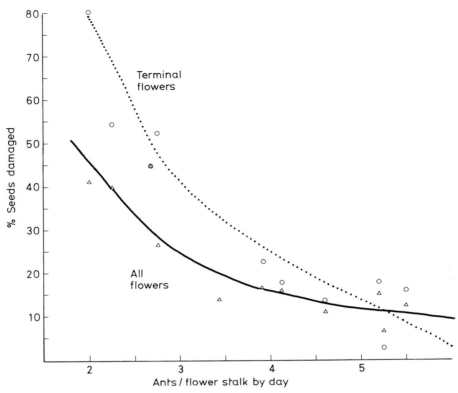

Fig. 2.4 Relation between the numbers of ants during the daytime on a *Helianthella* head and the percentage of seeds damaged. (After Inouye and Taylor, 1979.)

sugars and amino acids from bracts on the inflorescence. The ants *Campono-tus platanus* (Formicinae) and *Wasmannia auropunctata* (Myrmicinae) show a seasonal succession as dominants, the first in the dry, the second in the wet season (Fig. 2.5). Schemske (1980) showed that they protect the fruit from a fly which lays eggs under the bracts and whose larvae destroy the seeds and arils. The formicine drives ovipositing females off and the myrmicine, being smaller, can get between the bracts and eat larvae. Birds disperse the fruit that matures provided the oily aril has not been destroyed. Schemske points out that as *W. auropunctata* is a 'tramp' species in disturbed tropical habitats, it is unlikely to have been a factor in the evolution of the present ant/plant association. Though extrafloral nectaries by-pass the plant-sucking bug, this system of attracting help from ants still depends for its success on having a reasonable number of ant nests quite near the trees.

Some tropical trees have carried this to its logical conclusion by providing not only extrafloral nectaries to attract the ants with sugars but suitable nest

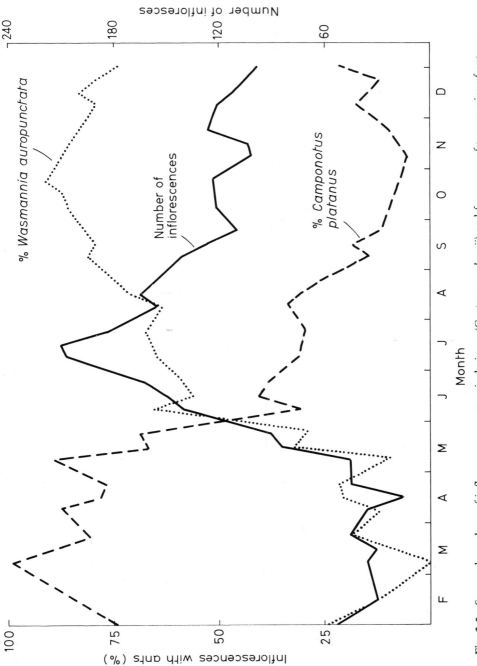

Fig. 2.5 Seasonal numbers of inflorescences on tropical ginger (*Costus woodsonii*) and frequency of two species of ant on these: *Camponotus platanus* and *Wasmannia auropunctata*. (After Schemske, 1980.)

spaces and even solid protein/oil food as well. Many species of *Acacia* (Papilionaceae) attract ants by means of nectaries on their stems but some also provide solid food, rich in protein and oil, in special 'Beltian' bodies on the tips of their leaflets; these contribute over 90% of the ant's solid food. In these *Acacia* plants the nectar from the stem nectaries is copious and can produce as much as 40 g glucose and fructose each day. Moreover, nesting space is provided in specially swollen hollow thorns in which a species of *Pseudomyrmex* (Pseudomyrmecinae) lives. These ants have strong stings and can bite, sting and kill phytophagous insects that fly on to the plant; some are then eaten, others are killed and dropped. It seems that the ants relate territorially to the nest tree and thus give it more effective protection. They not only attack mammals but clear herbs from an area around the base and kill branches of other trees as they encroach. This defence against other plants gives the tree space of course, but also prevents fire spreading to it (Janzen, 1966). Another ant subfamily, the Dolichoderinae, supplies defensive ants to plants: species of *Azteca* live in trees of the genus *Cecropia* which offer both board and lodging, and hairs, at the base of each petiole contain, amongst other nutrients, glycogen (Stradling, 1978b; Wheeler, 1942).

With these last examples the ant/plant mutualism reaches its ultimate refinement: in exchange for protection ants get a territory that includes all their requirements of space and food. In the following sections looser relationships will be described: between ants and seeds or green leaves and between bees or wasps and pollen.

2.4 Seed eaters

The use of seeds to supplement prey is common in myrmicine ants. Even the ponerine *Brachyponera* which is primarily a predator in savannah, collects and stores seeds during dry seasons; many are not eaten, and so contribute to the dispersal of the plants (Lévieux and Diomande, 1978b). *Atopomyrmex* (Myrmicinae), a mixed feeder living in dry Ethiopian regions, eats 48% prey, 28% seeds and honey-dew from Aleurodids; a quarter of the prey are Coccids (Lévieux, 1976). Seeds also eek out prey for *Tetramorium caespitum* of warm temperate regions. Workers collect, one by one, the minute seeds of ling (*Calluna vulgaris*) from the ground or directly from the capsules and carry them back to store in the upper galleries of the nest prior to crushing and eating. They prefer them to the larger seeds of *Erica cinerea* – even though these have twice the biomass (Brian *et al.*, 1965, 1967) – grass seeds are also collected and stored until spring when their embryo is cut out and fed to larvae. There is no doubt that they provide a valuable source of oil and protein, perhaps the staple diet for the sexual brood which must fly by June. Two *Novomessor* studied in the Chihuahuan Desert of North America are both facultative seed-eaters and predatory (Whitford *et al.*, 1980), whereas

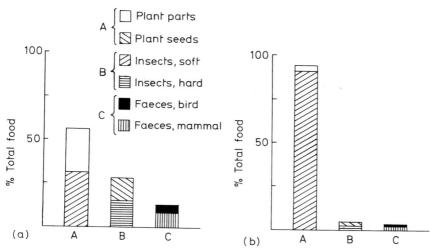

Fig. 2.6 The foods of *Pogonomyrmex* harvester ants. (After MacKay, 1981.) (a) *P. montanus*; (b) *P. rugosus*.

two species of *Pogonomyrmex* in California (Fig. 2.6) showed a contrast between a mountain one (*P. montanus*) that eats flowers, leaves, resins, insects and vertebrate faeces in spring and a plains one (*P. rugosus*) that eats almost nothing but seeds with a staple diet of *Erodium cicutarium* all year round (MacKay, 1981).

Seeds are, of course, not only rich in the nutrients needed by a plant embryo and hence valuable for ants, but are ideally designed for storage and may keep for years, provided they are dry and dark. The tribe Myrmicini includes many highly specialized seed-eaters such as *Messor* in Eurasia and Africa and *Veromessor* in America. In South-West France species of *Messor* prefer seeds to prey which they only eat when food is short; they even collect and eat green unripe seeds. *M. structor* uses grass seeds (especially *Poa annua*) as early as May and *M. capitatus* uses the seeds of the rock rose (*Helianthemum*). Unripe seeds do not keep and need to be eaten at once, but later in the year, ripe ones are first stored at the nest entrances and then put in special granary chambers. Food storage depends on the ants selecting well-drained soil to nest in, making large, well-ventilated chambers, harvesting in dry weather and drying damp seeds in the sunshine: any which start to germinate must be rejected (Delage, 1968). The embryo is eaten preferentially though this is not an essential act in the storage process. *Messor galla* in African savannah collects the seeds of various grasses in the dry season and in the wet, instead of hunting insects, collects seeds of a different kind, e.g. *Evolvulus* (Convolvulaceae), up which workers climb to cut peduncles each of which can yield 1455 seeds. This species will occasionally eat termites (Lévieux and Diomande, 1978a). *Messor arenarius* feeds mainly on the seeds of the grass

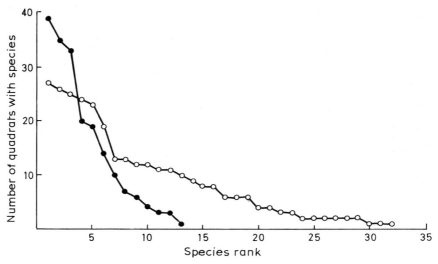

Fig. 2.7 The distribution of ant-dispersed plants (o) and non-ant-dispersed plants (●). (After Handel *et al.*, 1981.)

Aristida pungens which is common around their nests. They cut nearby at first but later 20 m or so from their nest, climbing up for whole inflorescences which they pile on the ground; in May they take grain from the soil surface where it is blown into depressions by the wind. This species also eats insects, mainly termites, and small corpses.

Veromessor pergandei studied in the Californian deserts by Tevis (1958) has a 93% seed diet collected from over 12 species of plant. Capsules are cut open whilst they are still green and, later in the year, seeds are collected from the ground. With a copious supply they choose seeds of *Oenothera*, *Malvastrum* and *Mentzelia*; in fact they are so attracted to the last that they queue up outside the capsules. After prolonged drought *V. pergandei* are prepared to take *Pectocarya* and *Plantago* as well as a lot of inedible rubbish. Both seeds and chaff are stored loosely in the surface chambers to be cleaned up later; seeds are then taken into the lower galleries and the chaff thrown away. In this way *V. pergandei* can survive many years in weather too dry to germinate annual plants. As many of the seeds collected are eaten the question arises: how mutual is the arrangement? From the plant point of view, does it pay to sacrifice some seeds to improve the dispersal of others?

One way round this problem is to enclose the seed in a fruity tissue which is eaten by ants and wasps or to provide it with an aril containing attractive oils. Although the simple *Myrmica* of more temperate regions will not take oils, other members of the subfamily will; thus *Solenopsis invicta*, the 'fire-ant', now living in North America from South Carolina to Texas, though primarily a predator on a range of small invertebrates, is fond of soy bean oil as well as carbohydrate or protein (Glunn *et al.*, 1981). In fact, this species, like *Tetramorium caespitum*, uses oil stored in the crop of queens to fuel new colonies; it is passed, bit by bit, to the pharyngeal gland and digested, for this is a caecum of the gut rather than a salivary gland (Vinson *et al.*, 1980). Two acids, linoleic and linolenic dissolved in the oils are appetizers (Lofgren *et al.*, 1975).

Lasius alienus (Formicinae) provides a good example of ants that collect seeds but do not eat them: it takes the seeds of dwarf gorse (*Ulex minor*) to its nest but only removes the aril. As this species nests in open patches between bushes and is continually ejecting soil, the buried seeds establish themselves successfully in an ideal habitat free from competition or ingestion by vertebrates; buried seeds also resist fire. On the mature plant of *Ulex* an aphid, *Aphis ulicis*, lives under soil canopies made by the ant which supplies it with honey-dew and possibly solid protein too. This is why the trio ant–plant–aphid show a strongly associated distribution in heathland (Brian *et al.*, 1976). Similar plant/ant relationships are widespread in dry scrub in Australia and Berg (1975) estimates that 1500 species in 24 families use ants to bury and to plant their seeds. Most are shrubs living in dry places with seeds that are first dispersed ballistically then picked up by ants because of the aril which they eat in the nest.

By contrast with Australia, only about 300 plant/ant species are known elsewhere in the world. Moreover, the plants are mainly herbs living in moist temperate forests. Their fruits open downwards to release seeds with soft collapsing arils instead of shooting them out. An excellent example is *Carex pedunculata* (Cyperaceae) whose seeds have a 'fat-body' that causes ants to collect them; later they are either thrown away or simply left in the nest, which is frequently in rotting logs on the forest floor. Since the seeds germinate at once, this provides the plant with a good start over its congeners whose seeds are dormant over-winter although the congeners make up later as they have competitive advantages (Handel, 1976, 1978). Thus, ants, particularly *Aphaenogaster rudis* which is common on the forest floor in temperate American woodland, will take *Carex pedunculata* seeds to new receptive spots as they move from nest to nest; 13/45 herbs (Fig. 2.7) are dispersed by ants (Handel *et al.*, 1981). In North America, *Sanguinaria canadensis* (Papaveraceae) also has seeds that can be carried up to 12 m by ants, especially in stable habitats. The distance the seed is transported is correlated with the size of the individual; thus *Camponotus pennsylvanicus* with a length of 9.3 mm transports the seeds much further than *Lasius alienus* with a length of 3.0 mm (Pudlo *et al.*, 1980).

Viola species disperse their seeds ballistically and six have seeds carried by ants, mainly by *Aphaenogaster*; again ants take the seed to their nest and eat the aril. The main gain to the plant seems to be that birds and mice find fewer of the buried seeds and as the ants sometimes move their nest (Culver and

Beattie, 1978), they may well help the *Viola* to establish in good spots.

Another, even more intricate relationship, centres around an advanced myrmicine ant, *Crematogaster longispina*, only 2–3 mm long, which makes fibrous carton nests in the epiphytic vine *Codonanthe crassifolia* (Geraniaceae) in the lowland rain forests and plantations of Costa Rica. The plant flowers all the year round and its floral nectaries give way after seedset to extrafloral ones on the fruit base; it also harbours bugs over which the ants build carton nests. The fruit has an attractive pulp and an aril which the ants eat, leaving the seed, which is somehow fixed or planted in the nest carton. As the ant scavenges for insect food as well and defecates freely, it no doubt provides the plant, which has roots at every node, with nutrients too. This many-sided microsystem is not obligate, as another ant genus *Azteca* (Dolichoderinae) can play the same part as the myrmicine (Kleinfeldt, 1978).

Since the ant appears to plant these seeds, the question can be asked: do grass-seed eaters in arid habitats plant seeds and practise horticulture? Are ants like *Messor* planting seeds near their nests when they take out germinating ones and pile rubbish on them? Do any ants actually dig holes or loosen earth before putting the seeds in? Undoubtedly many seeds germinate and form new plants conveniently near the nest in what is otherwise desert (Hermann and Leese, 1956). Do seeds establish in the nest and force the ants to move? Such a complementary relationship must enable the plant/ant pair to extend its range into very dry habitats.

2.5   Leaf eaters

Other myrmicine ants have approached horticulture differently. They culture a fungus on a vegetable substrate and eat its special nutrient bodies (see Plate 1). All are closely related and form a tribe, the Attini, which is found only in America between 44° S and 40° N (Weber, 1972). The earliest may have eaten yeasts nourished on caterpillar faeces; *Cyphomyrmex*, a small ant with small colonies still does this today. Later, biochemically more powerful, filamentous fungi took over and the ants began to collect plant litter, leaves and flowers for them to digest; today *Trachymyrmex smithi* collects leaflets of mesquite (*Ephedra*) in the Chihuahuan Desert, flowers of *Nama* and leguminous seeds; later in the year they use the buds of *Eriogonum trichopes* (Schumacher and Whitford, 1974). The most advanced genus is *Atta* with millions of workers of many sizes in each colony, and species that range from desert to tropical rain forest. Their fungus is fed with specially cut fresh leaves, flowers, fruits and other vegetable matter; it is thought to be a Basidiomycete, similar to that used by termites (R. J. Powell, personal communication). It produces 'staphylae' that are concentrations of nutritious material.

When the fresh leaves are brought in they are first rasped and chewed by workers to remove the surface waxes (which are antibiotic) and make the pulp accessible to the fungal hyphae; at the same time, workers extract oils for their own energy supply (Peregrine and Mudd, 1974). To this substrate the worker adds faeces that contain nitrogenous wastes and, surprisingly, some fungal enzymes: a proteinase and a chitinase, which together with the fungal cellulase, are capable of degrading pectin, protein, cellulose, starch, xylan, chitin and esters (Boyd and Martin, 1975; Martin *et al.*, 1975, both cited by Stradling, 1978b). The larval saliva which is rich in proteinase may be collected by workers and passed to the fungus too. Anyway, the staphylae produced are chewed by the workers and a juice of carbohydrates, mainly trehalose and mannitol (no polysaccharides are found), extracted. This is not rich enough in energy by itself but supplements the oils that the workers squeeze out of the fresh leaves (Quinlan and Cherrett, 1979). The residue of proteins and amino acids is given to the larvae, for whom it makes a good, sufficient, food (Martin *et al.*, 1969). The fungus comb of Attini is liable to infection with weed fungi and bacteria; to control these the workers lick all surfaces in the nest chamber regularly and possibly spread antibiotic substances such as phenylacetic acid and β-hydroxydecanoic acid from their metapleural glands. There is also a heteroauxin which may improve fungal growth.

The most striking difference apart from the type of fungus cultured between the Attini and the Macrotermitinae, is that the former use fresh, often green, plant tissues and the latter use dry, dead, woody or fibrous tissues. The living plant is richer in proteins and minerals than the dead tissues used by termites: in birch, dead leaf has only 35% of the protein, 53% of the phosphorus and 44% of the potassium present in fresh green leaves (literature cited by Quinlan and Cherrett, 1979); for oak the corresponding values are 31%, 75% and 78%. The termite fungus breaks down lignin and releases bound cellulose, which it digests and converts into sugars of various kinds, perhaps then synthesizing glycogen. The ant fungus digests the protein in the fresh oil-free leaf pulp and the fungal bodies later offer to the ant mixtures of carbohydrate and protein which are more easily digested than the original material, particularly by the larvae, the main recipients. There appears to be much more enzyme sharing in the ant but this is, as yet, uncertain.

Raw materials for ant fungus combs are sampled haphazardly but once a locality has been selected instead of a leaf here and a leaf there, whole bushes and trees are defoliated in a short time. *Atta cephalotes* in Guyana rain forest collects from 61 tree species yet favours *Terminalia amazonica* (11%) and *Emmotum fagifolium* (10%), whilst the commonest local tree *Marlierea montana* is avoided (Cherrett, 1968, 1972). It has been suggested that trees defend themselves against attine attack by means of the tannins in older leaves (Rockwood, 1976) or by means of a glutinous latex (Stradling, 1978b). *A. mexicana* in the Sonoran Desert, nevertheless, cuts creosote bush (*Larrea tridentata*) leaves and flowers; this aromatic plant is normally avoided by insects, and its use by these ants suggests that the creosote is quickly destroyed by their symbiotic fungus (Mintzer, 1979).

2.6 Pollen eaters

Apart from seeds and green leaves plants, of course, produce pollen. Originally this was probably passed from plant to plant by wind but on the 'flower' it must have attracted many different insects for its food value. Perhaps the plant first started to secrete nectar to attract prototype wasps who being predatory kept the flower clear of enemies; later the wasps may have started to eat pollen too. Some Vespoidea, (e.g. Masaridae) still collect pollen by mouth and regurgitate it into brood cells, and so do some Sphecoidea, to which the present bees are very closely related (Malyshev, 1966; Michener, 1974). The plant also switched to animal pollination (by flies, beetles, moths, birds, and others) and many settled to a co-evolution with bees. Whilst still solitary, bees evolved hairs to retain pollen grains, and converted their normal grooming behaviour into a way of brushing pollen into collecting baskets under their abdomens or on their legs (Jander, 1976). This enabled them to feed larvae on pollen/nectar balls. Flowers evolved scents and colours to help bees identify them and their shape changed from a radial to a bilateral symmetry to suit the bee structure and approach so that pollen could be shed on to that part of the bee that takes it quickly whilst still fresh to the stigma of the next flower (Fig. 2.8). Bee mouthparts lengthened to form a tongue and the flower corolla became effectively tubular; so long tongues and long tubes evolved at the same time (Free, 1970; Percival, 1965). Enclosing the flower and developing a corolla tube not only keeps flies and short tongued bees out but provides a steadier supply of nectar at a less variable concentration. It keeps out rain and sunshine and prevents excessive evaporation of water as may occur in quite humid atmospheres. Other methods of stopping evaporation are hairs that reduce the orifice (Fig. 2.9), lipid monolayers and chemical additives (Corbet et al., 1979).

Pollen is an excellent growth food containing 7–30% protein with all the necessary amino acids and vitamins, including cholesterol (Herbert et al., 1980). It also contains an appetizer, an unsaturated C18 alkane (Johansson and Johansson, 1977), and when stored contains honey and yeasts that produce lactic acid. Nectar from floral nectaries is a solution of sucrose with its two hexoses, glucose and fructose, and many amino acids too. Honeybees (Apis mellifera) have to evaporate a lot of water before storage and so it does not pay them to collect nectar with less than 20% sugar unless for immediate use. Bumble bees store only thin honey which they use within a few days. The sugars power flight, maintain hive activity and create warmth, and the pollen mixed with honey is used as larval food directly or to generate 'bee-milk' for young larvae and queens. According to Ribbands (1949) a load of pollen may

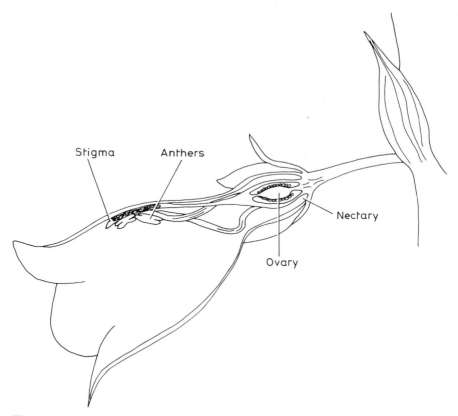

Stigma   Anthers

Nectary

Ovary

Fig. 2.8 Section of the flower of foxglove (*Digitalis purpurea*). (After Proctor and Yeo, 1973.)

involve between 7 and 120 visits and a crop full of nectar 250 to 1446 flower visits.

In an experimental study of honeybees, Free (1971) found that, although colour is important, scent is more so; this is interesting because young bees must come to appreciate the scent of flowers in dark hives before they go out to see them. He also found that radial symmetry is preferred to lateral symmetry, that a coarse disruptive outline is more attractive than a fine or plain one, and that nectar guides, especially if broken lines, are attractive as well as directive. Bees do not 'see' flowers as we do; their sensitivity to light lies in the ultraviolet, the blue, and the green wavelengths (Wehner, 1976). Honeybees have innate preferences for certain flowers, e.g. *Medicago sativa* (lucerne) (Mackensen and Nye, 1969). By importing queens from different countries and keeping them together in England, Free and Williams (1973) showed that some strains favoured blackberry (*Rubus fruticosus*) whereas

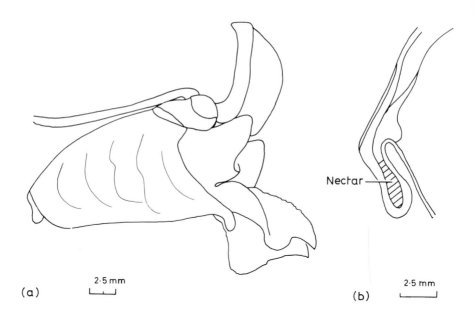

Nectar

(a)   2·5 mm

(b)   2·5 mm

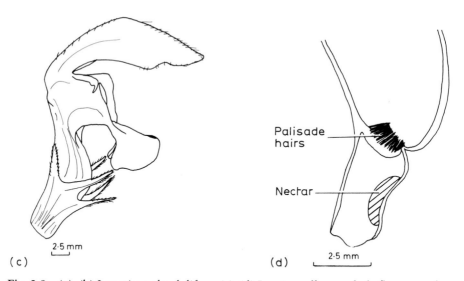

Palisade
hairs

Nectar

(c)   2·5 mm

(d)   2·5 mm

Fig. 2.9 (a), (b) *Impatiens glandulifera*; (c), (d) *Lamium album*; whole flowers and section of nectaries which shows the usual position of the nectar. (After Corbet *et al.*, 1979.)

other preferred lime (*Tilia*) or collected more pollen from white clover (*Trifolium repens*).

An interesting problem is how bees co-ordinate their seasonal activity with the flowering time of plant species. *Bombus*, after hibernating, may emerge in response to a temperature in their soil refuge (Prŷs-Jones, 1982). Solitary bees may have a generation geared to a single flower, e.g. *Andrena* and dandelion (*Taraxacum officinale*). Social bees are released from this restriction and go on through the season, though some *Bombus* have short and others long cycles to fit geographical high or low latitudes, respectively. Spring brood rearing by honeybees can make use of over-winter stored pollen/yeast mixtures ('bee-bread') but fresh pollen can usually be obtained from forest trees such as elm (*Ulmus*), poplar (*Populus*), and willow (*Salix*) which together constitute 11% of the total annual collection in England (Synge, 1947). After that, grassland herbs and crops, especially Papilionaceae are important whilst shrubs of the family Rosaceae provide most of the summer forage. Ivy (*Hedera helix*) yields a very late supply of pollen (October) that can only be useful for early spring breeding.

CHAPTER 3

Foraging by individuals

All individuals in an insect society share their food in a nest space but only some go out and get it; these foragers are usually older and larger than average. As a society grows or as new species with bigger societies evolve, more food has to be found and taken into the nest from further afield. This creates problems not only of navigation, which is highly developed even in non-social insects, but of search, choice, communication, efficiency, transport and defence. (For insect foraging see Hassell and Southwood, 1978.) Here the method and organization of food collection is considered in stages of complexity. In this chapter a system is explained in which little communication between individuals occurs, and the methods which each forager uses are of paramount importance. In the next chapter a system is discussed in which transfer of information about sources by conducting parties, using trials, signs and symbols can be superimposed on individual action, and the decision for the individual is whether to arouse nest mates or not. Finally, a system is described in which nest extensions into the surrounding habitat are developed and territories organized.

3.1 Foraging strategy

Undoubtedly the social bees of the Halictidae and Bombini (Apidae) rely on the food collected by individuals; they do not co-operate in this process (Michener, 1974). This does not mean that workers that are old enough and big enough cannot pick up useful hints from returning nest mates: a returning forager of *Bombus* sounds different even to human beings (Schneider, 1972, quoted Michener, 1974). There seems to be little doubt also that bees can identify flowers from the pollen on foragers as well as by tasting the liquids they regurgitate. The result can be a concentration of work on a few of the large number of possible forage plants in the vicinity, as Free (1970) reports for *B. lucorum*. Wasps, too, are able to inform nest mates about food, probably through smell, for wasps are notoriously sensitive to chemicals (Hölldobler, 1977). The coming and going of both bumblebees and wasps is, in fact, sporadic but analysis of *Bombus affinis* data by Plowright (1979) suggests that external changes due to weather rather than contagious behaviour cause this.

In the wild, species of *Bombus* use flowers with a corolla depth that suits their tongue length, though this correlation is never close because both bees and flowers vary a lot in size within a species and nectar can sometimes be reached even in long tubes. Thus, Viper's Bugloss (*Echium vulgare*) secretes nectar at a rate influenced by air temperature but the honey-bee with its relatively short tongue can only reach it in humid weather, when it is more abundant but less concentrated. Honey-bees were seen to collect from one patch of plants with unusually short corollas even in dry weather (Corbet *et al.*, 1979), and they only go to red clover (*Trifolium pratense*) when the nectar has risen far enough up the corolla tube for them to reach it (Free, 1970). Many *Bombus* have very long tongues (e.g. *B. hortorum*) and can be the sole users of plants with deep nectaries.

Different species of bees are probably attracted to different classes of flower that suit their tongue length, yet little is known about what cues they use. Before they leave the nest they cannot experience colour and shape, only scent and taste, and it may therefore be that these dominate their first reactions as foragers. Yet colour plays a part in flower selection, for short corolla flowers tend to be white and yellow whilst long corolla ones are more often blue and purple (Knuth, 1906–1909, in Proctor and Yeo, 1973). Three species of *Bombus* chose yellow or orange artificial flowers more often than blue or purple ones and white or red ones were ranked lowest (Brian, A.D., 1957). Moreover, bumblebees are more easily trained to blue than to white artificial flowers; white needs more reward and bees forget white training sooner (Heinrich *et al.*, 1977). However, shape may also be influential since long corolla types tend to be irregular or bilaterally symmetrical; they usually come singly and hang down so that a bee has to approach from below. In contrast, short corolla types are often clustered and form a 'head' or compound flower as in Compositae; *B. lapidarius* is attracted to clusters (Prŷs-Jones, 1982). Some bumblebees try to settle in flowers which are already occupied by other bees (Brian, A.D., 1957); clearly this behaviour, though perhaps primarily aggressive, could save a lot of research and lead new foragers to good types.

Bumblebees learn which flowers suit them and how best to approach and enter. Heinrich (1979) released young bees into a natural meadow from which other wild bees had been excluded by nets; they tried five species before settling for jewelweed (*Impatiens*), a flower with a deep nectary and bilateral symmetry. Even then their first visit was clumsy: they landed on top instead of on the lip and took 60 visits to become skilled enough to fill their crop from 20 flowers in 6.2 min. In contrast, the open radial flower heads of *Aster* and goldenrod (*Solidago*) needed no experience but gave least reward. The importance of the nectar supply showed up when all the bees were released at once: the nectar supply from jewelweed was reduced to a quarter, that from red clover and *Aster* to a half but the goldenrod not at all. Thus competition led

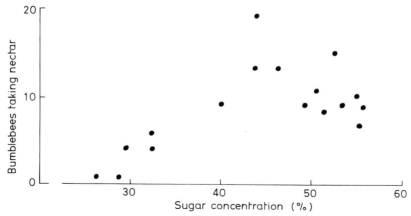

Fig. 3.1 The number of bumblebees taking nectar from lime flowers (*Tilia*) in relation to the concentration of sugar in the nectar. (After Corbet *et al.*, 1979.)

some bees to give the clover priority over jewelweed. Again, if white clover (*Trifolium repens*) and cow vetch (*Vicia cracca*) are netted from bees they build up their nectar so that *Bombus vagans* will later concentrate on them in preference to the same species elsewhere (Morse, 1980). Stronger nectar is usually preferred (Fig. 3.1) but the viscosity can interfere with suction in long-tongued bees at high concentrations and so gives an optimum which maximizes nett energy gain to the bee (Prŷs-Jones, 1982). Pleasants (1981) shows how bee abundance 'tracks' flower abundance through the season in the Rocky Mountains, Colorado (Figs. 3.2 and 3.3).

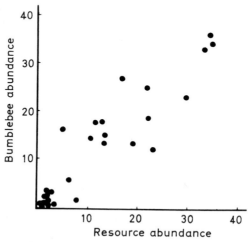

Fig. 3.2 The relation between bumblebee and nectar resource abundance. (From Pleasants, 1981.)

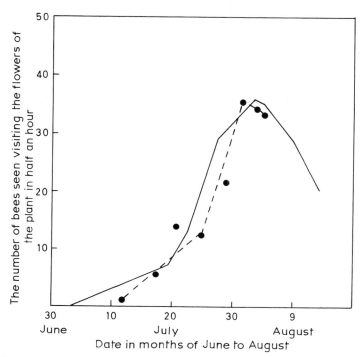

Fig. 3.3 Seasonal change in nectar available (————) and numbers of bees (*Bombus bifarus*) seen foraging (– – – – – – –) in a 30 min sample. (After Pleasants, 1981.)

So far the collection of nectar has been considered. It is doubtful whether long-tongued *Bombus* ever gather pollen without nectar; they belong to a very specialized flower/bee association. Other bees are free from this restraint and will visit flowers that yield only pollen (e.g. poppy, *Papaver*). On others such as sunflower (*Helianthus annuus*), which yield both nectar and pollen, they go only to the florets with freshly dehisced anthers if collecting pollen. Pollen is ejected after the bee vibrates these but in other types of flower bees 'milk', bite or glean pollen (Eickwort and Ginsberg, 1980). Nectar collectors may visit the same area and become powdered with pollen which they transmit to older flowers on the same sunflower head by walking, but they quite often comb it up and discard it. This cannot be because it is unsuitable; it is just not part of their immediate goal. In spite of this habit they may be better pollinators (Free, 1970). In socially advanced bees then, pollen and nectar are often collected on different trips from different flowers and pollen collectors prepare for a trip by drinking stored honey. Such species can also collect water to dilute stored honey and there is no doubt that the quantities of the main raw materials of food (nectar, water, pollen) collected are regulated by communication inside the nest.

From the plant point of view, pollination would be more effective if a bee visited the same species for a time instead of going randomly from one to another; have plants been able to impose this condition? The honey-bee nearly complies for only 6% of its pollen loads are mixed. *Bombus* are not so good: 34% for *Bombus lucorum*, and 63% for *B. pascuorum* (Free, 1970). Whilst this must help cross-pollination of herbaceous plants and small bushes, it cannot help trees, where a bee is more likely to move to another branch of the same plant rather than to another tree altogether. To avoid the resulting self-fertilization, many trees have evolved separate male and female flowers (Maynard Smith 1978a, p. 136). Yet the pollen of 'pollinator' trees planted in orchards is carried to nearby trees, and it is suggested that this transfer occurs between successive trips since pollen carried on the bee's hairs remains viable for some hours (Free and Williams, 1972).

Sugar and pollen supply vary throughout the day and depend on the internal rhythms of flower opening, pollen release and stigma sensitivity, all modified by weather conditions. Thus a bean (*Vicia faba*) flower opens on the first day to shed pollen, on the second day to receive pollen (the stigma is ripe) and on the third day it withers and ceases to attract bees. Such a group of flowers form an inflorescence which is commonly just a shoot that grows flowers instead of leaves, it has an apical growing point which generates young flower buds, each of which colours, opens, emits scent, sheds pollen, produces a receptive stigma and finally fades into inconspicuousness. This design is intended to ensure that bees land at the bottom of the spike and add fresh pollen to flowers with receptive stigmas, then, walk or fly up (depending on the distance apart of the flowers) until they reach the highest open flower; they then carry a new dose of fresh pollen to the oldest lowest flower on the next plant and so on (Free, 1970). Pyke (1978, 1979) investigated the way *Bombus appositus* feeds from, and pollinates, monkshood (*Aconitum columbianum*) and found that the bees do not move systematically up the age spiral. They fly between flowers, and tend to move up to the next nearest neighbour (90% of the time) rather than to the next youngest. This reduces flight time even though it means missing out some of the flowers in the spiral and Pyke worked out the rule of behaviour '. . . fly to the closest higher flower not just visited . . .'. There were irregularities of course: one in five times the bee actually moves down and two in three times leaves the spike between the bottom and the top visiting the youngest flower less than one in five times. Nevertheless, the system works and a flower is re-visited only 1.2% of times on a single trip. Pyke pointed out that the plan of going to the nearest rather than the next flower in the spiral is good as long as the spikes (plants) are not too far apart. If they are it is better to concentrate on each and visit every flower systematically.

Another case where a cursory, quick collection of nectar pays is in the flower of the desert willow (*Chilopsis liniaris*) visited by *Bombus sonorus*;

Plate 4

Above: Honeybee forager returning with pollen loads. The pollen is transferred to storage cells before conversion into bee-milk.

Below: Bumblebee collecting nectar from ragwort, becoming smothered in pollen grains as it does so. (Photographs courtesy of J.B. Free.)

each flower offers nectar in a pool at the confluence of five grooves. In the morning queens go round sucking up the pool and getting 1.7 μl in 2 sec then they go back and clean up the grooves getting only 2.4 μl in 5 sec. However, Whitham (1977) calculated that to take first the pools and then the grooves gives a 25% greater gain than being thorough all the time; from the plant's point of view the flower also gets twice as many visits.

Besides being stronger, bigger bees travel more quickly between inflorescences though not between flowers; yet energy consumption in flight increases in proportion to body weight and a big bee uses more energy than a small one. The best (most economical) size for B. appositus on Aconitum is 0.24 g which compares very well with the actual average weight of 0.28 g (Pyke, 1978). Pyke's model also predicts that the more nectar per flower the bigger the bee expected and, in fact, B. appositus and B. kirbyellus feed on delphinium and monkshood with 2–4 μl day^{-1} whilst B. bifarius and B. sylvicola feed on composites which have short corolla tubes in each floret and only a few open at once (Proctor and Yeo, 1973). Another model for Bombus foraging for nectar considers three states: bee in the nest, bee in flight and bee in the flower, and assumes return to the nest when the crop is full (Oster and Wilson, 1978). The time in a flower depends on its structure and nectar supply and the time between flowers depends on their distribution including their scatter and on bee flight velocity which takes into account take-off and landing energies as well as flying energy. This makes large bees more efficient for big flowers with a lot of nectar provided they are well spaced. The authors produce an expression for energy gain per forager which increases with flower density to a limit on the usual diminishing return function and predicts an optimum forager size for each resource size. Bigger bees are, in general, more efficient foragers but, of course, cost more to make.

The idea of optimal foraging has been a stimulus to research on factors which influence the exploitation of a 'patch' of food. The forager should reduce all equidistant patches to the same level of supply; this level is the rate of intake they expect, estimated by an integration of past experience (Charnov, cited Hassell and Southwood, 1978). The further the patch from the nest, the more deeply should it be extracted. Using Bombus appositus queens working for nectar on Delphinium nelsoni in the Rocky Mountains in Colorado, Hodges (1981) found support for Charnov's model. The amount of nectar in flowers on the same spike (a patch) was correlated so that a bee only had to sample one flower to decide whether to go on or go off to another spike. If nectar was boosted experimentally the bees stayed longer on each spike and went to all of the flowers (usually about three).

Though the honey-bee has a superb system of communication, there are times when an individual works alone. Their behaviour has been studied when foraging for nectar by Butler et al. (1943) and when collecting pollen by Ribbands (1949). In both cases individual bees were attached to a group of flowers in a small area even where the whole crop was extensive. They visited these flowers as long as the supply held out but left before it failed, even though other bees were still using it. A bee will have sampled other crops and '. . . the extent of her tendency to change is determined by her memory of past yields . . .' (Ribbands, 1949, p. 63). Ribbands stresses that they show '. . . continuous exercise of choice . . .'. In short, they are continuously assessing the food supply and exploiting the best available. Weaver (1957) suggests that they research after satiation on the main nectar source.

There is one complication: some bumblebees collect nectar by biting through the corolla tube even before the flower opens, and those that cannot do this use the holes made by others (and even by birds). In Europe, at least 300 species of plant are robbed in this way (Brian, A.D., 1954). In cultivars reproduction is assured but wild plants, unless they are self-sterile, must be able to get pollen transferred within the flower or by other insects, vibration, or wind. Heinrich and Raven (1972) argue that nectar robbery only causes the long-tongued pollinating bees to visit more flowers, and so increase outbreeding. This 'illicit' nectar gathering means that one flower, say bean, can energize the pollination of another unrelated flower, say poppy (*Papaver*) which produces no nectar. This works well and indeed probably depends upon the floristic jumble with which humans have replaced earlier stable, or at least slow-changing, ecosystems. Old-fashioned long-tongued bees that derive their complete sustenance from a single flower, e.g. *B. hortorum* on comfrey (*Symphytum officinale*) (in Prŷs-Jones, 1982) may well be relegated to nature reserves before long.

3.2 Worker variability

Workers vary in size, sensitivity and activity. This is used to create a useful mix of types in each colony. Work in the nest is light and intricate; outdoor work is heavy and dangerous though requiring sensitivity to environmental cues and intelligent flexibility. Ideally then foragers should be both older and larger than average: this is, in fact, achieved through a variable rate of development which enables initially bigger individuals to develop tough pigmented skins and an attraction to light more quickly than smaller ones and so to become foragers sooner (Lenoir, 1979a, summarizes the evidence for this in ants, bees and wasps). The period in the nest in darkness has only limited use for the development of foraging behaviour and can be curtailed without harm. This still leaves the question: how much size variation is genetic and how much phenotypic? Variation in adult size due to unequal feeding during the larval stage is common. In *Vespula*, for example, larvae near the nest entrance and in the centre of the comb are larger because they are nearer the food channel and in those *Bombus* whose larvae feed on a pollen mass, some get better access than others. Do the better fed ones become foragers earlier in adult life or is a genetic component needed to clinch this development?

In most eusocial ants, bees and wasps the frequency distribution of worker size is unimodal and approaches a normal curve distinct from that of the reproductives; such workers are 'monomorphic'. In a few ants, however, the shape is skewed towards a log normal and in extreme cases can be broken into two or even three modes (polymorphic); this must have a genetic basis. Associated with this increase in the size range is a greater variation in shape due to an ability to grow allometrically after the general worker caste characteristics have been determined during development. Allometric growth means that some parts, in this case the biting system and some associated head glands, grow faster in relation to the whole than other parts. A log scale graph of part/whole gives either one straight line, or several lines with different slopes. This implies the existence of different relative growth rates at different stages of development (polyphasic growth) but it depends on the metrics used (Brian, M.V., 1979a). Polynomials also fit some data well (Baroni Urbani, 1974).

The variation in size of monomorphic workers does not arise simply from inequality or 'errors' in food distribution though reduced food supply certainly reduces mean size. The workers, though all one sex in ants, bees and wasps, are not cloned and are, at most, only sisters. Food supply is a social and seasonal variable which interacts with this genetic variation. This produces workers with a range of size, development rate and behaviour that covers the requirements of the society, though how efficiently is unknown; some may be underemployed.

Chromosomally determined sex differences add to other genotypic effects in termites; there is no known rule as yet, but in the Macrotermitinae the large workers are female and do most of the foraging (Noirot, 1969, 1974). In *Hodotermes mossambicus* (Hodotermitidae), the workers which cut grass are older and stronger and darker than those which work in the nest (Watson, 1973). *Bombus morio* foragers are larger than average (Fig. 3.4) and in *B. pascuorum* large foragers collect more pollen than small ones; is this tendency acquired through success? Their greater size may give them more strength to enter flowers and their surface area help to collect pollen. In the European wood ants (*Formica rufa* group), the bigger foragers tend to hunt, whilst the smaller ones collect honey-dew; the bigger ones also go further afield, wander off the tracks more, and become involved in special activities such as fighting (Brian, M.V., 1979a; Mabelis, 1979a), but the development of these behaviour differences is obscure.

Workers of *Formica* will co-operate in the retrieval of large prey as described by Sudd (1965, 1967). So, too, will workers of *Oecophylla*, another formicine ant with relatively large foragers (Hölldobler and Wilson, 1977a). In the Myrmicinae this is true of *Myrmica* but in *Tetramorium*, with much

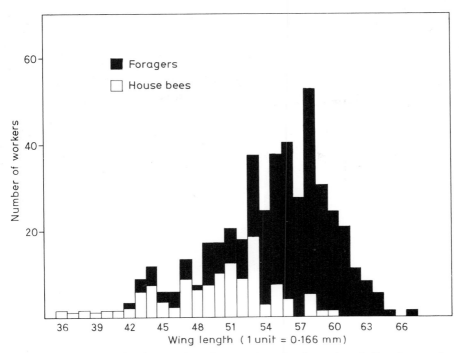

Fig. 3.4 To show that foragers have longer wings than house bees in *Bombus morio*. (After Garofalo, 1978.)

smaller workers, which are very effective individual collectors of small ling seeds, it is usual to cut up a big prey and carry it back in small pieces. In *Pheidole dentata* the worker population is dimorphic and consists of small, normal workers on the one hand and large ones with abnormally big heads and jaws on the other; whilst the normal workers cut up big prey they are defended by the big 'soldier' ones (Oster and Wilson, 1978).

Solenopsis invicta workers tend to 'match' prey size to their own size; bigger workers never collect small termites in the lab (Oster and Wilson, 1978). This 'resource matching', it is suggested, is an important principle in ant evolution: these authors compare ant size and the size of their invertebrate prey and suggest that new invaders such as *S. invicta* when they entered North America from the south acquired a whole range of new food species and enjoyed 'ecological release'. They suggest that evolution of a wider size range by a genetic shift could be an important adaptation which extends their prey capture capability, although efficient deployment must remain a problem.

Atta cephalotes in tropical rain forest shows, in samples taken from the nest, a log normal frequency distribution of worker size (Fig. 3.5) (Stradling, 1978a). On the trails the very small workers are missing except for a few that

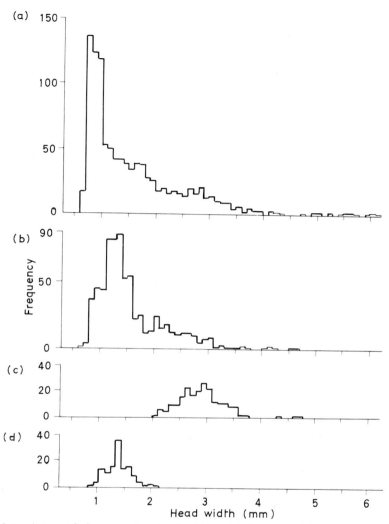

Fig. 3.5 *Atta cephalotes* worker head widths: (a) from nest; (b) from a trail; (c) carrying pieces of leaf; (d) riding on leaf pieces. (After Stradling, 1978a.)

Plate 5 *Opposite*

Above: A soldier of *Atta cephalotes* (Attini) in Trinidad. These defend the nest, not the foraging columns. They come out if the soil is vibrated and can bite strongly, though they do not cut leaves.

Below: Workers of *Atta cephalotes* in Trinidad, carrying leaf fragments that they have cut. Very small workers climb on to the leaf as it is being cut and lick up exudates; in this way they are often carried back to the nest. Wilson (1980b) has shown that bigger workers cut tougher leaf material. (Photographs courtesy of D.J. Stradling.)

ride on the leaf pieces and clean up juices. The cutting is done by medium sized workers, the bigger ones of which seem to attack tougher tissues. The biggest workers are defensive and come out of nests that are disturbed (Autuori, 1974; Lowenthal, 1974; Weber, 1972).

Atta sexdens is similar to *A. cephalotes* in its polymorphism and Wilson (1980) has assessed the efficiency with which its workers perform, by testing the cutting speed of different size classes and dividing this either by their weight or their respiration rate. He found that the workers which cut leaves were slightly longer (modally and on average) than those which cut softer tissues, such as petals. The energetic efficiency thus showed a sharper peak at a larger worker size for leaf cutters than for petal cutters (Fig. 3.6) and the best size corresponded well with the actual size for leaf-cutters. Wilson concluded: 'the species is accurate to within . . . about 10% of the energetic optimum . . .'. In short, *A. sexdens* has evolved a class of leaf-cutters too big to work in the nest and too small to be useful in defence. They appear to be able to select a class of leaf that suits their size and strength. Doryline ants show a similar range of size in their advanced genera (Fig. 3.7); hunting is done by the middle sized workers who each carry a prey of about their own size. The larger workers stand head out in defensive postures (Topoff, 1971).

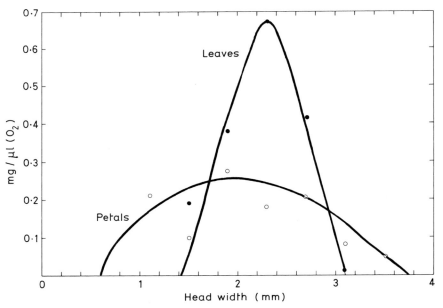

Fig. 3.6 *Atta sexdens* energetic efficiency of workers of different sizes when cutting soft petals (o) or hard leaves (●). The graph shows milligrams of vegetation cut divided by microlitres of O_2 consumed in the process, for a series of workers whose head widths (size) have been experimentally regulated. (After Wilson, 1980b.)

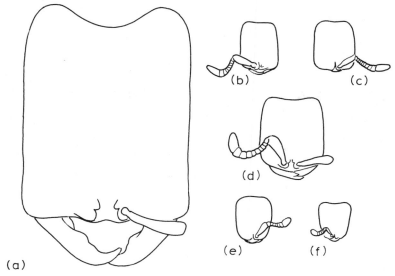

Fig. 3.7 Heads of workers of *Dorylus affinis* drawn to same scale to show differences in size and in number of antennal joints. (From Wheeler, 1910.)

Workers also vary in their responsiveness. Wallis (1962), using *Formica fusca*, noted that some workers would spontaneously go out and forage and suggested that this may be an important factor initiating foraging in ants. Quite recently, Lenoir (1979a) has come to a similar conclusion after a study of *Lasius niger*. Genetic variation can be considerable even in colonies developed from the progeny of a single queen, which provides for the existence of leaders or activists to enrol and guide their nest mates in any work that they perceive to be necessary. The effectiveness of the colony is then almost equal to the effectiveness of the most active and sensitive workers.

In a similar vein, Oster and Wilson (1978) have stressed the role of 'tempo' in societies: a disposition to be either calm or excitable, to be either deliberate or lively, to crash wildly into an action or go purposefully into it. Typical of the former type are ants in the subfamilies Ponerinae and Myrmicinae, and of the latter, ants of the subfamilies Dorylinae and Dolichoderinae. Rapid, unoriented motion may come close to being the requisite mode for hunting by dorylines. A tendency to be retiring and discreet can be a feature of small young colonies where big ones of the same species can afford to risk confrontation. A state of high tempo may be an important condition of colony division, but this is unlikely to be genetic and most likely to result from social conditions such as congestion and age structure.

To summarize, there are many societies of wasps and bees that collect their food through individuals that learn the resource pattern of their locality and keep it under frequent surveillance. They collect food which is rich and easy and have a staple with subsidiaries on which to fall back when supply declines temporarily or permanently, due to a seasonal change in food or to the arrival of effective competitor insects. The success of a society of this sort depends greatly on the quality of these individuals and many species show considerable variation in size, activity, sensitivity, and other useful attributes. Where one leads, others may follow, information, especially the adventitious clues of food smell, passing to others in the nest. From this, in ants and termites, there evolves a 'planned' size-class frequency that can produce exaggerated growth in the forebody giving extra strength to their jaws and so the ability to do heavier work than average. There is then a tendency for foragers to match food size (or leaf toughness in *Atta*) to their own body size, and this seems to be achieved by learning from trial what they can do most successfully.

CHAPTER 4

Foraging in groups

4.1 Communication about food

The simplest way to communicate about food is to display on return to the nest. A simultaneous secretion from the mandibular glands can activate any latent foragers and provide olfactory clues to the kind of food and its locality. If the nest has several entrances, that to which the forager returns can indicate the best sector in which to search. Further guidance may come from a marker spot, either a glandular secretion or faeces placed somewhere outside so that the combination of nest exit and spot if extrapolated indicate the direction to be followed. Stimulating chemicals can be present in the spot produced by yet other glands. For greater accuracy the finder can collect a group of workers and lead them out, using its visual topographic memory and the trail spots to find its way back and help others to keep in touch. The most refined system so far evolved is to direct by 'pointing' with a run in the nest; even distance can be indicated by means of a buzz whose frequency correlates with the time taken to arrive at the food.

In the majority of wasps, food communication is little more than an activation on return; this may give clues from acquired odours as in *Vespula* (Maschwitz *et al.*, 1974, cited Jeanne, 1980) but in *Vespa* helpers may follow a leader back to food. Melponine bees make simple trails, thus species of *Trigona* return from a food source making mandibular gland spots on vegetation. They do this slowly and methodically leaving a gap near the nest which is species characteristic: for *T. postica* a gap of 8 m follows spots at 1 m intervals (Lindauer and Kerr, 1958, 1960). This species, like all the others, uses a mixture of volatile compounds in which benzaldehyde predominates but 2-tridecanone and 2-pentadecanone are also present. *T. subterranea* uses monoterpene aldehydes (neral and geranial which are both stereoisomers of citral) and *Scaptotrigona* uses carbonyl compounds (Blum and Brand, 1972). In the nest these bees alert and activate others by zig-zag runs and buzzes that are species specific and carry no information. The informer then flies out to the food followed closely by recruits. There are thus three key elements in *Trigona* recruitment: a trail, a display and leadership.

The ponerine ants, like the Meliponini, attract help with their mandibular glands; *Odontomachus troglodytes* with a strong sting matches its prey to its own size, and takes driver ants by biting and stinging them whilst releasing a mixture of alkyl pyrazines which alert, attract and cause the recruits to attack (Longhurst *et al.*, 1978). When *Bothroponera soror* finds large prey it releases a mandibular gland pheromone which attracts other workers from a radius of 20 cm. At least 21 compounds are present, but a mixture of 2-undecanol 2-methyl-6-methylsalicylate and 2-undecanone in the ratio 40:10:1 elicits all the reactions of the full gland: porrect antennae, attraction, jaw grip and stinging. This species also conducts helpers to the site of very large prey (Longhurst *et al.*, 1980), but other species recruit help by going back to the nest and engaging another to follow them; they run out with the follower touching the gaster and legs of the leader who has an inter-segmental gland in the dorsal part of the gaster called the pygidial gland which helps this 'tandem' arrangement (Hölldobler and Traniello 1980a; Maschwitz *et al.*, 1974). This gland can be used to make a trail by bending the tip of the gaster down and rubbing it on the ground as the ant runs. In yet other species this enables several workers to be recruited at once along a single-file trail laid by scouts (Fletcher, 1973; Hölldobler and Traniello, 1980b).

Megaponera foetens, a famous ponerine, and an obligate termite eater, has three modes of worker size (van Boven, 1970) and uses canopy pattern to orientate (Hölldobler, 1980). After locating fresh termite workings in the soil, a scout lays a recruitment trail back to the nest. After about 5 min it comes out again followed closely by several big workers and many small ones; at this stage removal of the leader causes chaos even though there is a trail (Fig. 4.1) (Longhurst and Howse, 1978). When they get back to the termite zone the big workers remove the soil sheets and the little ones go in and pull out the termites. Piles are made outside until the operation is finished. The attack appears to be co-ordinated and concentrated by the emission of a mandibular gland pheromone comprising mainly dimethyl disulphide and dimethyl trisulphide. After 10 min or so the column reforms and returns in close formation three to five ants abreast, with big workers in the van ready to drive off robbers. It returns at 4.6 cm s^{-1} compared with 3.8 cm s^{-1} out (Fletcher, 1973; Longhurst and Howse, 1978, 1979a, Longhurst *et al.*, 1979). *Megaponera* like many other ponerines stridulates audibly (Markl, 1973) mainly when disturbed or trapped but it is uncertain that this helps in foraging. In Kenya the hunting programme involves three peaks a day and each ant nest eats 653 termites per day, mainly *Macrotermes subhyalinus* (Lepage, 1981a).

The subfamily Myrmicinae show more variation than the Ponerinae. *Myrmica* lives in temperate grass-scrub and feeds on honey-dew, carrion and small prey; it forages and navigates visually by reference to horizon and canopy shape and to the pattern of polarization of the sky or the position of the sun. Its response to food varies: for water or small prey it lays a trail of

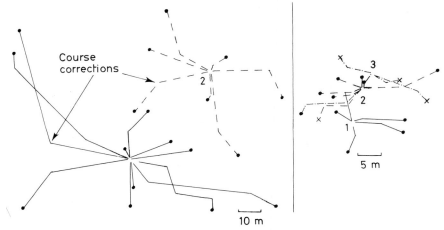

Fig. 4.1 Foraging trails of *Megaponera foetens* from different nest-sites. Two colonies, one from two and the other from three sites. The course may be controlled by scouts. Prey are *Macrotermes*. (After Longhurst and House, 1979a.)

venom with the sting as it returns laden (Kugler, 1979a, describes the evolution of the sting into a pen in Myrmicinae). Inside the nest it behaves calmly and stimulates only a few workers to go out and walk back on the trail. Large prey or a lot of small prey cause the discoverer first to assess the situation and then to return unladen, laying a trail as before but with the addition of a secretion from Dufour's gland; especially near the nest and around the food. When the worker enters it displays energetically and makes quick runs and rushes so that tens of workers return along the trail, accompanied or not by the discoverer (Cammaerts, 1978). Whilst the venom is mainly a solution of proteins and amines, it also contains a small volatile compound of low polarity which is the trail pheromone (Cammaerts-Tricot *et al.*, 1977). Dufour's gland synthesizes straight chain C13 to C19 hydrocarbons, three farnesenes (terpenoids), and four small oxygenated compounds: acetaldehyde, ethanol, acetone and butanone, the last four in the ratio 35:3:40:25 (Morgan *et al.*, 1979). The mixture evaporates progressively beginning with the small alcohols, aldehydes and ketones that arouse, and canalize the workers, then the terpenoids and finally, the hydrocarbons. This sequence enables the workers to assess the freshness of a trace. This is true of *Myrmica rubra*, and *M. scabrinodis* both of which can follow each other's trail. However, although Dufour's gland of *M. scabrinodis*, secretes the same volatile compounds as *M. rubra*, the latter has different heavier ones, used to mark territory, (Cammaerts *et al.*, 1977, 1978). To do this the ant walks slowly and places droplets.

Myrmica species also produce many chemicals in their mandibular glands which are released when prey is encountered, and attract help from nest mates in the vicinity (Morgan *et al.*, 1978; Parry and Morgan, 1979). *M. scabrinodis* contains 49% 3-octanone, 18% 3-octanol, 12% 3-decanone and at least 16 related compounds ranging from ethanal to 3-undecanone. *M. rubra* has a similar set in different proportions (Cammaerts *et al.*, 1981a).

Monomorium venustum workers give a motor display to get recruits out which is quite elaborate and includes wagging from side to side, head pushing, antennal beating and jerky rushes. This display diminishes as the source dries up, losing first the wagging then the antennal beating and last, the rushes (Szlep and Jacobi, 1967). *Crematogaster* took to the trees about mid-Tertiary times (Brown, 1973), and now lay trails with special tibial glands in the hind legs, which they drum quickly and vigorously on the substrate so that the tips of the tarsi spread pheromones (Fletcher and Brand, 1968; Leuthold, 1968). A *Novomessor* species can attract help from 2 m distance by releasing a volatile chemical from the tip of its sting, with its gaster raised above its body; it also guides nest mates out with a poison trail but its pygidial gland is not used to lay trails. Stridulation passes sound through soil and enhances reaction to pheromones (Hölldobler *et al.*, 1978; Markl and Hölldobler, 1978).

A similar series of steps exists in the Formicinae. Wallis (1964) could find no evidence of recruitment in *Formica fusca* but it may lay recruitment trails with hind gut material, and excite latent foragers with a waggle display (Möglich and Hölldobler, 1975). *Camponotus pennsylvanica* is like *F. fusca* except that it adds formic acid to its trail (Traniello, 1977). *Camponotus socius* shows more care in collecting helpers; the scout lays a hind gut trail (perhaps of faeces with the addition of rectal gland material) straight back to the nest, displays inside by wagging and offering food, collects a few tens of recruits and then leads them out along the trail; formic acid from its poison gland activates the followers (Hölldobler, 1971). In *C. sericeus* the discoverer goes out slowly with a single follower touching it; the leader stops and relinks using the trail as a guide if contact is lost (Hölldobler *et al.*, 1974; Maschwitz, 1975a).

Melipona, a genus of stingless bees, conducts parties to food after alerting nest mates by buzz-runs; the leader zig-zags in the required direction and is followed in flight by other bees. This only takes them 30–50 m (*M. quadrifasciata*) or 10–20 m (*M. seminigra*). Although in theory this is sufficient to set a direction, many repetitions are needed before the recruits search alone beyond the distance guided (Hölldobler, 1977). The duration of the buzz and the distance to the food correlate (Fig. 4.2), thus presenting the possibility, not yet confirmed, that distance can be communicated by sound (Esch, 1967; Esch *et al.*, 1965).

In *Apis mellifera* a buzz with a duration proportional to the distance of a food source is also made and is transmitted through the comb structure to receptive bees (Wenner, 1962). This buzz is made whilst the famous 'waggle'

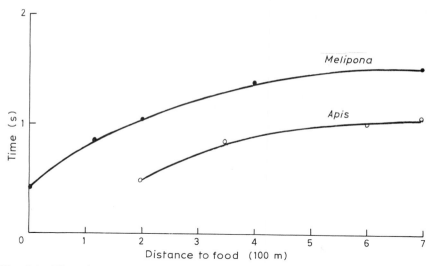

Fig. 4.2 The relation between food distance and sound signal duration for *Apis mellifera* and *Melipona quadrifasciata*. (From Esch *et al.*, 1965.)

display (essentially a straight run during which the abdomen is moved from side to side) is in progress. Also correlated with food distance is the depth in the hive where the display is given, the time taken to perform it, the length of straight run and the number of 'waggles' in it. All these pieces of information are redundant as is common in languages and it is uncertain which, if indeed any, is most important.

Apis florea often displays on a platform above its comb and is able to point towards the food; distance is, of course, scaled down (Free and Williams, 1979). *Apis dorsata* actually dances on its vertical comb, but transforms 'sunwards' into 'upwards': direction is then indicated as an angle to 'up' instead of to 'sun'; similarly with *A. mellifera*. This transformation is not unique to bees, for ants running horizontally on a plate at a fixed angle to the sun will re-orientate to gravity if the plate is tipped vertically (Markl, 1964). This system takes no account of the height of the food source above the ground. However, high sources and sources on the ground close to the hive induce 'round' dances which alert nest mates to search near the hive. The radial distance on the surface within which this display is used varies with the subspecies: in *A. mellifera ligustica* it is 100 m, in *A. m. carnica* it is 85 m, in *A. m. fasciata* 12 m and in *A. indica* it is only 2 m (Hölldobler, 1977). Beyond this surface limit there is a gradual transition to the waggle dance.

Honey-bees have a remarkable memory, a good time sense, a good sense of smell, a visual sense with colour and the intelligence to familiarize themselves with the procedure and routine of an experimenter (Wells, 1973; Wenner, 1974). Apart from the appearance and smell of an experimenter, there are the smells made when people trample vegetation, cut grass and use paints to mark the bees, as well as natural plant and locality differences. After training to a scented sucrose solution and then letting it dry up, it is only necessary to blow some of the scented air in through the hive entrance to obtain an outrush of bees to the old dish, or to any dish bearing that particular scent. Bees are also attracted by other bees, both visually and by the smell they leave as they walk about, probably the same smell that exists in the hive and at the entrance of the hive (Ferguson and Free, 1979). So there are many sources of error in experimentation and careful controls are essential.

To overcome these, Gould (1976) used the fact that bees will dance to a light in the hive as though it were the sun. He covered the ocelli of scouts and succeeded in reducing their sensitivity to this hive light. They then ignored it and danced normally up the comb and recruits who could sense the light because their ocelli were free, misinterpreted the direction indicated. A response to smell would not have had this effect. Parathion reduces the length of run in the waggle display which then falsely indicates a nearer distance than the true one; bees follow this and go to the wrong (near) place, rather than use cues from locality smells (Gould, 1976). There can now be no doubt that honey-bees communicate symbolically, but there is still some doubt about how much use is made of this. Quite often dances about food are ignored; only new, untried recruits respond, and then it takes them many dances to really get the message. Older bees, set in their ways, seem to have a repertoire of flowers from different times of day and wait in the hive in a quiet corner until they get a hint from a returning forager that 'their flower' is ready; this hint may only be a smell. In any case the bees learn roughly what time of day each flower is ready and wait for confirmation from returning scouts.

Small errors in their direction dances can be related to the presence of magnetic granules in the front of the abdomen that are sensitive to the earth's magnetic field. These errors can be eliminated by applying a counter field. Tomlinson *et al.* (1981), moreover, have shown that a magnet held near a bee dancing normally, reduces the time of the dance from 34 to 19 s; this confirms their sensitivity, and shows that a very heterogeneous field can interfere with their dance rate. Finally, there is even a trace of the 'party conduction' used by Meliponini: two lots of bees approached baits at different heights from different directions round an impenetrable hedge; a marked bee would return closely followed by a group of bees which it had apparently led (Wellington and Cmiralova, 1979). For more experiments that support this notion see Friesen (1973).

Apis mellifera does produce an odour from Nasanoff's glands (homologous perhaps with ant pygidial glands); it is fanned into the air and appears to be used when natural odours are scarce, as in the movement and clustering of swarms (Free, 1968). Also important is a less volatile material that is spread by their feet whilst walking or feeding at a source or whilst entering a hive (Ferguson and Free, 1979). This is colony specific and seems to come from intersegmental glands on the dorsal abdomen, not Nasanoff's glands; it is volatile and able to attract foragers to feeding stations.

4.2 Group slave-raiding

Leptothorax acervorum (Myrmicinae) is a small ant living in small colonies. In Europe it is preyed on by another ant of the same tribe: *Harpagoxenus sublaevis*. This lays siege to their nests and kills all the workers that try to escape by cutting off their antennae and legs, with their extra strong, though toothless, mandibles. Before starting a siege the *H. sublaevis* collect a group together by leading them out one at a time after a venom droplet 'call'. Brood is taken back and eaten but some pupae escape and hatch to give workers which help in the nest and are appropriately called 'slaves' because their labour benefits their hosts alone (Möglich, 1975; Winter, 1979).

In north-eastern America, *Harpagoxenus americanus* uses *Leptothorax longispinosus* and other species. Scouts search singly or in small groups which may even include a few slaves. If a single scout finds the entrance to a nest of the right prey species, it goes straight back, activates latent foragers and in less than a minute returns, dragging its sting along the substratum with a single file of nest mates following: they attack at once without a siege. If a group of scouts finds a nest some go back and recruit whilst others attack immediately. Apparently this species can vanquish its prey easily, probably by using a repellent from its mandibular glands. The prey workers and the reproductives run off and escape but *H. americanus* stops any workers with brood and takes it from them. Eggs and larvae are eaten but the pupae and sexual adults are not; clearly the pupae produce slaves and the sexuals have a chance to start more colonies: a mutually beneficial arrangement. Slave workers which take part in these raids cannot cause panic and never lead forays, though they fight their own species in the nest (Alloway, 1979).

Leptothorax duloticus in the same area, preys upon the same set of *Leptothorax* species but less skillfully; recruitment is slow, resistance strong and many ants die. They may move into the prey nest but instead of taking brood back remove the dead and consume much of the brood, including sexual pupae (Alloway, 1979). The behaviour of this species represents a stage in the evolution of predator/prey and master/slave relationships from an earlier state of interspecific competition (Wilson, 1975; also reviewed in Buschinger *et al.*, 1980).

Formica sanguinea preys upon and enslaves *F. fusca* and other related species of the subgenus *Serviformica* in Europe. The slave-makers are very specific about their slaves and it is unusual to find more than one species of slave in any one host colony; different races of slave may even exist (Dobrzańska, 1978). In the equivalent north American pair, *F. subintegra* (the master) and *F. pergandei* (the slave), scouts undoubtedly make trails, and lead recruits back to the target nest (Regnier and Wilson, 1971). Dufour's glands are unusually large in *F. subintegra* and they are used to cause panic amongst the prey workers who, very sensibly, disperse without resistance, frequently up the vegetation. There is no evidence of excessive use of this gland in *F. sanguinea* and the technique of attack is unknown, apart from the fact that they fight with their toothed jaws. In both species pupae are removed, and taken back; many are eaten but some emerge to provide more slaves. This is just a cull since larvae and eggs are left to mature.

The genus *Polyergus* has a cluster of species in Europe and north America that are obligate slave-makers. *P. rufescens* in Europe is not known to use trails for recruitment but *P. lucidus* in north America does and after arousal by a returning scout many workers move out along the trail. Hundreds of ants can take part in these raids. *Polyergus*, though they have sickle-shaped mandibles seem to avoid killing their prey and merely create panic whilst they lift prey pupae and take them back. Pheromones are probably used but the usual formicine poison (formic acid) is not; very little is, in fact, produced by *Polyergus* (Czechowski, 1977; Dobrzańska, 1978). The evolutionary convergence between the dulotic (slave-making) formicines and their myrmicine equivalents is striking.

4.3 Tunnels and tracks

Most successful ants and all termites, make tracks above or tunnels below ground from their nests into the surrounding land. These develop out of highly frequented trails and are kept in repair so that they last through the winter; where extraneous materials such as carton are used the term 'road' would be appropriate. They speed travel over normal rough terrain by smoothing the surface and removing obstacles and they are more easily defended against robbers. Where tracks exist, recruitment trails and displays connect to them rather than to the nest; they thus function as nest extensions and normally harbour a reserve of latent foragers. They are defended and the zones they serve tend to be exploited and monopolized likewise so that nest territory develops into food territory.

The examples which follow are arranged in order of complexity, first for ants then for termites. *Myrmica ruginodis* makes soil covers over its aphid clusters, and defends them. These spots of territory are not joined up structurally but the traffic between them encourages hunting in the same area so that the overlap of foraging zones is reduced. In this and other species of the genus the less volatile fractions of the Dufour's gland secretion are used for territory marking (Chapter 3). They comprise linear saturated and mono-unsaturated hydrocarbons in the taxonomically close pair *M. ruginodis* and *M. rubra*, whilst in another pair, *M. scabrinodis* and *M. sabuleti*, equally

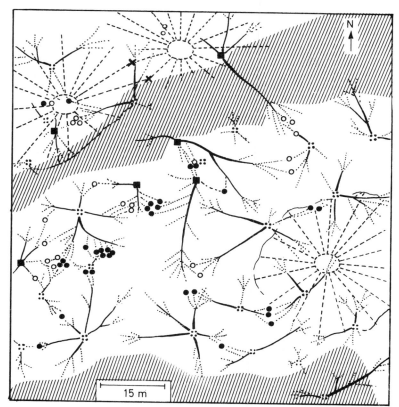

Fig. 4.3 Nest positions, tracks and trails or foraging areas for three species of *Pogonomyrmex*: *P. maricopa* (Pm), *P. rugosus* (Pr) and *P. barbatus* (Pb) in Arizona. (From Hölldobler, 1976a.)
Key: broken lines = *P. maricopa*, hollow squares = *P. rugosus*, black squares = *P. barbatus*, white circles = places where Pr and Pb fought, black circles = places where either Pr or Pb fought intraspecifically.

close to each other, they consist of farnesene and its homologues. The first two species cannot distinguish each other's territory as the trail pheromones in the sting gland are not specific (Cammaerts-Tricot *et al.*, 1977; Cammaerts *et al.*, 1981b). This does not mean that individuals cannot distinguish each other; they can, even at the colony level.

Monomorium pharaonis lays discontinuous trails with its poison gland and these, where they overlap near the nest, form recognizable tracks (Sudd, 1960). The trails are made of a mixture of alkaloids (at least five kinds) and a bicyclic unsaturated hydrocarbon: $C_{18}H_{30}$ (Ritter *et al.*, 1975). Venom is used to repel other ants from food and recruitment involves both display and Dufour's gland (Hölldobler, 1973). *Pogonomyrmex rugosus* and *P. barbatus* use their sting to lay recruitment trails of poison and Dufour's gland secretion: alkanes in the C12–C15 range both straight and methyl-branched predominate in the latter (Regnier *et al.*, 1973); the small oxygenated molecules of *Myrmica* appear to be absent; so too are the terpenoids. These trails develop into tracks which interdigitate (Fig. 4.3) with those of nearby colonies but do not cross or merge with them (Hölldobler, 1976a). Similarly with *P. badius*, this species has several tracks usually less than 5 m long of which one is a principal; from these they forage into a territory which is not entirely defended but clearly avoided by adjacent colonies (Harrison and Gentry, 1981).

Veromessor pergandei, an ant of the North American desert appears to have incipient tracks to seed collecting areas up to 45 m away; workers go out suddenly in the morning using the same route as on the previous day (Tevis, 1958). At first all traffic flows out from the nest to the food but when the temperature reaches 32° C, no more workers emerge and at 44° C all are underground. Many return without food. Such a collective foray during the brief phase when light and temperature are suitable argues for a systematic exploitation rather than a scout directed one but the nature of the organization is still uncertain (Hölldobler and Möglich, 1980). A regular rotation of about 15° each day has been suggested by Bernstein (1975) who thinks that higher up the mountains, where seeds are more abundant, the foraging axis is shorter and rotates more slowly.

Pheidole militicida, an ant with two modal worker sizes, makes 2 cm wide tracks which run for 40 m or so but only last a few weeks except near the nest (Hölldobler and Möglich, 1980). They structure a territory, for at the point where tracks of different colonies meet, fighting sooner or later breaks out, especially if workers get back to the wrong nest by mistake. Tracks are made by workers which collect near one of the many nest exits, and move to and fro slowly, laying sting trails of venom and Dufour's gland secretion; after several hours the group moves off along the new route. These tracks are made into new country, they do not develop out of well-used trails, as do those of *Monomorium*, they imply group exploration.

Tunnels are made by *Solenopsis invicta* in North American grassland; about six from each nest mound run some 1 cm below the soil surface, and branch repeatedly for some 20 m. Shafts rise to the surface at intervals. Their scouts lay Dufour trails that recruit without the need to display (Wilson, 1962) and appear to be mixtures of unsaturated hydrocarbons that are mostly species specific. *S. geminata*, and *S. xyloni* are similar to each other but different from *S. invicta* and *S. richteri* (Barlin *et al.*, 1976). Fresh trails are more attractive than stale ones, probably on account of volatile compounds. The related *Diplorhoptrum fugax* lives in nests of much bigger ants (commonly *Formica*) as a specialized predator and makes tunnels to their brood chambers to obtain larvae for food; those in use are marked with Dufour. The

Formica larvae are plastered with repellent venom that dispels their worker guards and contains an alkaloid, closely related to that of *Monomorium pharaonis* (Blum *et al.*, 1980; Hölldobler, 1973).

The most advanced Attini may travel hundreds of metres to obtain food for their fungi. Different species of *Acromyrmex* have trackways that vary in width and length and number. Workers of a desert species, *A. versicolor*, studied in Arizona by Gamboa (1975) forage singly around the nest for green vegetation after rain but otherwise build several trackways 10–14 cm wide and up to 17 m long which they clear of vegetation. Foragers leave these to cut leaflets by grasping the stalk with their serrate mandibles and moving around from side to side in a sawing motion. As usual the proximal part of a track is better marked than the distal part. Species of *Atta* with millions of workers in each colony build even larger trackways about 30 cm wide and up to 250 m long, which are cleared sporadically and cut skillfully as well as worn down by incessant traffic (Weber, 1972). To these trackways scouts return drawing trails with their poison gland which contains yet another alkaloid (Tumlinson *et al.*, 1971), able to recruit workers to new forage. The trail following of four species is described by Robinson *et al.* (1974).

Atta cephalotes lives in rain forest and workers walk over 200 m to cut leaf pieces; each fragment takes 2–3 min to cut. Unladen foragers will travel 2 m min^{-1} on uncrowded trails but only half as fast on crowded ones. In spite of this there is no evidence of traffic control. Daily foraging in Trinidad starts decisively at any time though usually at night and is in full swing after 2 h; it then goes on for an average of 7.5 h by day or 12 h by night. Forays are often unsynchronized between different nests and even between different arms of the same nest perhaps due to the vagaries of food discovery. Trails from different colonies do not overlap; presumably during their construction ants from different colonies just avoid each other for no fighting is seen. These tracks are thus the structural basis of a system maintaining dispersal and territory (Lewis *et al.*, 1974; see also Stradling, 1978a).

Under laboratory conditions, this species will lay a trail from green food to the nest; scouts run back and forth several times without cutting or carrying food themselves, though leaf-carrying workers will also lay trails (Jaffé and Howse, 1979). The number involved after a phase of rapid increase settles to a steady state because scouts start to collect instead of laying trails. This must be a response to the density of workers on the food or to the accumulation of leaf material in the outer nest chambers. More food is taken after deprivation and definite preferences are shown; in fact foragers are prepared to travel further for a more appetising food. Workers more active and intelligent than others may study forage availability, quality and location and attempt to guide others to the best patches. In addition to penning trails, *A. cephalotes* marks unfamiliar areas with the secretion of its 'valve' gland which lies at the base of the sting valves. They lay the scent by moving slowly to and fro with the tip of their abdomen on the ground, (cf. *Pheidole militicida*). Once an area has been scented with this the workers no longer show a characteristic gesture of insecurity called 'abdomen lifting'. This territorial secretion is both species and colony specific but rather astonishingly it only lasts an hour at 27° C and needs constant reinforcement (Jaffé *et al.*, 1979). In the driest parts of its range in east Texas, *Atta texana* makes tunnels from its nest which must protect foragers from desiccation. *A. mexicana* is even more of a desert species and in the Sonoran Desert makes tunnels 30 cm deep and 10–60 m long; when these surface they became visible tracks, which run for a further 70 m or so before their traffic disperses (Mintzer, 1979). Presumably they grow gradually with the population and are periodically orientated by shafts to the surface.

Only a few examples to illustrate the diversity and usage of ant tracks can be taken from the other sub-families. The interested reader can consult originals. In Ceylon, *Leptogenys occilifera* (Ponerinae) hunts worms and termites from permanent guarded tracks; on finding a worm a scout goes back to the nearest track and recruits help to pull it out and cut it up. Termites are piled up and taken away by workers recruited from the tracks. No display is needed; the smell of a recruitment trail brings hundreds of auxiliary workers to the scene (Maschwitz and Schörnegge, 1977).

The tracks of wood ants, *Formica rufa, F. opaciventris* and others, conduct workers as quickly as traffic permits to and from their food trees. Though pre-existing linear hard surfaces, such as human paths and tree trunks are used, workers also create their own tracks by a combination of incessant wear and deliberate clearance. As they use these right into the autumn it is not surprising that the tracks near the base nest survive the winter and can be re-developed in spring (Mabelis, 1979b). Foragers that find prey do not rush with it to the nearest track; they assess its quantity and quality first by prospecting in the locality and then lay a hind gut trail. On the track they belabour workers with a vigorous display, and these leave to start collecting. At dusk, workers rest in the nest but numbers rise again next morning, showing that memory of the locality and event is stored overnight (de Bruyn and Mabelis, 1972). Individual foragers stay on the same route day after day like honey-bees. The older ones, which are more attracted to light, tend to start work earlier in the day and stimulate the younger ones which are more ductile and will go where the traffic is dense (Rosengren, 1977). Old foragers tarry on the nest dome and guard it but they are a store of information and a reserve of labour, for if the foragers on one trackway are taken away the guards lead new young ones out to replace them. To block a track it is necessary to remove both the foragers and the guards and put a physical barrier across it (Fig. 4.4); then a new track will be made along the barrier until it reaches the next radial trackway (Zakharov, 1980). These wood ant tracks are defended against other colonies and species and comprise a truly

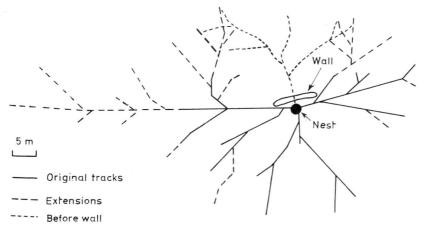

5 m

—— Original tracks

— — — Extensions

----- Before wall

Fig. 4.4 *Formica rufa* trackways from nest to trees with aphids. After blocking of one track with a wall, extensions were made to others and the blocked track withered. (After Zakharov, 1980.)

organized territory. They do not cross, and normally there is only one per tree; if two, they go on opposite sides of the trunk (Fig. 4.5).

Lasius fuliginosus is jet black and epigaeic though virtually blind. It has tunnels that run out from the nest for many metres, often just underneath a layer of moss but usually deeper, even in the roots of birch trees (Dobrzańska, 1966). These tunnels are lined with carton, like their nests of which they are thus linear extensions. At the bases of trees, 'stations' (which serve simply as shelters from sunshine, in which ants may stay for up to 2 h) are connected by tunnels and lined with carton (Fig. 4.6). This species also raids and collects pupae from the nests of other ants, e.g. *Formica pratensis*. Pupae are dumped in the stations or on the tracks rather like the piles of termites that ponerine ants make; other workers take them back in a sort of relay collection system. Dobrzańska suggests that, compared with *Formica rufa*, each individual shows more flexibility of task but less of locality, so that each area has its collection of multipurpose workers. The trail pheromone is species specific, stored in the rectal fluid and laid from a prey source to a track. It contains six normal aliphatic acids in the series hexanoic to dodecanoic acid (Huwyler *et al.*, 1975); workers respond to this without a display.

An entirely arboreal formicine genus *Oecophylla* in Africa and the Orient, hunts and collects honey-dew in a territory composed of several tree canopies (Fig. 4.7). It lays trails from a sugar solution bait with its rectal gland (Hölldobler and Wilson, 1978) and after display recruits major workers for defence with its sternal gland (Fig. 4.8). Most prey up to the size of a honey-bee can be caught and collected by one forager but if help is needed it does not lay a trail but uses its mandibular glands like some ponerines. These

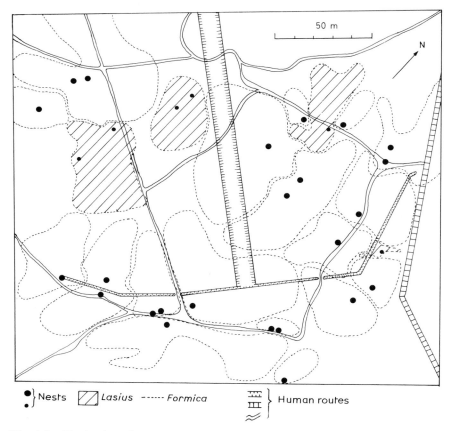

Fig. 4.5 Territories of *Formica polyctena* and *Lasius fulginosus* in woodland at Bierlap, Netherlands, 1968. (After Mabelis, 1979a.)

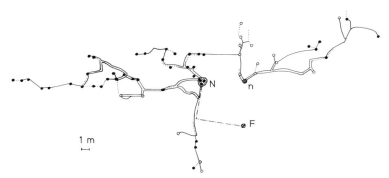

Fig. 4.6 Nest, tunnels and tracks of *Lasius fulginosus*. N, main nest; n, bud nest; F, nest of *Formica pratensis*. Pine trees are solid circles, birch trees open circles, carton-lined tunnels double line, tracks single line, trails dotted. (After Dobrzańska, 1966.)

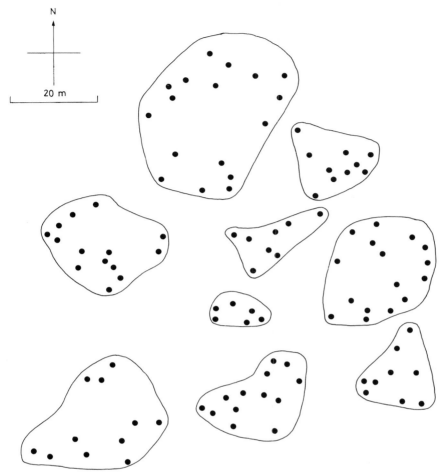

Fig. 4.7 Tree bases (black circles) and territories of *Oecophylla longinoda* in a reserve in Kenya. (After Hölldobler, 1979.)

produce a fragrant, musty odour from a liquid which exudes on to the mandible base when closed over the prey (Bradshaw *et al.*, 1979a). Is the use of this 'call' an arboreal adaptation or a primitive survival? Over 30 volatiles exist but hexanal and 1-hexanol are the principal ones. The aldehyde alerts, the alcohol attracts, and the minor constituents direct and focus worker behaviour. In fact, this is the order of their volatility and a recipient will first be alerted then attracted towards the site of action.

Iridomyrmex humilis (Dolichoderinae), the 'Argentine' ant, has tracks and makes trails to recruit workers with a gland near the anus called the 'ventral' gland (or Pavan's gland) which is found only in some species of this subfamily

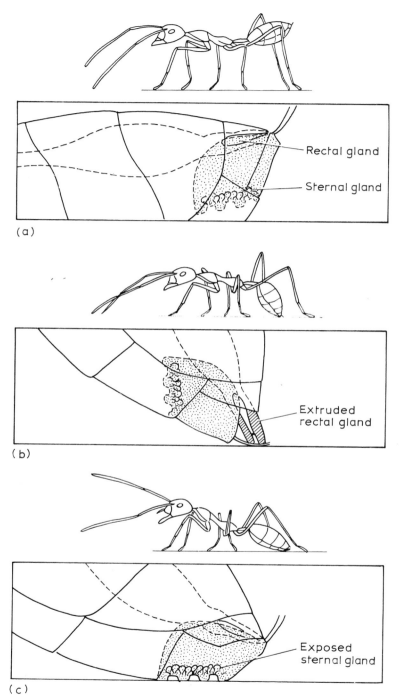

Fig. 4.8 *Oecophylla* workers mark new terrain with their rectal gland ((a), (b)) and recruit major workers to help defend it using their sternal gland (c). (After Hölldobler and Wilson, 1978, Vol. 1.)

(Robertson *et al.*, 1980). The mandibular glands also help in trail following as in *Oecophylla*. Another gland, the 'anal', homologous with the pygidial glands which are present in all ants except the Formicinae, contains a repellent which, if daubed on food, repels rivals, and may also be used to close worked-out avenues, an idea with great potential in traffic control.

Termites have probably always lived in tunnels and today Calotermitidae and Termopsinae still do. Others build them in soil to connect their nest with large immovable food items, e.g. logs. These tunnels have strong supporting walls inaccessible to predators and are smoothly lined to speed traffic flow. Some *Coptotermes* both compress porous soil and excavate compact soil, and often line their tunnels with faeces and a silt/saliva mixture (Lee and Wood, 1971). They are by no means direct links and appear to grow from one food item to another.

Stuart (1969) using a *Nasutitermes* (Termitidae) describes how tracks develop on a sheet of glass. The first termite to encounter the sheet 'jerks' and runs off to alert others by contact. Soldiers come out slowly and a column is formed which appears to lay a trail, each individual slowly advancing a little and then giving way to others as it runs back quickly, in the to and fro motion just described in ants. This is an exploratory trail which establishes a claim to new territory; it is a nest extension though it will never be used for brood. The sternal gland (Fig. 4.9), an intersegmental gland like the pygidial but ventral, provides the trail material and is present in all termites examined (Stuart,

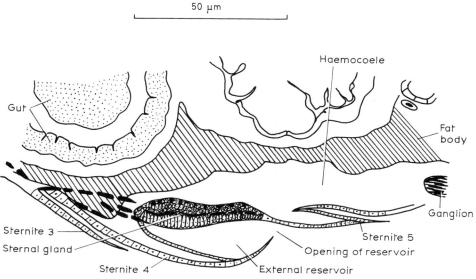

Fig. 4.9 Nymph of *Zootermopsis nevadensis*, longitudinal section through the sternal gland. (After Stuart, 1964, in Krishna and Weesner, 1970, p. 205.)

1969). If scouts discover food, more trails are added and if the food lasts long enough, faecal spots appear on the trail and form a visible track which soldiers line facing outwards; these tracks in nature are normally 1 cm wide and can be 50 m or more long. Then with big pieces of food to break up and collect, the track is slowly covered by pieces of chewed wood reinforced with anal cement, first near the food then near the nest and finally in between; soldiers stand on the walls until they are roofed over, and later reinforcement makes the walls opaque. So an exploratory trail buds recruitment trails and develops into a track.

The interesting thing is that a tentacle of the colony is protruded into new space which branches towards food items near enough for scouts to find. Of course it is difficult to know how much prior local knowledge is put into the choice of the direction taken, but the method resembles that of ants like *Oecophylla* and *Pheidole* which establish a familiar track from which they later hunt or collect.

Nasutitermes costalis will cover an artificial trail made with an alcoholic extract of faeces (Jones, 1980); the cover cuts off corners and straightens the trail out a bit. First, the workers make a skeleton with bits of solid material softened with saliva, then the cracks are filled in with faecal cement. A worker that finds a crevice turns round and forces its faeces in with a side to side movement of the abdominal tip. Having filled up these gaps, smaller irregularities in the surface are then polished whilst wet using the mouth parts and saliva; this leaves in the end a smooth, tough material. Though this sounds well planned it relies a lot on chance; a collision in traffic between two workers causes both to stop and remain still long enough to get a stimulus to build. Workers of different sex and caste and age respond differently.

Trinervitermes geminatus makes small mounds that function as shelters or 'stations' for mounting foraging expeditions. Tunnels are made underground to a food supply of dry grass and the termites open a hole in the surface. First, small soldiers emerge which stand facing out round the hole, then workers collect and move in and out until a compact mobile column, three to four individuals wide flanked by small soldiers is formed (this takes 30–45 min). This travels 2–3 m from the hole and then disperses over a grass tussock which may carry 500–800 workers with 40–50 cutting the same stem or blade of grass which is carried back in pieces 2–20 mm long. When foraging is over after about 2 h the workers return in a stream to the hole unladen whilst minor soldiers move rapidly around the column partly to repel ants and partly to round up stragglers. The hole is then plugged and any termites outside left to the ants. Foraging by different colonies appears to be synchronized and all holes are open within half an hour of each other (Ohiagu and Wood, 1976).

Some grass-eating termites are blind, e.g. *Trinervitermes bettonianus*. When they go out to reconnoitre in columns of workers and soldiers they lay

Plate 6
Foraging cover built by *Macrotermes* on twigs of *Balanites* in Kajiado district, Kenya. (Photograph courtesy of J.P.E.C. Darlington.)

trails that branch, and that are later reinforced with faeces to become tracks. These tracks enable foragers to return to base (a surface opening from a tunnel) since each bifurcation reduces the strength of the trail or track and gives a polarity to the information. When food is found a recruitment trail is made and attention drawn to it with head jerks; this is at least five times as effective as the exploratory trail by itself and recruits many new individuals (Oloo and Leuthold, 1979). Of course, the exploratory trail may be mainly faecal and the recruitment trail laid with the sternal gland which gives a qualitative difference between the two but this question is still open. In *T. trinervoides*, up to five workers can lay a sternal gland trail that lasts about 20 min; these attract a 100 or more new workers accompanied by soldiers that follow the scouts (Tschinkel and Close, 1973).

The genus *Hodotermes* (Hodotermitinae, Termitidae), unlike the *Trinervitermes* just described are both black and have eyes and are able to navigate celestially by day. They emerge from their tunnels irrespective of light and spend about a quarter of an hour searching, then establish a foraging direction to a determined and presumably 'agreed' area, usually within 3 m (Leuthold *et al.*, 1976). Along this trail they make pheromone marks that diminish in intensity away from the exit hole. At their destination they ascend grasses and cut pieces 2–6 cm long, usually dead, and pile them up; other workers carry them back. Thus, in spite of some visual potentiality they rely on an exploratory trail for navigation. On the whole it looks as though termites and blind ants with tunnels have converged behaviourally: both make exploratory chemical trails from nests or tunnels and use a special method of recruitment to food as well. Ants with celestial navigation may still make tracks but these are not needed for navigation, only to extend and determine the bounds of the territory and to speed and facilitate movement.

The way in which hypogaeic termites find their food is of interest. One would expect it to be through a chemical either dispersing, from the wood or litter or from a fungus decomposing it, but this is apparently secondary at least in Chihuahuan desert species (Ettershank *et al.*, 1980). Species of both *Gnathamitermes* and *Amitermes* go for the shade and coolness by day or the warmth retention by night, and then search further.

4.4 Nomadic foraging

The Dorylinae have evolved a technique of predation on invertebrates of about their own size that depends on changing the position of their nest frequently. Most are hypogaeic and blind, especially in the Orient and in Africa and they make temporary tunnels from their nests, in fact one could almost say that they spend their time in a mobile tunnel or track (if above ground) that links successive lodgings. From this base they search for food, make recruitment trails, collect food co-operatively, cache it temporarily on or near the track and then, after a spell of raiding, bring their juveniles, chiefly brood, and the queen, along to a new nest instead of back to the old. Workers are able to go out when only a few days old and take part in raids as 'beaters'; they flush prey for the older ones to capture. The track is made, to quote Schneirla (1971) by '. . . a relay trail laying process . . . ants in the van are not scouts but pioneers . . .'. Each ant, when it reaches the end of the trail, stops suddenly, slows down, and then advances slowly only a few centimetres at a time rubbing its abdomen against the ground. These tracks serve primarily for navigation and recall the exploratory tracks made by *Veromessor* and *Pheidole* and some termites. A track-laying column probably starts in response to a rising food demand from new larvae; it tends to move towards areas of good food supply, probably through the coalescence of recruitment trails. *Eciton hamatum* hunts in exploratory columns 100 m or more long that stretch in various directions from their shelter. Exploratory tracks are used for navigation and are made with hind gut material, as in termites, in short segments by a succession of workers, each proceeding for a few centimetres slowly and then returning. Recruitment trails made by scouts returning from the food, cause workers to react in a qualitatively different way and are probably made by secretion of the pygidial gland (Chadab and Rettenmeyer, 1975; Hölldobler and Engel, 1978). However, to avoid over-simplification one should note that the track not only guides but enhances response to the recruitment pheromones (Blum and Portocarrero, 1964). This 'taking the nest out to the food' and using all individuals co-operatively in a flush/capture process enables them to consume large prey such as whole nests of wasps or small vertebrates.

A raid by *Neivamyrmex nigrescens*, a nocturnal forager of this group, in Central America, begins with the renewed activity of peripheral old workers that make a raiding front which moves away and breaks up in response to scout recruitment to food. When growing larvae are present they move to a new nest site each day following an exploratory track that develops from the coalescence of food trails; the older workers outside and the younger inside (Topoff and Mirenda, 1978). These exploratory trails may depend on a tactile element too, for if chemical cues are weak, workers tend to go along the bases of rocks, tree trunks and other objects (Topoff and Lawson, 1979). However, the chemical recruitment trail is sufficient to start recruitment and bring in secondary recruiters that run faster. In this way a 'wave of arousal' spreads quickly as on the surface of a wood ant nest (Topoff *et al.*, 1980a, b). Moreover, laboratory studies have indicated, and field studies confirmed, that emigration is towards areas of a suitable nest site and a good food supply. Overfed colonies still emigrate but much less than underfed ones (Topoff and Mirenda, 1980; Topoff *et al.*, 1981).

African dorylines are more hypogaeic. *Anomma wilverthi* sends out columns that move under the cover of soil (Fig. 4.10), especially in sunny patches and on bright days (Raigner and van Boven, 1955). These are 2–6 cm wide with about 13 ants cm^{-2} walking over each other; they never run straight and tracks may bifurcate and rejoin a little later on. In this species, 36% of columns are underground but in *A. nigricans* 52% are. When an unorganized dispersal swarm leaves the nest it appears to start with a demand for food from larvae and the excitation spreads to the bigger workers on the outside, but the direction taken may be a matter of chance except that apparently they rarely go in an easterly direction! Individual ants run all ways sensing and feeling over a surface of hundreds of m^2 whilst presumably laying a rectal trail. This swarm gradually moves away and is supplied with new ants from the nest in a few columns for some time but eventually ants begin to come back with food and the direction is reversed so that any worker trying to leave the nest is pushed back by weight of numbers. *Anomma kohli* attacks nests of *Cubitermes* underground and slowly fights its way in over a period of several days (Williams, 1959). A species of *Dorylus* in the subgenus *Typhlopone* which makes undergound tunnels that only break the surface in an active hunt, is given to eating earthworms and other soil insects but also attacks other driver ants from time to time. Leroux (1979b) describes a long battle

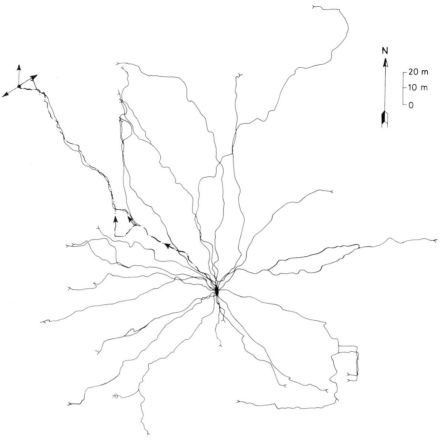

Fig. 4.10 The paths of raids by one colony of *Anomma wilverthi* in the Congo from one nest-site during June and July (complete lines). On 12–14 July it moved NW to a new nest (broken but thicker line). (After Raignier and van Boven, 1955, p. 128.)

undergound that began against a mobile column of *A. nigricans* and ended in their nest as they dispersed in all directions. The attacking *Typhlopone*, were inside for 5 h before they started to surface and make piles of *Anomma* brood which were later taken back to base in a column. Their workers have short, flat, cutting jaws that can bisect driver ants easily. After a 12 h raid, the relict colony of *Anomma* hid in a crevice: the queen is decapitated when found so the colony lasts only the few weeks taken for the workers to die; Leroux found no evidence that they could requeen.

In brief, communication enables the more active and sensitive individuals that have located a good food source to recruit help and exploit it before others from different colonies or species arrive. It is noticeable, though, that they do not rush at once back to base, they assess the quantity and quality of the food and then show a graded response. They activate, lay chemical direction trails, accentuate parts of the trails with other materials, lead nest mates out or allow them to go alone. In the case of the honey-bees the direction and distance can be communicated from the nest by dances; even in this case, however, the dance may only alert workers that are already fixed on a set of plants. Once a recruitment has built up a good population of collectors, the stimulus to communicate wanes and the number of foragers stabilizes.

The exploratory trail is used by both ants and termites and is probably built up after a series of scouts have returned from a profitable direction. It is a tentacle of territory, laid slowly and systematically, along which recruits move freely. It may, on the other hand, be a systematic means of exploring local terrain as when it is rotated from day to day. In doryline ants this has evolved, apparently, into a permanent nomadic way of life, in which food collection is largely concentrated into periods when larvae are growing. The use of communication reaches a bizarre extreme in slave-making where parties arrive at previously located slave nests and attack chemically so that they are able to carry off selected brood stages and leave the colony in a state to regenerate quickly.

CHAPTER 5

Cavity nests and soil mounds

All social insects are able to cut out a chamber in soft wood or soil or the pith of a plant and cover the access hole with a screen. Even species that build elaborate nests often pass through a stage when a single queen or a pair of sexuals makes a shaft and chamber and constricts the entrance. From this simple nest, made originally to provide shelter from weather and enemies and give a useful work surface, vast air-conditioned concentrations of population production have evolved. This has necessitated careful selection of raw materials, the evolution of glandular cements and in a few cases of the building material itself. Parallel with this increase in precision and complexity there has been an increase in co-operativeness during building. Though many quite complex nests, e.g. wasp comb, can be made by a single founder individual, others need group co-operation to give strength or warmth, and the use of various age groups with different contributions to make is the culmination of nest-making evolution and leads to the creation of large, subtly designed, structurally adaptable and independently supported nests.

There are two main styles of architecture: in one, many pieces of brood are kept in each chamber and are either free to move (termite larvae) or can be moved by workers (ants), which may sort them into life-history stages and keep them in different chambers; in the other, a separate cell is made for each piece of brood and though these can move about a little inside their own cell the workers do not move them from one cell to another (bees and wasps). In this chapter the structure and construction of a few selected nests is described; discussion of their micro-climate and their method of defence against enemies follows later.

5.1 Cavities and burrows

The tiphiid wasp progenitor of ants doubtless paralysed its prey and laid eggs on it in its burrow like *Methoca* does today (Wilson and Farish, 1973). Even the ponerine ant *Amblyopone pallipes* sometimes transfers its brood to its centipede prey (Traniello, 1978). Many ants still use ready made cavities in rocks, stones, galls and snailshells, e.g. *Cyphomyrmex*, and other primitive Attini (Weber, 1972). They are usually cleared out and their access hole is

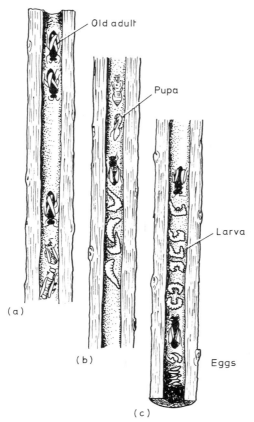

Old adult

Pupa

Larva

(a)

(b)

Eggs

(c)

Fig. 5.1 The bee *Braunsapis* (Ceratini) nesting in a dry pithy stem. (a) Constricted entrance; (b) mid section; (c) bottom of nest sealed. Young bees are paler than the old ones. (From Michener, 1974.)

often blocked with a guard-worker's head or a capping of debris. The primitively eusocial bees of the tribe Ceratini are the only ones to make chamber nests (Fig. 5.1); they enter hollow stems and clean and extend the cavity and constrict the entrance (Michener, 1974).

A very simple excavation is that made by queen ants or termite pairs after the nuptial flight and after they have removed their wings. Bulldog ant queens (*Myrmecia*) make an oblique shaft down to a brood chamber with a vestibule just below the surface from which they emerge to hunt at night, (Haskins and Haskins, 1950, 1964). *Manica rubida* (Myrmicinae) queens also come out to hunt; each starts a shaft against a stone and on leaving scratches a bit of rubbish over as a disguise. There is no vestibule as in *Myrmecia* and several narrow shafts go down into stony sand and end in small chambers (LeMasne and Bonavita, 1969). Attine queens often start to dig under a fallen leaf, digging quickly down some 20 cm depending on the soil stratification and moisture and then make a lateral chamber where they eject their infra-buccal pellet of fungal spores. The next day hyphae are present. Queens dart out quickly to collect food for these and reblock the hole afterwards. Two months later young workers break out and build a cone or turret of fresh soil (Weber, 1972). Vestibules exist even where queens no longer forage; *Solenopsis invicta* (Myrmicinae) queens do not forage yet the nest has a vestibule at the top and a chamber at the bottom (Lofgren *et al.*, 1975; Markin *et al.*, 1972). This gives the queen a choice of two temperatures. The termite queen is never alone and her mate helps dig and defend the nest against other males. Termites not only block the entrance with soil, they seal it with a quick setting proctodeal, and partly faecal, fluid (Nutting, 1969, p. 263); the walls may also be lined with faeces. *Allodontotermes* pairs kept in tubes in the laboratory dig co-operatively to make a vertical shaft 5–8 mm diameter, 8–10 cm long, down to a flat-floored horizontal chamber 2 cm long and 0.5 cm wide, that has smooth, shiny sides and is closed with a granular septum (Bodot, 1967). Even at this stage of evolution termites finish off their nests better than ants.

Ants frequently develop mature colonies in nests that are no more complex than those made by the queen alone, just replicated. This is the case with a large ponerine, nearly 2 cm long, *Paltothyreus tarsatus*, which lives in African savannah and cuts soil grains out of the subsoil and piles them on the surface. Their shaft descends obliquely to chambers no more than 50 cm deep (Lévieux, 1976). Many nests of unspecialized Myrmicinae are as simple as this but the harvester ants living in dry steppe/savannah build quite well differentiated nests. A seed harvester living in dry, but not loose, dune sand in North America, *Pogonomyrmex occidentalis*, makes about five almost vertical shafts, which branch several times and descend at least 2 m (Lavigne, 1969). The nest has a number of openings that vary seasonally in position and number: most in June, least in September. Oval chambers at various levels connect with each shaft; the top ones are large and used as nurseries and the bottom ones are small and used for hibernation. *Messor ebeninus* in alluvial plains in the Sahel region of Africa has a network of ducts (both horizontal and oblique) 3–5 mm diameter in the top 0–55 cm of soil that are used for seed storage. Between 60 and 100 cm down there are larger chambers (4–6 cm diameter) and at a depth of 180 to over 220 cm a set of brood/queen chambers (Tohmé, 1972). The quantity of galleries reflects the size of the colony and ranges from the single shelter cut by a queen up to vast excavations under areas of 50–100 m² that descend several metres. The workers clean out the top galleries in spring and excavate more deeply whenever it rains, as the wet rainwash soaks into the soil, softens it and makes it easier to carry clumps of soil grains up and away. In this they are helped by a beard of hairs. A relative, *Messor arenarius* is unusual in that it lives in dunes of blown

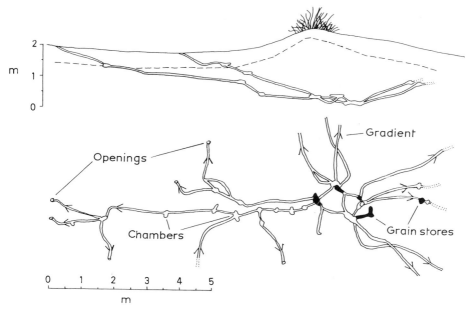

Fig. 5.2 Section and plan of nest of *Messor arenarius* in sand. (After Délye, 1971.)

Within the figure: 2, 1, 0 (m scale); Gradient; Openings; Chambers; Grain stores; 0 1 2 3 4 5 m

sand (Fig. 5.2). In the Sahara, over short distances, it digs with its feet or by holding its mandibles close together and travelling backwards like a mechanical excavator; over greater distances it makes a ball of sand with its forefeet, jaws and undertucked gaster and carries it out in a lump adhering to its beard. It dumps the lumps to form a striking semicircular crater. The shafts are very oblique and pass through 10 m to descend only 2 m. Side galleries, 1 cm high, and 3–5 cm diameter are usually empty but the workers gather in them before going out to forage. They also remove blown sand from their surface galleries each evening. Their granaries, lower down at 2 m, are large chambers 40 cm × 10 cm but only 2 cm high, often full of grass seeds. Brood is kept in nearby chambers (Délye, 1971). Attini excavate nests of very variable complexity. The *Cyphomyrmex* lives in a snail shell and has already been mentioned; many *Acromyrmex* live in soil pits roofed over with thatch and *Atta*, at the other extreme, excavate huge subspherical chambers 20–30 cm diameter, several meters below the surface (Weber, 1972). The spoil is thrown out over a wide area and in the end a hillock develops through which new ducts are passed. The fungus comb has finely malaxated vegetation added on top and waste removed below.

In the simple formicine ants, (Sudd, 1969, 1970a, b) digging consists mainly of making a grab at a soil surface using the jaws and then either carrying large grains away one at a time or letting small grains accumulate until a pile can be raked together with the forelegs. *Formica fusca* starts against a vertical surface such as a stone or in a hollow and digs a shaft straight down altering its direction only if large pebbles get in the way, and removing large, loose projecting grains; finally gaps and crevices are filled with medium-sized sand grains. Trowelling with closed mandibles is rare. In European *F. fusca* the opening on the surface is single and covered with debris thought to delay discovery by the dulotic *Formica sanguinea* and other social parasites (Dobrzanska, 1978). *F. japonica*, one of several sibling species in Japan plugs its nest opening with soil in winter, opens it again in May and increases the number of openings until July, when the sexuals are let out from a large subsurface vestibule. In the top 20 cm, shafts and chambers 2.5–5.5 cm long and 1–2 cm high are numerous and complicated; some shafts descend into the subsoil as deep as 3 m in a grassland form but are much shallower in woodland forms (Kondoh, 1968). In Europe the northern *F. lemani* makes superficial galleries whilst the southern *F. fusca* digs deep shafts and chambers, though whether this is a response to differences in latitudinal insolation or an adaptive species characteristic is uncertain.

The interaction between *F. fusca* and two of its slave makers whilst building has been studied indoors by Sakagami and Hayashida (1962). Soil was packed into one end of a tube and the amount moved measured. *Formica fusca*, a slave, showed a declining individual output as density rose; *Formica sanguinea*, a facultative slave-maker, could work by itself, though it was less effective than *F. fusca* and its rate declined more with density; finally *Polyergus samurai*, an obligate slave-maker, was unable to work at all and neither stimulated nor reduced the output of *F. fusca*. *F. fusca* with a high individual output and a relay system of moving soil particles is a good worker and a desirable slave. The number of idle workers increases with density, but in a limited space there is an optimum density for work and for some to stand aside rather than interfere or obstruct others is efficient. Individuals of above average activity, able to work at above average density may be called 'leaders' (Chen, 1937) or 'activists' (Wallis, 1964).

Camponotus is a very diverse genus; it has evolved from a single species in the oligocene (Baltic amber) to become the commonest genus today, (Brown, 1973; Wilson, 1976d). In Ivory Coast, West Africa, *C. acvapimensis* in herbaceous savannah, excavates to a depth of 45 cm to avoid dry seasons but has a widely dispersed nest underlying at least 15 m² of surface and Lévieux (1971) has calculated that 0.5% of the soil volume is mined in the level 0–25 cm by this ant; clearly a species with ecological impact. In cooler climates carpenter ants excavate nests in dead, but hard, dry wood. *C. vagus* workers tear strips of wood out along the grain but the ants also make partitions of wood spoil to form irregular chambers leaving the bark and the hard centre intact (Benois, 1972). Very thin wood septa are left in some species, and their skills as carpenters are formidable.

Having a fixed nest does not suit the life style of all ants; some are opportunistic. *Anoplolepis longipes* (Formicinae), which is well able to cut itself a nest out of soil, frequently lives in cavities under rocks, in drains or just under fallen logs and leaves. They even climb coconut palms and lodge in the crowns. As their colonies are polygyne (40 queens and 3800 workers on average in the Seychelles) this enables them to move to new sites when necessary and even to divide (Haines and Haines, 1978). The army or driver ants are another example: Ecitonini in American tropics camp by day under logs and rocks where they are shaded from the sun (Schnierla, 1971).

Termites that excavate hard wood (Calotermitidae) or soft wet wood (Termopsinae) cut out tunnels and then plaster them with faeces which are more fibrous and plastic than ant faeces as they do not filter their food. They often make partitions by placing a grain of sand and cementing it with faeces to create a lenticular fabric up to 2 mm thick (Lee and Wood, 1971). *Zootermopsis*, when making tunnels in wood, cuts pieces out, drops them and pushes them backwards with its legs; often a line of several individual works in series. Eventually these pieces can be used to fill crevices or even be eaten. In building, a common technique is for a worker to take a piece of wood or a dry faecal pellet and go to the construction site; there it first deposits a drop of liquid faeces with its abdomen and then turns around and adds its solid to the wet mass; and so on repeatedly. Stuart (1969) suggests that builders recruit help by banging their heads on the substrate. Several individuals work together responding to situations created by others as though they had created them themselves, a procedure called 'stigmergy' by Grassé.

The Macrotermitinae (Termitidae) mix a secretion of the salivary glands with clay and then apply it to the lining of galleries and the construction of partitions or use it as a matrix in which to embed sand grains (Lee and Wood, 1971). Many live underground making chambers connected by a maze of galleries; the chambers are either for reproductives (the male and female pair) or for fungus combs or are nurseries. Josens (1972) found that the pair usually continue to live in their foundation chamber and the fungus combs though widely scattered, are concentrated in the food area often quite a way from the reproductive chamber. In mature nests of *Ancistrotermes cavithorax* the reproductive cell is 5 cm long, 3 cm broad and 2 cm high, has a flat, horizontal floor and a vaulted roof. The walls are smoothly plastered with a reddish, clayey soil obviously selected for its fine grains and coated with faeces to give a thickness of 3 mm and no corners. Access to this cell is by two narrow galleries through which the reproductives cannot escape: all food that they need is carried in and all eggs that they lay are carried out by the workers. Chambers with fungal combs are only a little bigger and are likewise plastered with clay but have a concave floor which gives access to the underside and allows air to circulate; in *Pseudacanthotermes militaris* the combs stand on conical legs or pillars (Fig. 5.3). This species piles vegetable debris in

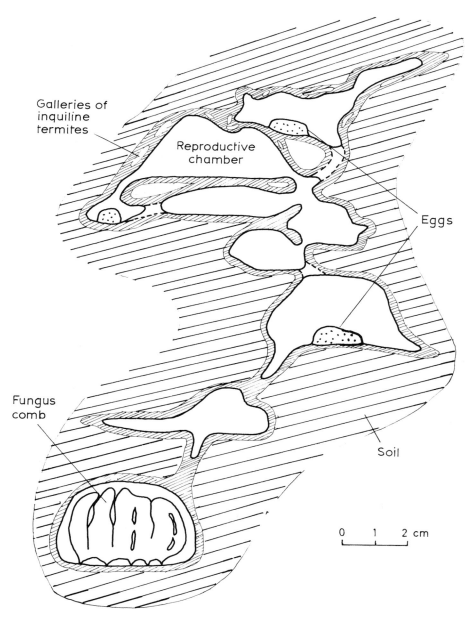

Galleries of
inquiline
termites

Reproductive
chamber

Eggs

Fungus
comb

Soil

0 1 2 cm

Fig. 5.3 Vertical section of a soil excavation of *Pseudacanthotermes militaris*. The reproductive cell has galleries of an inquiline *Adaiphrotermes* in its walls. (After Josens, 1972.)

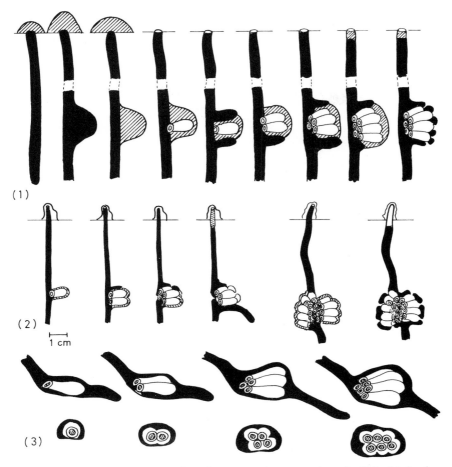

Fig. 5.4 Stages in excavation of nest burrow and construction of cells in this by three species of *Lasioglossum*. 1. *L. malachurus*; 2. *L. linearis*; 3. *L. nigripes*. Black, cavity made; hatched, loose excavated soil; white, the cells themselves. (After Plateaux-Quénu, 1972.)

chambers only 1–2 cm under the surface but like other termites and, unlike the Attini ants, it makes its combs of faeces. Bees in the tribe Halictini whether social or not make long shafts skilfully furnished with cells (Michener, 1974; Plateaux-Quénu, 1972). The bees smooth and compact each cell inside with their mandibles and then line them with a secretion of the salivary and Dufour glands; there is no wax. The access hole is usually constricted to fit a single bee. *Lasioglossum malachurum* makes cell clusters and then cuts a space out around them so that they give the appearance of having been constructed in a chamber (Fig. 5.4). Parasocial bees share a main shaft but have their own side shafts and cell groups.

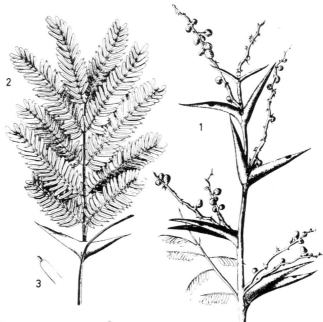

Fig. 5.5 A species of *Acacia* inhabited by a species of *Pseudomyrmex*. 1, swollen thorns with holes cut by ants as entrances; 2, a leaf with an extrafloral nectary at its base and 'Beltian' bodies at the tip of each leaflet; 3, these enlarged. (After Wheeler, 1910.)

The use made of twig cavities as nests is astonishing in tropical savannah/ forest. One tree, *Cordia alliodora*, (Boraginaceae) harbours 54 species of ant (Wheeler, 1942) and one ant, *Azteca longiceps* (Dolichoderinae) occurs in 85% of stems of various plant species. It constructs carton partitions to make chambers. This ant/plant relation has become very specific and obligatory in a few cases; species of *Pseudomyrmex* (Pseudomyrmeciinae) inhabit *Acacia* plants (Leguminosae), some of which have evolved hollow, swollen stipular thorns (Fig. 5.5). The ant queen cuts her way in and founds a colony that grows eventually to fill the whole bush. Likewise *Tetraponera aethiops* (same sub-family) lives in *Barteria fistulosa*, an understorey shrub in tropical rain-forest. Queens enter slightly swollen hollow branches 10–20 cm long and the growing colony moves into new ones as these appear; they find coccids and fungus already inside (Janzen, 1972). All the queen does is identify the tree

---

Plate 7 *Opposite*
The stem of *Cecropia peltata* has a weak spot through which ants of genus *Azteca* (Dolichoderinae) penetrate into the hollow pith (*above*). They clear this out and make horizontal partitions with the debris (*below*). (Photograph courtesy of D.J. Stradling.)

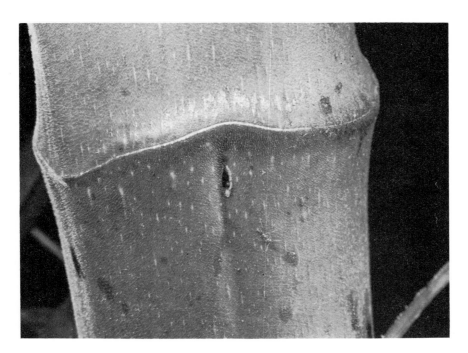

species and cut in, an action which is well within the scope of most presocial insects.

Although this section has only considered nest excavation it is clear that a great variation exists in the skill involved. Some make a simple shaft and chamber, others have a vestibule as well and others add camouflage on the hole or even seal it. Further development of this system enables the chambers to differentiate into granaries, fungus gardens, nurseries or hibernation chambers. Termites add a reproductive chamber and furnishings such as fungus combs. These all vary in the extent to which walls are thickened, corners eliminated and smooth linings given. Partitions can be made too.

5.2 Soil mounds

Spoil heaps acquire a skin under rain impact in which plants establish; after stabilization these heaps are re-excavated to form mound nests. Many ants and termites by-pass this stage and build a mound directly using stones, logs, plant shoots or branches as a scaffolding or more lasting mechanical support. Others rely less on these and build a cellular foam of soil or wood chewings that is self-supporting. The advantage of such a purpose-built mound is that it can avoid stones and branches and be easily sealed against insect enemies.

Simple myrmicine ants build small mounds up to 10 cm high in herbs and either leave the inside fairly free of partitions (*Myrmica ruginodis*) or make many small mud-plastered chambers inside (*M. scabrinodis*). The latter system makes the nest more resistant to desiccation or invasion. *Myrmica* certainly defecate a brown liquid into crevices and there is some evidence that this sets into a flaky, brittle cement, but how it is used is in doubt. The more advanced *Solenopsis invicta* forms mounds with some 40 litres of chamber space in them. The first small worker to arise from the founding queen cuts chambers and after 3 months the mound can be 7 cm high with the same diameter. Additional shafts, about 10, are made and foraging tunnels radiate from them so that by 5 months the mound is sponge-like and invaded by vegetation. After 2–3 years it is dome-shaped and overlies a set of chambers and shafts that get fewer deeper down until they finally send shafts to the water table. In clay soils steeper sides are possible (Markin *et al.*, 1973). *Tetramorium caespitum* in temperate Europe forms a similar mound and tunnel system, though in sandy soils this is very easily eroded and most nests lie just beneath the surface crust between bushes (Brian and Downing, 1958).

Lasius is variable: *L. alienus* never makes mounds, *L. niger* either uses the mounds of *L. flavus* or nests under stones. *L. flavus* rejects stones as nest covers and piles soil over them. In European meadows at least, this species builds mounds into which the roots and rhizomes of living plants invade, stabilizing and strengthening the structure. Clay-rich minerals are taken selectively from the subsoil to produce steeper and firmer mounds than sandy material. Many herbs exist only in or on these mounds either because they grow up from soil-covered shoots, (e.g. *Helianthemum* or *Thymus*) or because they seed on to the loose new top soil (e.g. the grass *Festuca ovina* and various annual crucifers). These plants consolidate the mound by binding the soil granules with rootlets, but some grow too fast and make too much shade, e.g. the grass *Brachypodium pinnatum*. Rabbits (*Oryctolagus cuniculosus*) are attracted to the mounds probably because they feel warm, give a clear view around and offer a chance to graze attractive grasses (such as *Festuca*); even large grazers walk between hills and graze from the tops (King, 1977). *Lasius* ant-hills do not grow as fast as those of *Solenopsis invicta*; Waloff and Blackith (1962) give the height (m) of *Lasius* as 0.9 times the age in years. This means 22 cm after 10 years, when the maximum diameter is about 84 cm. For the first 2 or 3 years the growth rate is high and the ants excavate a vast system of galleries and chambers down to 2 m in well-drained deep soils. *Lasius flavus* is unable to make mounds on slopes or shallow soil; the north American form does not build mounds at all (Wilson, 1955) yet its nests are water resistant and can be found along stream edges (Talbot, 1965). In European salt marshes *L. flavus* mounds rise above water level and carry *Festuca* and rabbits and on the southside even have a Mediterranean shrub (*Frankenia laevis*) (King, 1977; Woodell, 1974). Even more remarkable are the dolichoderine ants that live between tide levels in a Mexican halophyte-covered mud flat (Yensen *et al.*, 1980); as the tide rises, foragers retreat and other workers bring up soil particles to the rim of the nest crater. They cease as the tide floods over and reappear when it retreats, surviving submersion in the air-locks inside.

Termites, like ants, can be arranged in a sequence that broadly represents their evolution from simple excavators to mound builders. Unlike ants they have evolved enormously complicated, enduring and gigantic buildings (a better word than mounds in many cases) which may have lasting effects on the ecosystem (Fig. 5.6). Small mounds of soil, no bigger than those made by many ants, are frequent in tropical mountain grassland – the humivore *Cubitermes* in Kenya makes regularly dispersed flat-topped mounds of mineral grains, fine soil particles and digestive residues laid down in layers, and covered in herbs and grass (Harris, 1955). *Amitermes vitiosus* in northern Australia builds mounds 0.5 m high on shallow or sandy soils but in soils with more clay they reach 1.5 m high and are conical and free of vegetation. Of *Nasutitermes exitiosus* and *Coptotermes lacteus*, Lee and Wood (1971, p. 30) say that the workers place sand grains in position and cement them with clay and saliva giving an open, spongy structure that is later filled in with transported soil particles to give an amorphous appearance; the more this is done the harder the resulting product. The outer casing of *Macrotermes bellicosus* mounds has a similar appearance, and again the intensity of re-packing affects the final hardness. In *Microcerotermes*

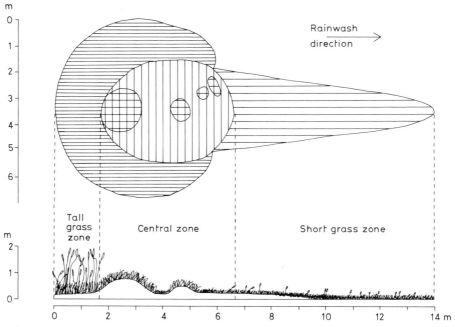

Fig. 5.6 Vegetation zones around a rainwash eroding nest of *Odontotermes* on sloping ground in the Loita plains, Kenya. The break up of the nest is started by Aard-varks (*Orycteropus afer*). (After Glover *et al.*, 1964.)

arboreus, faeces are also added, as they are in the nursery and wall linings of *N. exitiosus* and *C. lacteus* just mentioned. Others have nest fabrics predominantly of fibrous faeces without soil, e.g. *Mastotermes darwiniensis* whose pellets are dry and give a granular wall structure when pressed together (Lee and Wood, 1971, p. 35).

Nest construction by *Nasutitermes costalis* (Nasutitermitinae, Termitidae) has been studied in the laboratory by Jones (1979). Expansion of the nest is episodic and not to do with apparent need; as much as 10% can be added in a single operation. Clusters of holes are first cut in the outer envelope, soldiers and workers come out to explore and within minutes the workers start building on the surface bumps. Where workers have access from all around they make pillars 3 mm diameter rising to about 6 mm; then a roof disc is started. Walls are made from one side only and are virtually metamerized pillars. They spread out on top, and meet others and either fuse after height adjustment or else divide. Finally, an envelope which may contain spines and ventilation chimneys is made to cover the whole structure and enclose some space as well. Soldiers stand defensively; they do not build. In *Drepanotermes* (Termitinae) only late instar foragers build but soldiers come outside for defence (Watson and McMahan, 1978).

Plate 8

Above: Soil mounds of *Lasius flavus* (Formicinae) supported by living herbs and grasses in Dorset, England. They tend to develop in lines along the margins of cattle tracks in grazed limestone grassland, and form vertical sides where the soil is rich in clay.

Below: Soil mounds of *Lasius flavus* in sandy soil in Dorset, England, join the soil surface at an obtuse angle due to poor cohesion between the soil particles with consequent erosive slip. They frequently carry ericaceous shrubs on top.

Cavity nests and soil mounds 87

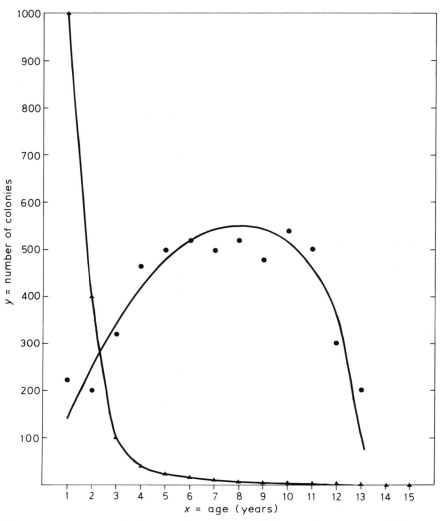

Fig. 5.7 *Macrotermes bellicosus* colonies; survival (l*x*), ⌂, and expectation of life (e*x*), ●. (From Collins, 1981a, Table 2.)

The most spectacular mounds are made by Macrotermitinae. They can be either round mounds 20–30 m diameter at the base and over 6 m high with a covering of vegetation or monolithic spires with steep sides; the latter may reach 8–9 m high on a base no more than 2–3 m diameter. The survival curve of one, *Macrotermes bellicosus* is shown in Fig. 5.7. They start underground where a new couple have dug a shaft and chamber about 20 cm deep and reared a few small workers (Grassé and Noirot, 1955, 1961). These workers take down wood and build a miniature fungus comb seeded with fungi from

Plate 9
Overdispersed 'Disc' mounds of *Odontotermes* sp. in Savanna near Kajiado. The soil between mounds is a deep-cracking clay with thin patchy vegetation, that on mounds is loose friable and crumbly and covered with a dense grass turf. (Photograph courtesy of J.P.E.C. Darlington.)

adult faeces. This is enclosed in an envelope of clay 1 mm thick with a hole at the top and takes only a night to make; in fact, the whole microcosm can be established within 100 days. The nascent mounds break the bare surface soil after rain and grow at a linear rate (Pomeroy, 1976). Grassé (1944, 1945, 1949) has described stages in this growth and development (Fig. 5.8) into the final form which may, according to Collins (1979) take either of two designs on the same soil with the same drainage. This difference in design may be due to different species for one lives over 1700 m and the other below 1400 m (Werff, 1981). In one type, the brood nest with the reproductive cell, the fungus comb and galleries for young, rests on conical pillars which stand in a cavity, called the cellar, enlarged out of the original soil excavation; open chimneys arise from this (Fig. 5.9). In the other type built below 1400 m and called 'cathedral' from its pointed hollow spires which can be 6 m high and 3–4 m diameter at the base, the reproductive cell, fungus comb and brood galleries together form a spherical assemblage at about ground level resting on a plate 3.5 m diameter. This plate is supported by a single pillar, though not actually attached to it. From the cellar very large shafts up to 20 cm diameter descend into the soil as deep as 3 m and are probably used for

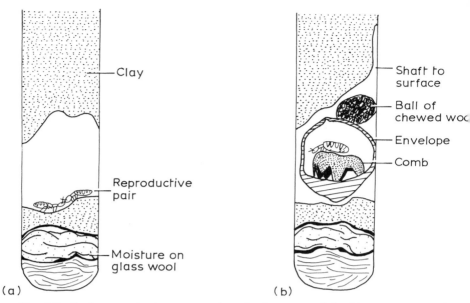

Fig. 5.8 Early stages in the construction of a fungus comb by *Macrotermes natal-ensis*. (a) Reproductive pair in chamber excavated in clay; (b) Fungus comb surrounded by envelope and ball of chewed wood all made by workers that have broken through to the surface. (After Grassé and Noirot, 1955.)

mining clay. Valves or flat spirals which are up to 2.5 cm thick and covered with salt crystals hang from the plate into the cellar. The mound wall is of clay and carton, perhaps faecal in origin and the inside is mostly clay mixed with saliva but incorporating some lime (Grassé and Noirot, 1961). Even in sandy deserts like the Kalahari the termites collect these materials selectively (Bouillon, 1970). *Macrotermes michaelseni* uses more superficial material than the deep-mined clay of *M. bellicosus* (Leprun and Roy-Noël, 1977). Their workers will build a cover over their queen if she is exposed in the laboratory (Bruinsma and Leuthold, 1977): first, they make pillars 1.5–2.0 cm high around her and then a nearly horizontal lamella to create a vault with only one or two passages in it. In a slow stream of air, to which they are very sensitive they build nearer to her on the upwind side which suggests that a part of their recognition is chemical.

Much remains to be discovered about how termites organize their building. Apart from the great pyramidal mounds, they make spheroidal nests, e.g. *Apicotermes* (Termitinae) in cavities underground. These are regularly and intimately designed with floors connected by ramps and a cortex of galleries under a perforated skin the pores of which are species characteristic (Noirot, 1970).

The theme of this chapter has been that in the beginning social insects lived

Plate 10

Above: Mound nest made by *Macrotermes michaelseni* in African savannah. Large ventilation holes (a), and smaller holes from which sexuals have flown recently (b), are both visible.

Below: A section through a mound nest showing three zones. The central zone normally houses brood rearing chambers (the nursery). Around this are distributed the chambers for fungus cultivation (see also Plate 1). The outermost layer contains ventilation shafts. (Photographs courtesy of J.P.E.C. Darlington.)

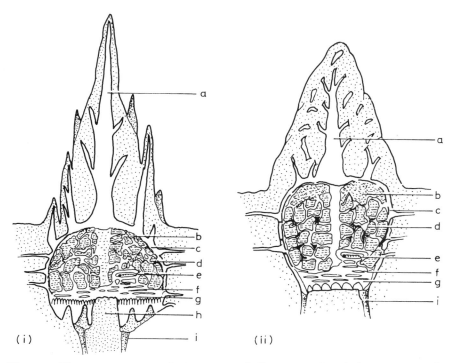

Fig. 5.9 The two nest types of *Macrotermes bellicosus*. (i) Spiral plate; (ii) normal type. a, mainshaft; b, food stores; c, envelope; d, fungus combs; e, reproductive chamber; f, galleries with larvae; g, base plate with either spiral (1) or supporting cones (2); h, pillar (in 1); i, caves. (After Collins, 1979.)

humbly in cavities either ready-made or self-made and that they adapted these for coming and going, paying special attention to the entrance hole and smoothing the walls. Living chambers were placed at various levels where possible and came to be differentiated into those for brood or forage collection with, in termites, the addition of a cell for the reproductive pair. Wasps and bees started social life with much more developed building skills than ants and termites, and queens of annual wasps can still build, unaided, quite elaborate miniature nests.

Co-operative nest building at its simplest perhaps involved relay chains of workers, some cutting at the faces, others carrying away debris, and from the pile that resulted mounds grew into the surrounding vegetation which provided support. These mounds were, in turn, excavated, then purpose-built and came to provide a micro-ecosystem of higher metabolic activity than average for the area. In termites, very intricately designed nests were produced, often of enormous size and durability.

Plate 11
Above: Mounds of *Odontotermes* sp. in the Loita Plains, Kenya, to show chimneys.
Below: The fungus comb of *Odontotermes monodon in situ* with air ducts painted white. (Photographs courtesy of J.P.E.C. Darlington.)

Cavity nests and soil mounds   93

CHAPTER 6

Nests of fibre, silk and wax

6.1 Mounds of vegetation and tree nests

Soil is not an easy medium to use in trees and in both termites and ants the proportion of vegetable fibre or masticated wood (carton) has been increased. In the former group, only genera of the subfamilies Amitermitinae and Nasutitermitinae (both Termitidae) have achieved this. Nests often start as galleries in dead, dry branches and retain the usual shape with a thick-walled reproductive chamber, and an envelope with reduced access. Some species of *Bulbitermes* and *Lacessitermes* make spheroidal carton nests in bushes within 1 m of the ground.

Both main subfamilies of ant have succeeded in making their own nests in trees as distinct from using prefabricated hollow stems and thorns. The genus *Tetramorium* (Myrmicinae) does this in the tropics (Bolton, 1980); *T. aculeatum* in West Africa, where foliage is dense, fills out the spaces between leaves with 'a rough-looking light mass of decomposed vegetable matter' (Wheeler, 1922, p. 188) that often acquires a binding fungal mycelium. *Crematogaster* (also a myrmicine genus) ranges from soil to trees. The first stage is when soil nesters build against trunks and ascend trees to plant bugs which they cover with a mixture of puddled clay and vegetable fibre, perhaps softened with saliva (Soulié, 1961). Arboreal species enlarge these covers and move in with the bugs. *C. impressa* nest in twigs and grass stems in regularly burnt savannah (Delage-Darchen 1974), but when leaves develop they make small nests (up to 8 cm wide) with partitions out of a mixture of loosely held vegetable fibres and mineral fragments. Improved carton is obtained by using less clay, and chewing the vegetable fibre more whilst adding saliva which may include mandibular gland secretion (Delage, 1968). The nests are usually slung around branches with a main entrance underneath and are divided into

---

Plate 12 *Opposite*
Above: A carton nest of the dolichoderine ant *Azteca chartifex* in Trinidad.
Below: A carton nest of *Nasutitermes* sp. in a citrus tree in Trinidad. Both species use wood fibre as the basis of the nest, but the cement used in the construction of the nest originates in the salivary gland of the ant and in the faeces of the termite. (Photographs courtesy of D.J. Stradling.)

chambers by thin partitions that give a fairly constant floor to ceiling distance but show no systematic replication, and create the impression of a randomly partitioned space. Chambers become more like galleries towards the outside and form a distinct cortex. The envelope is toughened and sometimes covered by overlapping water-shedding scales.

C. longispina mentioned earlier makes carton nests in which the epiphytic vine *Codonanthe* fastens its roots; this helps, probably, to bind the nest material together, for if the *Crematogaster* is replaced in natural circumstances by *Solenopsis picta* the nest disintegrates, the vine's roots are exposed and it falls off.

The formicine *Lasius fuliginosus* cuts galleries in rotting wood and makes carton partitions which extend out into foraging tunnels and stations at the base of the trees in which they forage. Carton is made by softening wood fragments with honey-dew or nectar: mandibular gland fluid is not used though this was once suspected (Maschwitz and Hölldobler, 1970). The wood ant, genus *Formica* is famous for its huge mounds up to 1 m high with a base 2 m diameter; *F. ulkei*, one of this group uses a mixture of soil and vegetation and forms a hard crust covered in vegetable debris. The dome is asymmetric since it is enlarged on the sun-facing side. European wood ants, (*F. rufa* and sibspecies) build even larger mounds of twigs, leaf petioles, bud scales, conifer leaves and other plant remains. Some workers dump them at the top of the mound, whilst others arrange them so that they lie flat. This, presumably, reduces the tendency to slide down as well as filling out any hollows, yet rain and wind erode the mound top and pieces have to be continually replaced (Chauvin, 1970). It is not an aligned 'thatch' but it makes a convex surface that sheds water; as no cement is used skill is needed in placing the pieces stably and Dobrzański (1971) has shown that workers improve with experience.

Genera in the tribe Camponotini have evolved diverse techniques. *Camponotus* excavates in soil, in hard dry wood or makes carton nests in trees, and several, e.g. *C. senex*, use larval saliva for binding nest material together. It is thought (Sudd, 1967) that this behaviour derives from the common formicine practice of burying fully developed larvae before they spin a cocoon, since they need an external attachment on which to fix the silk threads. The silk is synthesized in the same labial glands that earlier in larval life secrete proteases. *Dendromyrmex* mixes silk with the vegetable fibre used to make carton and some is used by itself to line the nest and veil the entrance. Larvae often spin unaided by workers and make a communal web in which each pupates in a thin cocoon but sometimes workers hold them and move them about whilst they spin a fabric which strengthens the nest carton. In this way the workers come to take an active part in guiding the application of silk (Wilson, 1981).

Species of the genus *Polyrachis*, mainly tropical and oriental, vary in the degree to which they use larval silk. *P. simplex* of Israel binds soil and vegetable debris with silk and makes a very inconspicuous nest; only the movement of workers outside gives it away (Ofer, 1970). Nests are built in rock crevices, tree trunks, under bark, under stones or in leaf litter on the ground. Shelters are also made for aphids on tamarisk. Ofer describes workers mending a tear in the nest fabric: some dozen workers pull the torn pieces together from outside whilst others bring out large larvae ready to secrete silk. They hold these with their jaws so that their bodies and mouth face away from the worker and their head and thorax are free to move. The weaving worker then 'moves slowly and cautiously' along the rent antennating the larva which bends its body periodically and touches the fabric with its mouth parts. The fine silk it secretes is invisible. Fewer than 10 workers weave whilst others collect sand grains and plant material and fasten it on with their own mandibular gland secretion. Weaving is not just a response to damage; it seems to go on most of the time, even at night and normal nests have as many as nine overlapping layers of silk. Each fibre is 0.7–1.0 μm thick and about 2 mm long. The silk protein resembles that of the silkworm but is not quite the same: it has alanine 30%, serine 13%, glutamic acid 13%, glycine 9% compared to the silkworm's glycine 45%, alanine 30%, serine 12%, tyrosine 5%, and lacks methionine and cystine, both sulphur-containing amino-acids (Ofer, 1970).

Lastly *Oecophylla*, the 'weaver ant' with only two species is the supreme exponent of silk nest-making. Way (1954) lists 89 trees in 35 families in which it nests in eastern Africa. Workers are more constructive; they co-operate to bind leaves together with larval silk and so make chambers; they use no soil or dead vegetation at all. In fact, if a leaf dies, they make a new nest. Way says that about a day before starting to build on cloves the larger workers crawl slowly over the nest area. After this large workers often in chains of up to 12, begin to draw leaves together: whilst the first grips the leaf edge each ant behind grasps the petiole of the one in front with its jaw and the last hangs on to the leaf edge with its tarsal claws. Several chains co-operate in a most striking way sometimes re-arranging leaves into more convenient positions. Once the leaves are together, they are held by workers while others, carrying medium-sized larvae held in a particular way and antennating continually (unlike *Dendromyrmex*) guide the larvae who bind the edges with salivary silk.

Worker larvae contribute about 10 times as much silk as male larvae. This is remarkable since, in formicine ants as a rule, male cocoons are about the same size as female ones. Wilson and Hölldobler (1980) conclude that the difference must have evolved since communal nest spinning started and represents a substantially greater investment by females in the nest fabric. Once the silk tension is released the workers move off. In addition to the larger workers outside, smaller ones spin inside (Hemmingsen, 1973); both work outside at night. These nests are started by single queens who do not feed but settle in crevices and nourish larvae until they are old enough to spin webs of silk across their entrance. Sudd (1967) gives a useful summary of work by Ledoux and Chauvin on *Oecophylla* behaviour.

6.2   Combs of cells

Halictine bees (Halictidae) excavate shafts and cut out horizontal, flat-bottomed cells from the sides; the interior of each cell is smoothed with clay and then lined with a substance from Dufour's gland and the mandibular gland. Blind burrows drain off excess water. Some species make vertical cells of the same shape and sometimes sets of cells are excavated and supported by very few pillars or plant rootlets and the earthen walls between cells can be reduced to very thin partitions. Other cells are built by adding material to the walls, and they may be used again after relining though most Halictinae just fill used cells up with soil. Although co-operation takes place in construction it is still possible for a single female to build the whole nest by herself. Social evolution is not related to the capacity to build elaborate groups of cells (Michener, 1974; Plateaux-Quénu, 1972).

Bumblebees save excavation by using rodent burrows and nests which they tidy up and re-lag; they may even use old birds' nests in bushes (Alford, 1975). Eventually, they make a canopy of wax secreted by inter-segmental abdominal glands (Fig. 6.1). This wax melts between 35° C and 40° C and is very heterogeneous. Hydrocarbons comprise 37%, monoesters in the range C40–C50 comprise 29% and the rest (34%) is a mix of gum, ethyl esters and alcohols. Tulloch (1970) summarizes the main differences between this wax and beeswax as follows: there are diols and there are more unsaturated and branched compounds with four to six carbons less; it melts 25° C lower than beeswax.

The rough agglomeration or pyramidal pile of cells which bumblebees make does not constitute any symmetrical or organized comb structure at all yet is well adapted for moulding into the shape of a rodent's nest. It begins as a single wax chamber into which the queen lays several eggs; these hatch and continue as larvae to share space around a pollen store usually called a 'pocket'. Sometime later, the chamber is partitioned by larval silk and just before the larvae pupate each spins a cocoon which forms a new individual wall. Wax is scraped together from the old wall to make a new egg chamber and restart the cycle. No doubt because this is a material secreted by inte-gumental glands, it is used sparingly and plasticized with a watery ingredient probably from the mandibular glands. It is then re-used repeatedly. Honey and often pollen is stored in old brood cells that may have their rims extended with wax (Alford, 1975).

The stingless bees (Meliponinae) also nest in natural cavities; they clean

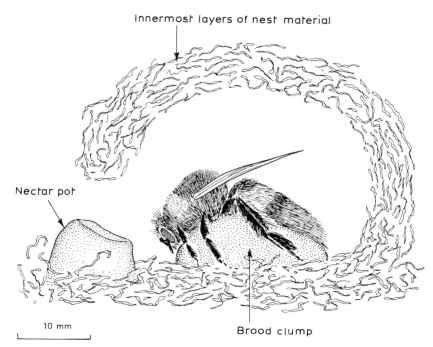

Innermost layers of nest material

Nectar pot

10 mm

Brood clump

Fig. 6.1 The nest of *Bombus terrestris* in an early stage (a section) that shows the queen incubating her brood in a wax cell. (From Alford, 1975.)

and seal them with cerumen, a mixture of wax from dorsal glands and resin collected from trees (Kerr *et al.*, 1967). Some very small bees, (e.g. *Trigona portoi*) use twigs with hollow centres, others use branches, tree trunks, rock crevices, buildings or even the old comb chambers of Attini. Some frequently nest inside occupied termite or ant nests. In setting up in an occupied *Nasutitermes* nest one species begins with a flask-shaped tube containing a few bees which is gradually enlarged and extended inside, lined with cement, furnished with combs and supplied with a queen and more workers (Wille and Michener, 1973). In a *Crematogaster* nest hollowed out by a pangolin, a colony of *Trigona oyani* established themselves by occupying and blocking off a central store and brood chamber, with a thick tough partition (the 'batumen') made of resin, wax, or mud, or all of these mixed (Darchen, 1971).

No matter where the nest is placed it is usually strongly zoned. Just inside the protective envelope are the storage cells (Fig. 6.2), clusters of large pots with honey or pollen in them opening upwards until sealed (one species makes cones for pollen). Unlike *Bombus* which uses refurbished old brood cells for storage, they are specially designed. As far as the cavity allows they are arranged to make a second shell, a sort of lagging or second line of defence

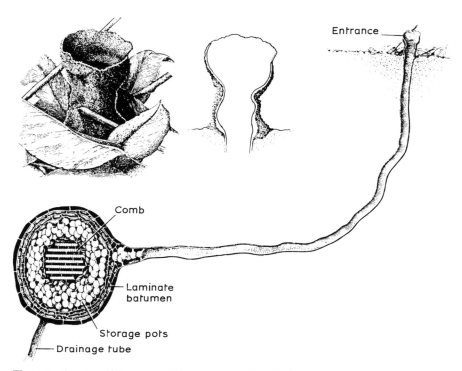

Fig. 6.2 Section of the nest of *Trigona recursa* in soil; there is no involucrum between the comb and the storage pots. (From Michener, 1974.) The entrance, enlarged, is also shown. (No scale given.)

between the envelope and the brood chamber. This itself is surrounded by a perforated shell that supports brood cells arranged in irregular piles, clusters, horizontal plates or even spirals. Each spheroidal cell made of soft cerumen opens upwards; it must, for it is filled with liquid nutrient, receives an egg and is sealed and left alone. Cells are made independently, not in a comb: only part of their walls are shared and their relative position is maintained by struts and pillars of cerumen (Roubik, 1979). Since each is structurally independent it can be deprived of its cerumen as soon as the larva has spun a cocoon; furthermore once the adult has emerged the cell can be destroyed and a new one started in a group higher up, as in *Bombus*, but spaced well away on struts which serve as temporary scaffolding.

Often there is a long and elaborate access tube made primarily of mud or resin; it may have diaphragms to close at night, sticky zones to trap invading insects on or a spongework of struts in a vestibule to carry guard bees. Drainage tubes are built by soil-nesting species. One bee (*Dactylurina staudingeri*) uses a new design (Fig. 6.3); it builds its combs vertically and back to back with cells that lie on their side. New cells are made at the bottom

Plate 13
Above: A nest of the bumblebee (*Bombus lucorum*) with the cells arranged in a cluster surrounded by vegetation fragments from a mouse nest. On the right are 5 cocoons from which the wax has been cleared for the construction of new egg cells.
Below: Honeybee nest in a natural cavity in a tree. Nine irregular combs are visible, with the drone cells on the lower margin. (Photographs courtesy of J.B. Free.)

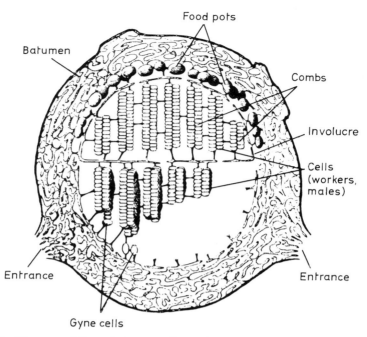

Batumen

Food pots

Combs

Involucre

Cells
(workers,
males)

Entrance

Entrance

Gyne cells

Fig. 6.3   The nest of *Dactylurina staudingeri* in section. (From Michener, 1974.)

of the combs so that these grow down as in *Apis* combs but do not have the same mechanical rigidity and permanence. Several double combs occur in each nest and gyne cells are placed near the bottom rim again as with *Apis* (Darchen, 1972; Michener, 1974, p. 341; Wille and Michener, 1973).

Stingless bees can be said to have developed the pyramidal pile of *Bombus* cells into a stratified and suspended tower which grows up as it is destroyed below until a new cycle begins at the bottom again. The rigidity is given by a skeleton of struts and more regular wall sharing by cells, especially after cocoon spinning so that hexagonal cross-sections form a comb. The logical extension of this, to build tiers back to back so that space is more fully used and greater strength achieved, necessitates making cells that lie on their side and thus produce vertical combs. How this rotation of 90° took place is quite unknown.

Stingless bees are social all the time and nest building is wholly co-operative. In bumblebees, by contrast, the solitary queen can start a nest which is a miniature of the final one. In wasps, too, the founder female can construct a pedicel, a set of combs and even an envelope by herself; workers which appear subsequently only add to this foundation. In many solitary species great skill is shown in nest making; Evans and West-Eberhard (1973, p. 76) who list wasps in the order of the complexity of their nests, include five families of excavators that are not social at all; others make clusters of mud cells and show incipient sociality, e.g. some Stenogastrinae and a few *Polybia* (Polistinae).

All Polistinae are social and have evolved a great variety of nest form. This, perhaps, follows the use of pulp made by chewing together vegetable fibres and mandibular gland secretion and letting it set to a hard shiny lacquer that is waterproof. *Polistes* wasps construct a pendant stalk (Fig. 6.4) which carries a comb of cells: *P. goeldi* cells are cylindrical and attached to the stalk so that only a few walls are shared; *P. canadensis* cells are hexagonal and extended together with common walls but the stalk is asymmetric to the comb; *P. fuscatus* or *P. gallicus* with similar cells have a comb that is symmetrical to the stalk (West-Eberhard, 1969). The nest of *Ropalidia cincta* (Polistinae) is suspended by a specially flexible stalk from the underside of a palm leaf. The cells are arranged eccentrically and descend in several irregular

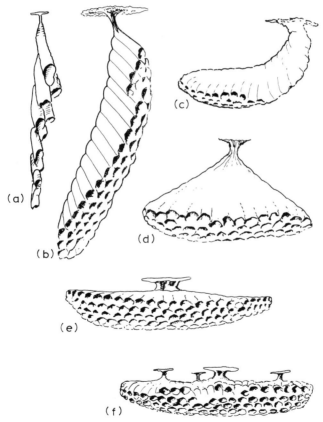

Fig. 6.4 Nest shapes of various *Polistes*. (From Evans and West-Eberhard, 1973.) a, *P. goeldii*; b, *P. canadensis*; c, *P. annularis*; d, *P. major*; e, *P. flavus*; f, *P. fuscatus*.

groups giving 200 or so in all (Darchen, 1976). Cell wall sharing is extensive and this, with the substantial carton, makes a strong nest.

After what has been said of the stingless bees, it may seem odd that wasps can rear brood in downwards opening cells. The answer is that the cell is not provisioned, the egg is fixed on the paper wall with adhesive, the larva fits the cell so tightly that it can easily stay in (the cell walls are extended, lengthening as it grows) and food is given a bit at a time on the mouth.

The tribe Polybiini are all tropical and have evolved (probably from a basic *Polistes* type) a remarkable variety of nest form (Evans and West-Eberhard, 1973; Jeanne, 1975; Richards, 1971, 1978) (Fig. 6.5). Layers of comb may be fixed together and to the cavity by struts, or the combs may be cylindrical and concentric. Another nest is cylindrical, with a tough wall to which combs are fixed all round and the wasps move up and down through a central hole. Yet another is stalkless and the comb is made flat on a branch and covered with an envelope. In spite of this complexity no special storage cells are made as in bees.

Of special interest is a wasp in this tribe, *Polybioides tabidus* which makes

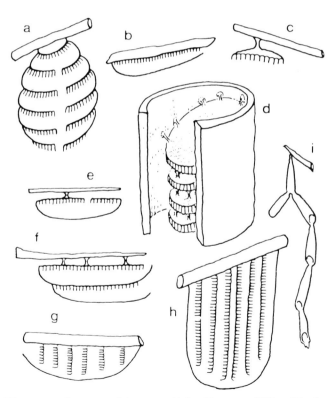

Fig. 6.5 The nests of some social wasps. (After Evans and West-Eberhard, 1973.)

a number of leaf-shaped combs which hang down from a small branch and are enclosed on each side by envelopes (Richards, 1969). Each comb can be conceived as evolving from a very eccentrically fixed pedicel, and the whole nest composed of several such units started close together. Combs do not all face the same way, in fact the two central ones hang back to back but the baseplates are not fused and there is a wasp space between them. Outside these, there are a few others that face outwards and are not back to back. A curious feature is that the central cells (all but an outer fringe of 3–10 cells) have their bottoms cut out once the larvae are large, with the result that the faecal pellet is easy to remove and may fall free of its own accord. The larvae spin cocoon caps at both ends, but that at the hind end is small and workers later take it away. After the new workers come out the other caps are removed and more eggs laid even though the cells are open at both ends. In *Belono-gaster griseus* workers also cut holes to take away larval faecal residues and the hole is then sealed with a transparent setting secretion perhaps of their mandibular glands (Jeanne, 1980).

The horizontal down-facing *Polistes* comb has been developed with great success by the Vespinae (Fig. 6.6). They make several combs linked by struts and surrounded by thin, overlapping envelopes in a branch or soil cavity which is enlarged and shaped as necessary. The strength of these nests lies in their endoskeletal cells and their supporting struts not in the envelopes (as in many Polybiini) which are broken down inside and rebuilt outside as the nest expands.

As already pointed out, the founding queen can construct a miniature nest complete in all respects except for the bigger cells used to rear gynes. Worker progeny continue the process and it is clear that however they measure the size of a cell, it cannot be related to their general body size which is very

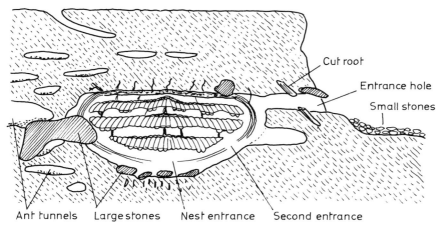

Ant tunnels   Large stones   Nest entrance   Second entrance

Fig. 6.6 Section of the nest of *Vespula rufa*. (From Edwards, 1980, after Janet.)

different from that of a queen. The finished cell is hexagonal in section and opens downwards but it begins in the 120° angle made by two earlier contiguous cells (Edwards, 1980) and has an outer edge which is at first circular but later straight. The cells are made by applying pulp roughly to the rim and then thinning it out several times whilst moving backwards and antennating neighbouring cells (West-Eberhard, 1969). As long as both antennae can feel a wall, the new wall is made straight but if one antenna cannot make a contact the wall on that side is circular. The normal straight line of cell intersection is thus a constraint imposed by neighbouring cells.

This much is known, but a little speculation is interesting. The ideal shape for a cell is cylindrical since it best fits the body in it. To save space cylinders have to be packed in staggered rows so that any three contiguous cells form an equilateral triangle rather than a square. To save material each should share walls with its neighbours and since each circle has six neighbours this means changing to a hexagonal shape. This eliminates waste space between cells and confers maximum mechanical strength.

So, the hexagon gives strength and saves space and material. It is, incidentally, one of only three figures that fit together perfectly; the other two, the triangle and the square are hardly right for a cylindrical larva! But how the cells are measured out is still a puzzle: is there a 'wasp-unit'? This could be the radius of the circle into which each hexagon fits and conveniently, a hexagon can be made by marking off chords of radius length round the circumference of a circle. If wasps make use of this they only need one measuring device, and an obvious suggestion is that they use some body part (as we would stride it out) but wasps vary in size and the size difference between the queen who starts a nest and the workers she makes it for, is substantial.

A further difficulty (even if a size length or radius 'stick' is found) will be how to account for the large hexagonal cells that are used to rear gynes. Here I can only guess that it may be to do with whether a cell is built just inside or just outside an imaginary circle with the standard radius marked out round the periphery. If chords are made from point to point a small cell results but if tangents are made at each point a large cell results. The ratio of area of the big to the small cell would then be 0.75; in *Vespula vulgaris* it is 0.763 (Spradbery, 1971).

Whether wasps build big or small cells depends upon their physiological state; this includes their age, if they are queens and whether they have an old or a young queen if they are workers. Young queens of *Vespa orientalis* build big cells before winter and small ones after and Ishay (1976) has shown that this behavioural development depends on the need for a short spell at temperatures around 5° C. He also found that pentobarbital causes them to make bigger and bigger cells as the dose is increased, which leads to the idea that an unreactive nervous system is necessary for big cell building. Other curious effects are also clearly related to the development of their behaviour. Thus, individuals that emerge from the pupal stage in the dark cannot make co-ordinated combs of cells until they have experienced a brief period of light; then they can build normally even in the dark! Again, solitary workers can build if taken away from their society very early, and they improve with practice. Yet if moved later on, after a social induction phase has passed, they become dependent on social stimuli for building to a degree that overrides their actual building experience and necessitates the presence of a queen (Kugler *et al.*, 1979; Motro *et al.*, 1979).

In the honey bee, each comb is composed of two layers of hexagonal cells aranged back to back, but staggered so that a cell in one comb backs on to three contiguous cells on the other comb, and shares a third of its bottom with each! The cell base is, in fact, composed of three rhomboid-shaped planes each shared with one cell on the opposite comb. Thus, a system of hexagons on one comb is mirrored with a vertical shift of one cell radius on the other. No space is wasted at all! Another detail is that with few exceptions the hexagons have two opposite sides vertical, not horizontal. Each cell also has a slight uplift from base to opening making an angle of about 13° with the horizontal – this perfection prevents liquid spill. The combs are self-supporting and able to carry heavy loads of honey, brood and bees. *Apis mellifera* makes several such double combs arranged usually in parallel and placed in a cavity that is sealed with vegetable gum (propolis) not with cerumen. *Apis florea* and *A. dorsata* both make one double comb suspended from a branch without any cover other than the bees themselves which cluster systematically head up, tail down. They are restricted to the tropics but *A. mellifera* and is sibspecies *A. cerana* have extended into temperate forest zones thanks partly to their metameric nest but also to the evolution of hibernation procedures and cavity nesting which enables them to remain with their stores instead of migrating as do the other *Apis* species (Fig. 6.7).

As in Vespinae, all *Apis* species have two sizes of hexagonal cells, but the bigger is used to rear males, not gynes. Cells for gynes are specially made in spring and are ovate, fixed at the broad end and tapering downwards; normally they are placed on the lower rim of combs but in emergency they can be extended out of the hexagonal comb and turned in a right angle to point down; this must be important but why is not clear. In *A. mellifera* the drone cells usually occur in blocks but in *A. dorsata* they are dispersed amongst the worker cells. There is no special food store; pollen is packed in normal cells conveniently near the brood and honey is further away in outside combs so that it blankets the brood and serves as a heat store as well as a sugar store. Cells are not broken down and recycled as in the stingless bees but are used over and over again. Their normal solitary state 'instinctive' behaviour develops differently in a social context in which queen stimuli are vital.

A general correlation is known to exist between constructive activity, social status and degree of ovary development in wasp workers (*Polistes* and

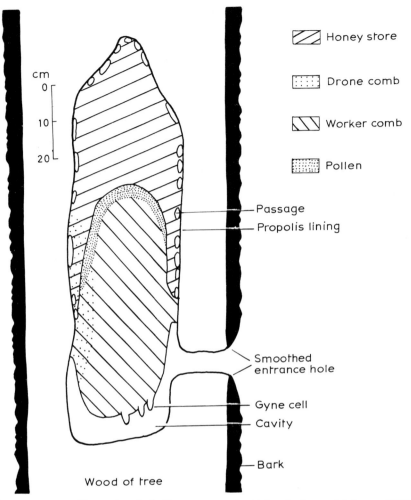

Honey store

Drone comb

Worker comb

Pollen

cm
0
10
20

Passage
Propolis lining

Smoothed
entrance hole

Gyne cell
Cavity

Bark

Wood of tree

Fig. 6.7 Nest of honeybee in hollow tree trunk. (After Seeley and Morse, 1976.)

Vespinae), and bumblebees (Montagner, 1966; West-Eberhard, 1969); this is, in fact, a widespread maternity syndrome, and the corpora allata are also enlarged, (Roseler *et al.*, 1980).

Honey-bees (Apini) all make combs of wax which is largely miricyl palmitate but contains some other acids, higher alcohols and hydrocarbons. Tulloch (1970) found that the main acid is 15-hydroxypalmitic acid and that the hydrocarbons consist of monoesters (35%), hydroxyesters (24%), diesters (12%) and free acids (8%). The alcohols lie in the range C30–C32. Thus about half beeswax is difunctional hydroxy acids and diols, both absent in the wax of bumblebees. Straight-chain saturated compounds are also more common in *Apis* than in *Bombus*. These features provide a harder, stronger wax workable at the hive temperature of 35° C, and which does not melt until 62° C. The combs are vertical, attached to the superstructure all along their length without stalks or struts, and hang with a catena-shaped margin along which new growth is added. In time they blacken and lose their attraction for the queen even though they are specially lined with some unknown material by workers. There are no major wax depots; it is deposited briefly on the rims of growing cells and as a capping over pupae and honey-filled cells. Thus, in the interests of obtaining a uniform repeatable design honey-bees have given up specially designed storage cells. Perhaps only the peculiar ovoid gyne cell remains as a vestige of some earlier evolutionary stage.

Wax is secreted from ventral intersegmental glands by bees that hang in festoons in the centre of a cluster that keeps the temperature around 35° C. From these it is passed to a set of chewers who mix it with a watery secretion of the mandibular glands and change it from a brittle to a plastic state. It may then be immediately incorporated into the comb structure (Darchen, 1980). This wax is synthesized in the worker body from sugars which like clay or plant fibre cost energy to collect; there is also a metabolic cost involved in its manufacture. Nevertheless, sugar is a widespread raw material with other uses in the hive but wax once made cannot be re-assimilated and is in no sense a food store.

The way in which *Apis mellifera* makes these cells, measures them out, adjusts angles to within a few degrees, polishes and planes the walls to a remarkable thinness is still only partly understood. Ribbands (1953) gives a good summary of early work by Huber, Darwin, and others. Unlike the situation in wasps, the process is necessarily co-operative, not merely because some secrete the wax, some emulsify it and put it on the cell rims and some fashion and polish it, but because the cell makers adjust their distance apart to suit the type of comb (worker or drone) being made, and arrange the combs with a space between sufficient for bee movement and access.

Darwin commented on the extent to which, in difficult situations, bees will make, break and re-make cells; especially in man-made hives when the comb foundation is put in irregularly. Darchen (1954) tried various deformations such as supplying a curved spine of foundation instead of the normal plane one, and found that the bees adjusted the depth of the cells to give a roughly equal distance between the two faces of a comb. In straw 'skeps' which are never given foundations, the combs are not always parallel; they may curve or even change direction but they tend to fill out the space, just leaving a passage for bee movement. Any obstacles put into comb are either taken out or incorporated with the minimum of distortion. Bees readily fill in holes in walls or cell bases but crushed cells are not easily rectified; for some reason they are reluctant to break and remake them.

Normally small 'worker' cells on one side of a comb necessitate the same on the other side because of the need for the bottoms to fit; similarly with male cells. So Darchen (1958) fixed worker cells and their foundation back to back with male cells without a foundation. Then the bees left the worker cells alone but converted the male cells into smaller ones, not by destroying them to ground level and starting them again but by splitting their walls from a point about half way up; more were created but they were shallow. When instead he left the bases on the male cells, they were accepted and the worker cell bases on the other side were re-organized into male cells. When Darchen stuck small cells and cell bases on to large ones the bees found a solution very difficult and broke through the bases to thin the wax out. Honey-bees are evidently loth to scrap and start again in the way that stingless bees, with their more free style of building, find easy. Can it be that their wax loses its plasticity? They also find angles other than the 60 and 120° very difficult to deal with (Naulleau and Montagner, 1961). One last question remains unanswered: how is the 13° slope engineered?

Although bees always work in a co-operative group it is possible for one bee alone to construct a very thin cell wall: 73 μm for worker and 92 μm for drone cells (1 μm = 10^{-6} m). As a wall is progressively thinned by a planing action of the mandibles it restores itself more quickly after deformation by a push outwards; without oscillating it just moves back into the equilibrium position and stops, fully damped (Martin and Lindauer, 1966). This assumes a constant temperature of 35° C and a fixed mixture of wax and mandibular gland fluid but it also varies with the position of the cell wall. Sense organs exist on the mandible and on the antennae which are probably used in the process of cell wall thinning, and if the antennal tips are removed or desensitized, though the workers can still build orientated cells, they have thick double walls with each lamilla separated by a filling of unworked wax nodules.

Hair-plates in the bee's neck are essential for equilibrium and without their free use the bees cannot even start to build. The antennae, on the other hand, are not essential though it is possible that they are used to measure distance and angle. Centrifugal forces caused by rotation can affect the orientation of cells (Gontarski, 1949, cited Martin and Lindauer, 1966) and so can the Earth's magnetic field. When deprived of visual and tactile cues bees build combs with an orientation to the Earth's magnetic field that is colony specific (Pickard, 1976). Using fields of magnetism of the same strength as a bee is likely to experience, experiment showed that the normal rate of walking in darkness can be increased by the periodic imposition of a magnetic field (Fig. 6.8). The latent period is considerable: 40–60 min, which rules out a simple disturbance of their abdominal magnetite. In addition, the time of a dance can be reduced from 34 to 19 s by applying a magnetic field (Tomlinson et al., 1981).

In this chapter the complexity of ant nests has been traced in stages from

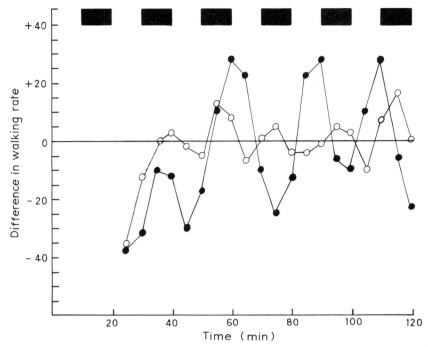

Fig. 6.8 The change in walking activity when a magnetic field is switched on (black bars) intermittently. o, no field; •, 10 min intermittent field. (After Hepworth *et al.* 1980.)

the use of a thatch of dead vegetation over soil mounds by wood ants through the increasing use of processed vegetable fibre (carton) in tree nesting species to a stage culminating with one tribe (Camponotini), some species of which use larval silk to make a communal cocoon nest. This last is a highly co-operative activity and involves gland sharing, between larvae and adults, some workers holding and some spinning with the bigger ones on the outside.

Termites, though they have evolved very complex and highly differentiated mound and tree nests, made of selected minerals and carton, do not spin cocoons and never have silk nests as a result. Although both ants and termites have stuck to the chamber design, wasps and bees have developed an extremely effective cellular design using either carton or a special body wax softened with mandibular secretion and mixed with plant gums.

Why is the one cell/one juvenile system so effective in bees and wasps? It provides, of course, a tightly-packed, regularly spaced, easily controlled, brood chamber and is economically made. The only raw materials needed are: in bees, sugar and gums; in wasps, wood. Can the more flexible system of stingless bees be thought of as a compromise between concentric chambers and regular hexagonal cells, a compromise which allows both tight-packing of individual 'closets' and access for cleansing and reconstruction; as well as keeping stores separate from brood yet using them as lagging?

CHAPTER 7

Microclimate

Diurnal variation is usually greatest on the soil surface because of the properties of electromagnetic radiation and is buffered by water whether as vapour, droplets or liquid in vegetation. Any insect seeks a suitable microclimate and selects out of the range available to it. Social insects respond in this way when foraging unless they make shelters for bugs or rest stations. These and the true nests are the special features by which social insects distinguish themselves and modify the microclimate in which they live. The nest is intended as a refuge from weather and a steady environment for the conversion of food into population. The aim is to buffer natural disturbance and hold the microclimate as near as possible to the best state for biological activity. Stages in the evolution of perfect control start with social insects moving with their young up and down in the natural daily microclimate strata, then using their nest to collect more of whatever is in short supply, then lagging their nest to conserve metabolic products and finally sensing deviations and actively producing what is needed to correct the error (homeostasis). These do not preclude each other and the most advanced social insects with a regulatory response both lag their nests and use them to pick up supplies from the environment.

An interesting point is that very few parts of the biosphere actually have the right microclimate. Even in the tropics it is not hot enough and in desert it can be too hot for parts of the day and too dry most of the year. So there is a problem of some sort for social insects to solve, everywhere.

7.1 Environmental regulation

In rain forest, temperature fluctuates around 25 to 30° C all the year for heat is only pumped in by insolation 12 hours a day and is largely stored in the biomass, water and soil. Yet even 25° C is not high enough for many social insects and they have to collect some sunshine. Ants living in the leaf zones show species characteristic zonation that will be described in the chapter on communities. In a search for sunshine they can run into wind and *Oecophylla* in Tanzania (6° south) build on whichever side is sheltered from the monsoon; this blows from the north-east between December and April and from the south-east between May and November (Way, 1954). Rain is heavy and tree-nesting ants need protection. *Crematogaster* nests can be severely damaged by mechanical impact if the rain hits directly but dripping from leaves can usually be deflected, by overlapping hydrofuge scales of carton (*C. skonnensis*) or by using a thatch (*C. ledouxi*) or just by allowing limited temporary absorption (*C. vandeli*) (Soulié, 1961).

Doryline ants avoid direct sunshine and will wait for cloud before crossing a gap in the forest; many are nocturnal. Soil nesters, however, seek the heat it gives and drive shafts down deep across the temperature gradient to avoid direct sunshine, wind and rain though not flooding. *Anoplolepis custodiens*, in East Africa, at 6°S selects bare soil surfaces with lethal temperatures of 60° C in sunshine (Way, 1953). The nests have several shafts that go down to 2 m or more with chambers at 10–15 cm intervals, and at 25 cm, temperature can be steady at near 30° C optimum like the surface at night (Fig. 7.1). The same species in the Seychelles nests quite superficially under logs and stones and in drains, possibly because the soil is too hard for excavation (Haines and Haines, 1978).

Savannah ants move up and down in the temperature wave and avoid foraging when it is either too hot or too cold on the surface. Thus in dry savannah in the Ivory Coast at about 10°N the day goes from 06.00–18.00 h and the temperature at 1 cm depth rises from 16°C to 30°C by 10.00 h and on to 34°C just before sunset. *Messor regalis* collects seeds in the morning between 08.00 and 11.00 h and again in the afternoon between 15.00 and 18.00 h and *M. galla* works at night. *Brachyponera senaarensis* in the same area works in the morning from 08.00–12.00 h and afternoon from 15.00–17.00 h (Lévieux and Diomande, 1978a, b). *Solenopsis invicta* from the Matto Grosso (12°S) in South America (the imported fire ant of North America) arranges its mound nests in Florida with a long axis running north/south (Hubbard and Cunningham, 1977); this presumably moderates insolation.

Surfaces can be too hot or too cold in the Arizona desert (33°N). *Acromyrmex versicolor* does not forage during the heat of the afternoon with a surface temperature over 44°C or at night when it falls below 8°C (Gamboa, 1975, 1976). July is a difficult month between 10 and 20 h and most surface work is done at night. Workers caught outside the nest in heat try to find shade and then press their bodies against the relatively cool surface soil around 32°C. Rain in July and August is just enough to germinate seeds and green up the vegetation. *Trachymyrmex smithi* only forages between 10 and 37°C in the Chihuahuan Desert (33°N), but *Formica perpilosa* works in the shade of bushes and nests at their base thus avoiding hot sand (Schumacher and Whitford, 1974), and *Atta sexdens* is nocturnal in Paraguay (22°S) when trail temperatures are more than 30°C (Fowler and Robinson, 1979).

In dry African savannah the diffuse nature of the nest of hypogaeic Macro-

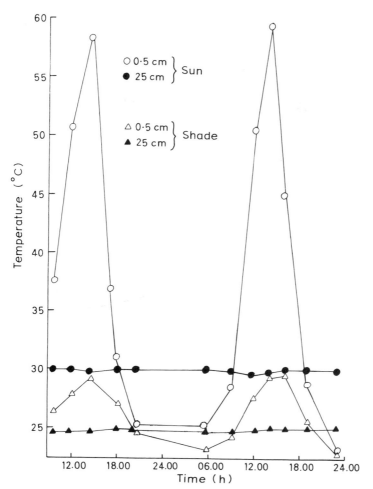

Fig. 7.1 Temperature fluctuation in sun and under the shade of a Mango tree in Tanzania in an area inhabited by *Anoplolepis custodiens*. (After Way, 1953.)

termitinae makes water conservation unlikely, but soil in general has a high gaseous humidity in its pore spaces even when 'dry'. That these subterranean species do not have vertical shaft and chamber nests is no doubt due to their solidly constructed fixed reproductive and fungus chambers but the species do differ in the strata they occupy (Josens, 1972). In very dry desert conditions shafts are sunk to water sources; *Psammotermes* does this in the Sahara.

The microclimate in the mound of *Trinervitermes geminatus* (not big by termite standards: 40 cm high and 50 cm basal diameter) has been studied in the Ivory Coast savannah at 6°N (Josens, 1971b). Beneath the crust a gallery system is filled with a lagging of cut pieces of grass later to be used as food, then come more empty galleries and finally, in the interior, brood galleries which extend as far below ground as above; there are also shafts that go down several metres for mining clay and water. The mound surface can be 50°C with an atmospheric shade temperature of 35°C but in the brood galleries the temperature is constant at 28°C, probably all the year round, especially in big nests. Mounds without termites but in good condition, show more fluctuation but average the same as 'living' mounds and Josens concludes that the termite lodge stores heat coming from outside and that their metabolism does not contribute much. The temperature wave generated by daytime sunshine passes more quickly through the mound than through the soil nearby, which has a lower heat capacity but still peaks an hour later in the soil immediately under the mound. Thus *T. geminatus* mounds collect solar heat and store it, only passing it gradually to the brood galleries. There is no doubt that termite mounds are often designed to capture and store radiated heat; this leads, in the case of *Amitermes meridionalis* and other species to a famous elliptic cross-section with the long axis north to south whereby the intake of heat from the sun is moderated by a reduced angle of incidence and so spread more evenly through the day (Lee and Wood, 1971). How termites orientate and build such a shape is unknown.

The tree carton nests of termites are waterproof and have special scales for shedding rain, or ridges on trunks to divert rainwash from their nests. Mounds are quite common in tropical rain forests (Abe and Matsumoto, 1979), and like tree nests are waterproof so long as the inhabitants are alive but quickly break up when they die, presumably from lack of repair. Mounds also tend to be built higher in areas liable to water logging, e.g. *Macrotermes* and various grass feeders in northern Australia. A humivore, *Cubitermes fungifaber*, which builds 'hats' looking a bit like a toadstool has them better developed in wetter parts of its range. Again, in Australia, more species of the genera *Amitermes* and *Nasutitermes* build mounds in the warmer, tropical areas than in the cooler, temperate ones (Lee and Wood, 1971, p. 81). This at first appears to run counter to ants which use mounds to extend their range into cool climates but these mounds may be cooling towers not solaria.

Further away from the Equator, summers can be long and hot but winters cold. *Pogonomyrmex occidentalis*, a harvester with a shaft and chamber nest in dry grassland, and a clear area of some 3 m diameter which it chews around its nest entrance, moves up and down with the season (Lavigne, 1969). In March zero temperatures extend to 2 m and the workers and queen are tightly packed, broodless in the 1–2 m zone. By May the workers are up on the surface with the queen, where the temperature at 10 cm is still only 10°C but rising towards a peak of 30°C in late July; it is then only 21–24°C below 1 m. Temperature inverts in November dropping to 2°C at 10 cm and 5–6°C below 1 m. *P. occidentalis* follows the seasonal wave of heat down and gets trapped by the cold.

Plate 14

An excavated soil nest of *Acantholepis custodiens*. A fine example of chambers at different levels connected by vertical shafts, allowing the ants to select an area of optimum temperature for their brood. (Photograph from M.O. Way (1953).)

Temperate forests generally, and mountain forests in the tropics, are low in ant species. Brown (1973) comments on sparsity at 2300–2500 m in the Andes, Madagascar and South India and says '. . . even at 2100 m in tropical mountain forest ants are exceedingly scarce . . .' yet higher up in the treeless zone they can be locally abundant. He points out that radiation is the vital factor. No species has solved the problem of nesting in the dense canopy of temperate trees; they are forced, with a few exceptions, to start their nests in glades kept treeless by grazing mammals or where old trees have died and collapsed. In such spots the majority of ants live in dead branches or build mounds. They prepare for winter by a brood-rearing programme that stops all larvae in a limited set of stages and forces a food storage syndrome on the adults. Physiological adaptations usually reduce the temperature thresholds of egg laying, brood growth and metamorphosis. This is certainly so in *Myrmica rubra* where the optimum lies near 21°C instead of over 30°C. Placed in a temperature gradient in saturated air, a colony settles at 20°C by first collecting brood from higher temperatures then more slowly from cooler zones, down to about 5°C. Queens may move themselves or be carried or dragged (Brian, 1973b). Many ants abscond when the environment cools locally as when tree growth shades their nest. The movement of a colony of *Myrmica rubra* from one nest to another has been analysed: activist foragers locate and examine new sites for suitability, lay trails to indicate direction then go back inside the nest and display in a specific way. This means jerking towards workers which soon fold up and allow themselves to be lifted and carried away. Other workers follow the leader and then brood is taken over. Most of the moving is done by a few responsive workers (Abraham and Pasteels, 1980). A dolichoderine *Tapinoma erraticum*, that reaches into temperate latitudes, makes small mounds of vegetation and soil; it readily moves to a new nest site after exploration by activists, a trail is laid and a few workers transport brood. Meudec (1977) showed that when groups of either active or inactive workers were singled out they developed a new inter-individual arrangement generating new leaders and followers. The activity of some must inhibit that of others.

The subgenus *Serviformica* comprises a very adaptable group of formicine species living in north temperate forest/scrub. In Japan at 36°N, *F. japonica* makes a typical multiple shaft and chamber nest. Ants are found below 80 cm in March but come up with eggs in early May to 10 cm; by June, the brood, mainly pupal, is widely distributed from the top down to 140 cm. The total annual fluctuation on the surface is 25°C (3°C in January to 28°C in August), but in the nest where it never falls below 13°C in April or rises above 29°C in October, the range is only 16° (Kondoh, 1968). This shows a delay of about 2 months in the annual temperature wave, similar to normal soil. *Formica fusca* in the British Isles has a similar nest structure even as far north as 54° but beyond this is replaced by *F. lemani*, a sibling species that develops more sub-surface galleries and probably makes better use of fleeting sunshine. *F. fusca* readily abscond; after finding a better site, a scout runs back and forth laying a hind gut trail; this done, it comes in and performs a special display that involves fast, short runs, waggles and to and fro movements. It may grasp a nest mate and drag it fowards a little and lift it up; this worker responds by folding its legs and antennae and bringing its gaster forwards and is carried to the new place (Möglich and Hölldobler, 1975). Many young workers, and all the brood, are carried but the queen walks with a group of workers protecting her. Even founder queens choose and defend warm places against competitors.

Lasius niger excavates chambers under flat stones which then store heat as well as protecting the ants from trampling. In complete contrast, *L. flavus* builds its mounds on south facing slopes that are well drained (Waloff and Blackith, 1962). The mounds are said to be triangular with the apex pointing north (Hubbard and Cunningham, 1977), though this is not always so; there is no doubt these spongy mounds have vast heat collection and conservation potential, however. The ants reduce vegetation growth by nibbling at roots.

7.2 Metabolic regulation

Large colonies may produce a lot of heat, water and carbon dioxide. Their problem is to dissipate the last and conserve the other two according to their needs: in cool climates to retain heat and get rid of water and in dry climates to do the opposite. Carbon dioxide is able to diffuse out of soil remarkably effectively and this, no doubt, is how it escapes from mound nests though in highly reactive populations such as honey-bees it has to be actively driven out during ventilation. Heat is normally retained by lagging, that is, the use of spaces in which air is stagnant and its poor thermal conductivity utilized. Water, if it will not evaporate sufficiently, is ventilated away in air currents. Nevertheless, the correct adjustment of these three variables is a difficult and delicate matter.

Retention of metabolic heat is well attested in the nests of some termites; in *Nasutitermes exitiosus* for example, occupied mounds are nearly 10°C warmer than 'dead' ones and nursery temperatures are always higher than in soil or air nearby (Holdaway and Gay, 1948) and *Coptotermes* nests in the centres of trees in Australia range from 33 to 38°C, which is between 13 and 20°C higher than in similar uninhabited trees. In the huge and elaborately structured mounds of *Macrotermes*, fungus combs augment the metabolism of the termites themselves. Conservation of the heat from this must be responsible for the 5°C hotter centres; living mounds average 30°C with a 4°C amplitude of variation. Evidently any radiant heat captured is not stored and perhaps the ducts and chimneys which contain circulating air are, in fact, cooling it rather than collecting heat. Air rises from the fungus into the spires or chimneys and falls back into the cave or cellar. This movement may be

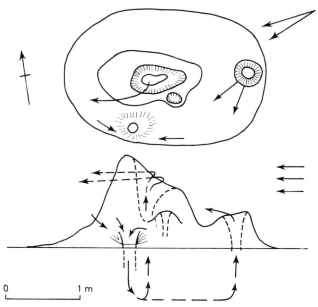

Fig. 7.2 Wind direction and air circulation in a nest mound of *Macrotermes subhyalinus*. (From Weir, 1973.)

convectional but it may also be due to wind over the chimneys sucking air out and wind at ground level blowing it in. Evidence that a *Macrotermes* adjusts the aperture at ground level has been provided by Weir (1973) who blocked the openings up and caused the nest materials to become so wet they collapsed. He suggests that dry afternoon winds both cool and dry out the nest. The salts in the cellars might be a result of crystallization of soil solutions after such evaporation. A ventilation current almost certainly helps to dispose of the carbon dioxide though much may diffuse through the crust (Fig. 7.2).

Termites adapt their nest structure to the habitat. In dry areas they need to supplement metabolic water with that collected from deep soil. *Cubitermes exiguus* lines its nest with more faecal material and some *Macrotermes* set their nests deeper in the soil in dry areas. *Macrotermes natalensis* builds mounds in areas with cool winters, e.g. plateaux in southern Africa. *Nasutitermes exitiosus* in dry areas of Australia conserves water so effectively that the relative humidity of its chambers rarely falls below 92% and for this it is thought that metabolic water is adequate because of the 1–4 cm thick outer skin of the mound made of clay cemented with faeces and, perhaps, saliva (Lee and Wood, 1971, p. 63). In southern areas it builds mounds on south-facing slopes and the inhabitants move away from the periphery in cool weather but there is no strict hibernation.

Wood ants of the *Formica rufa* group owe much of their success in temperate forests to their mounds of litter and the habit which some have of clearing the herbs and bushes and tree seedlings for several metres around, e.g. *F. exsectoides* in north-eastern America (Janzen, 1969, p. 152). The mounds of *F. ulkei* near Chicago (41°N) in oak-hickory forest are mostly started in forest margins and grow asymmetrically with a large surface towards the light. They are mainly a response to direction of insolation not to the Earth's magnetic field. Each mound is waterproof on account of the 5 cm thick roof of litter and soil which keeps the nest interior drier than nearby soil all the year round, and each clearly absorbs and concentrates heat giving 47°C at a depth of 5 cm in sunshine (Scherba, 1958, 1959, 1962). The mounds of the European *F. polyctena* are slightly larger and looser and seem to incorporate less mineral matter in the symmetrical dome. These ants forage on the ground both in shade and sun as well as in the tree canopy but nest on the forest margins, moving away when shaded. In a study of the foragers of *F. rufa* at the northern limit of its range in the British Isles, Skinner (1980a) found that they avoid rain and darkness but otherwise forage in proportion to temperature.

In a thermal diffusivity study on 2-year old nests of *F. polyctena*, Brandt (1980) found that the nest structure was such that the daily heat wave travelled in more quickly than it did into unstructured piles of nest material or even normally structured soil. In a nest it takes 20–30 cm for a heat wave to damp out completely, compared with 10 cm in soil or an unstructured nest. This, of course, gives a deep warming effect, maximal heat storage overnight and ample choice for brood maturation. Brandt, in fact, argues the thermal equivalence of a nest mound to a stone and points out that the ants actually benefit if the insolation is partly occluded by shrubs so that too high a surface temperature (say 50°C) can be avoided. In larger nests the role of metabolic heat and of heat release by aerobic decay of the nest material may be more important, and these have now been measured and compared in a shaded nest. Coenen-Stass *et al.* (1980) found that, although the heat emanating from ants of all categories exceeds that from the same weight of nest material, the greater amount of the latter gives it a greater heating power. Thus 82 kg nest material gives more than seven times the heat production of 1 kg ant material. Their summary figures are: 16.21 W for the nest and 2.62 W for the ant population; together about 19 W (Fig. 7.3).

It remains to consider what part worker behaviour plays in this: they select insolated sites, structure the nest material, open and close the surface vents, move into the sun and back again. But as far as composting vegetation is concerned do the ants go for the right sort of materials? Are conifer leaves, bud scales, and twigs particularly suitable or just the easiest to collect? To summarize: wood ants make mounds that trap and store sunshine and decompose slowly to give a substantial background heat. They also retain their own metabolic heat but the question still remains: do they increase

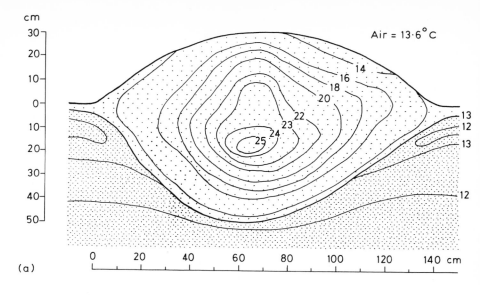

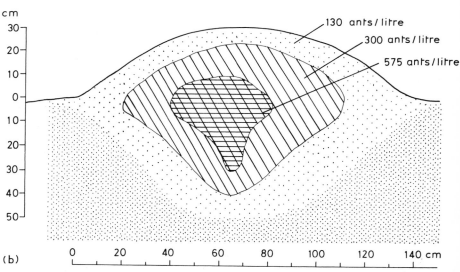

Fig. 7.3 (a) Temperature (°C) distribution on a mound of *Formica polyctena*. (b) Ant density at the same time. (After Coenen-Stass *et al.*, 1980.)

their heat production in response to, or in anticipation of, a drop in nest temperature?

The comparative fixity of wasp and bee nests means that their microclimate depends more on choice of site and on an ability to correct 'errors' in microclimate behaviourally. This includes the use of their bodies to screen sunshine or rain if envelopes do not exist and the creation of a draught by wing fanning as well as spreading water for evaporation.

Polistine nests that hang from a pedicel are often built under cover but on ones in the open the wasps orientate 'tail' down so that their wings and gasters drip rain water off. Envelopes, if they are made, protect brood from direct sunshine and fanning can regulate temperature, humidity and carbon dioxide inside, although the muscle contractions during ventilation may raise the body temperature by 3°C (Evans and West-Eberhard, 1973). In dry climates *Polistes* collect water to cool their nests.

The soil burrows of Halictidae presumably depend on the site chosen and the burrow depth to give, on average, appropriate environment as in non-social bees. However, bees may ventilate at the burrow mouth (Sakagami and Michener, 1962). Meliponine bees may either avoid or regulate their microclimate. *Leurotrigona mülleri* is one of the former kind with no control at all but a habit of nesting in termite nests; similarly, *Trigona (Apotrigona) nebulata* nests in the globular tree nests, 30–50 cm diameter, of a *Nasutitermes*. Bees that nest in holes which they line with cerumen are often unable to regulate their temperature if kept in artificial hives (Darchen, 1973).

Meliponines often build in the derelict mounds of *Macrotermes* or *Trinervitermes* in savannah. Old mounds are first eroded by vertebrates and recolonized by other termites, e.g. *Amitermes*, *Ancistrotermes*, or by ants, e.g. *Camponotus*, *Crematogaster* or other genera; bees later nest in the centre of the complex making a long, winding tube to the outside.

Other Meliponini are very effective temperature regulators. *Trigona spinifer* in Brazil keeps its nests at 33–36°C with an ambient temperature between 8 and 30°C (Zucchi and Sakagami, 1972). So too, does *Dactylurina* which chooses to live in dense vegetation and makes a ball-shaped nest with a 9 cm thick foam lagging of galleries under a thin, resinous skin 0.5 cm thick. The nest is ventilated by fanning which can be made more effective by opening a hole at the opposite side to the entrance tube or making smaller holes in the pellicle. At mid-day on hot days the noise of fanning can be incredible. Small bees (*Hypotrigona*) which live in cavities in branches ventilate in files with their head towards the nest centre (Darchen, 1973). Thermoregulation enables a few stingless bees to extend down into S. America as far as a zone of cold winters where they stop breeding but make no special preparations for hibernation.

The open comb nest of the oriental bee *Apis dorsata*, is curtained by three or more layers of bees hanging by their claws, gaster downwards. This sheds rain very effectively, shades the brood from sunshine and holds warmth in. If the temperature drops below 30°C, bees pack together to reduce heat loss; if they become too hot they fan or even collect water for evaporation. In an ambient from 20 to 34°C they can keep their comb at 30 to 31°C (Morse and

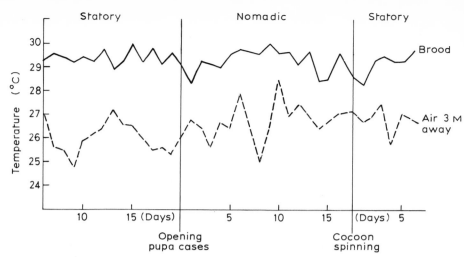

Fig. 7.4 Temperature in and around a cluster of *Eciton hamatum* measured daily during a complete activity cycle. (After Schneirla, 1971, p. 65.)

Laigo, 1969, cited Michener, 1974). *A. florea* in east Africa (20°N) does the same and can regulate at 33 to 37°C in an ambient between 18°C and 33°C (Free and Williams, 1979). Neither of these species of *Apis* appears to need to generate heat by day; the warm climate, normal metabolism and layer of bee lagging suffice. This applies equally to *Eciton hamatum* (Fig. 7.4).

Such a reaction for space heating might evolve out of incubation behaviour, i.e. close contact of the adult body with the young to impress their microclimate. There is evidence of this in those halictine bees that make cell clusters underground. Hornets and wasps certainly press against brood and pump their abdomen, thereby both circulating heat generated in the thorax and increasing gaseous exchange at the same time. They do not brood empty cocoons but do brood pieces of filter paper which have pupal extract on them – no doubt a chemical recognition (Ishay, 1972). The tight curling of the young *Vespula* queen around her first brood cells may well raise their temperature and, since she makes an envelope too, that of the air space as well. However, *Bombus* is the supreme brooder. The queen, who starts a nest, makes a shallow groove in the first cell cluster into which her body fits closely (Alford, 1975). Honey stored in a specially made pot within reach, supplies fuel for respiration and the temperature of the brood rises well above ambient (Michener, 1974). Heinrich (1974) was able to show that between 3°C and 33°C queens can regulate their thorax at 35 to 38°C and their abdomen at 31 to 36°C, a little lower. This enables them to keep the brood at 25°C with an ambient of 5°C and at 33°C with an ambient of 30°C; clearly regulatory. The queen's oxygen consumption increases as the outside temperature falls and it is evident that she measures deviation from optimum and responds by increasing her metabolism via her flight muscle activity in the way that simple homeostats do. The success of this reaction depends on the lagging as well as the active response and there is clearly a premium on those queens that choose, or are lucky enough to find, a good site to start in. Should this be insolated or shaded? The one saves fuel, the other gives a steadier regime. As workers accumulate thermoregulation improves, especially if a canopy is constructed. The nest temperature then varies by less than 3°C in the range of 30°C and overheating only has to be prevented by fanning cool air in; no water need be collected.

The temperate Vespinae with their paper-lagged underground nests, are also excellent regulators especially during the summer when populations peak; over winter they are not social and their queens diapause like most insects. Their optimal brood temperature is in the lower 30s and they can maintain this even at 10°C outside; they cool by ventilation and, when necessary, the collection and evaporation of water (Edwards, 1980).

Space conditioning nears perfection in the honey-bee, *Apis mellifera* and *A. cerana*, its sibspecies. They range all through temperate regions without the help of man by living in sealed and lagged cavities, because they store a vast amount of honey and burn it up slowly during winter. They must prevent hive temperature becoming too low since workers die at 8°C. Preparations begin when daylength diminishes, well before winter cold actually sets in: the queen stops laying and young larvae are destroyed so that only adults exist during winter.

A cluster forms as the temperature drops below 18°C; this cluster has a cortex of tightly packed immobile bees for insulation and a medulla of actively moving and 'shivering' bees to generate heat (the wing muscles can vibrate without moving the wings). Bees do not swap places between cortex and medulla very much. The cluster can maintain a temperature of over 20°C inside with an outside temperature around freezing. Curiously enough the combs interfere very little with cluster structure and mobility; only in *A. cerana* do the bees cut holes to allow sideways movement. During this stage the syrup consumed is very thick but humidity can be adequately maintained from metabolic water. Bees retain their (liquid) faeces and never fly outside until spring sunshine and warm air return.

In spring the bees raise the temperature in a small part of their brood nest to 35°C. Possibly, they respond more to daylength than to ambient temperature which varies a lot. In summer, with a large population, brood temperature can be regulated to nearly half a degree (Büdel, 1968). This stability is aided by the bulk of brood and honey which are effective heat stores though not heat generators (even the brood is negligible in this respect). Moisture is usually adequate and the bees ventilate not only to dispel hot air but to bring in dry air to evaporate water from their nectar during the ripening of honey, itself a cooling device. They do better in dry atmospheres than moist ones as a result (Michener, 1974). If the weather is exceptionally hot and dry water is collected and spread on the combs. Lensky (1963, cited Darchen, 1973) has shown how, in Israel, *A. mellifera ligustica* (the Italian race) can maintain a brood nest at 34 to 38°C in an ambient that varies from 22 to 50°C. This adjustment seems to be outside the scope of an African subspecies *A. mellifera adansonii* which tends to avoid overheating by living in cavities amongst the roots of trees, by limiting its population by constant swarming and by absconding (Darchen, 1973). This is not only a response to dry season overheating but also to food shortage in the wet season (as in *Apis cerana indica*, Woyke, 1976) or to disturbance by mammals, ants or other arthropods such as wax moths (Fletcher, 1975). Preparations are made, sealed brood is allowed to emerge, no more eggs are laid, and stores of honey are eaten. Recently swarmed colonies are more likely to abscond than others and some subspecies are more prone than others (Winston *et al.*, 1979). Absconding is clearly an irregular form of the migration shown by *A. florea* and *A. dorsata*.

Carbon dioxide is, of course, removed during ventilation though quite high concentrations are tolerated: up to 3%. *A. mellifera* will ventilate in response to carbon dioxide even when temperatures are below optimum and very large colonies are more efficient at controlling concentration than small ones (Seeley, 1974).

To summarize, ants select a suitable place, make a multiple chambered nest which provides a wide range of microclimates into which brood can be spread. In cool climates they use natural heat stores such as flat stones and construct heat capturing and storing mounds in insolated spots; some are even capable of using the heat of aerobic decay in vegetation which they collect into a pile and some perhaps of generating metabolic heat in response to an ambient temperature drop. In hot, dry climates they avoid exposure and collect water from deep soil, dew or plants. Termites are less mobile than ants because they often have rigid reproductive cells but they select sites carefully, build water-shedding envelopes and regulate air flow by adjusting the size of openings into ducts and caves. They have heat-insulated mounds in which they conserve or disperse metabolic heat as appropriate. Wasps and bees can incubate their brood and lag their nests or form a tight living curtain against heat fluctuation and they can generate heat by metabolic activity in response to a temperature deficit, or disperse heat (along with waste gases and water vapour) by ventilation and evaporation of water in response to a surplus. Honey-bees are capable of very sensitive thermoregulation in summer and of keeping out the cold in winter. This active regulation of microclimate involves the expenditure of energy and is necessitated by the immobility of the young stages. After special preparation in summer, whole colonies may abscond with much of their stores. Probably bees and wasps, but less so ants, abscond in response to a serious climatic or biotic threat and if this happens regularly, migration may evolve. For more details, see Seeley and Heinrich (1981).

CHAPTER 8

Defence

Many organisms are attracted by the huge concentrations of population that social insects often build up; their brood, flying sexuals, stored food and nest material can be used as food and the steady warm moist nest space also attracts many animals. Bacteria and fungi that cause disease, nematode and helminth parasites, mites that scavenge, steal food or are carnivorous, spiders and many other arthropods all attack and exploit in various subtle ways. Vertebrates even include fish that eat sexuals trapped on the water surface and toads that enter the nest. Other social insects, their own and related species may compete for the same resources or attempt social parasitism. In this chapter defence against arthropods and vertebrates is considered; there are two aspects: the weapons used and the organization behind their use. There is no possibility of a complete account of the enemies of social insects but many references to them will occur throughout this book.

8.1 Painful and paralysing injections

The first weapon available to the Hymenoptera was the sting (Fig. 8.1). Originally derived from a parasite's ovipositor, it was probably used to paralyse hosts by an ectoparasite and dissociated from egg laying altogether (Malyshev, 1966). It is still used to paralyse prey in several primitive ant subfamilies: Ponerinae and Myrmeciinae. As they sting, workers attract help by releasing alarm pheromones from their mandibular glands. *Odontomachus* give out a chocolate smell due to alkyl pyrazines and workers gather and snap at anything that moves, including their sisters. Many ants including dorylines can be killed by biting alone but once *Odontomachus* secures a grip they inject venom too (Wheeler and Blum, 1973). Another ponerine, *Paltothyreus tarsatus*, releases alkyl sulphides whilst defending its returning booty-laden columns; *Megaponera foetens* also does this. The Australian red bull-ant (*Myrmecia gulosa*) uses its sting both to kill prey and in defence, generally with good effect; the venom contains histamine, a hyaluronidase and proteinaceous components (Cavill and Robertson, 1965).

In the Pseudomyrmeciinae the sting is used more for defence and less for predation. This subfamily contains the two famous genera *Pseudomyrmex*

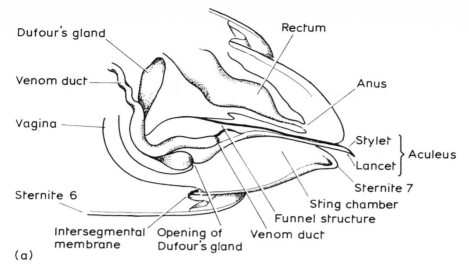

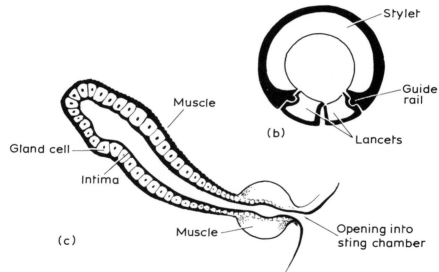

Fig. 8.1 The sting of *Vespula*. (From Edwards, 1980 after Schlusche). (a) Median section of gaster and sting; (b) cross-section of aculeus; (c) longitudinal section of Dufour's gland.

and *Tetraponera* that live in plants and obtain food as well as shelter. A species of *Tetraponera* that inhabits the small tree called *Barteria* uses its sting with great effect even against elephants. Janzen (1972, p. 890) says that they drop on to humans from their tree a few at a time, search out bare skin, insert their sting and pump in venom; pain is not immediate but when it comes it is deep and throbbing, lasts several days and even makes the muscles sore. Humans and grazing vertebrates quickly learn not to work under these trees. The procedure and weapon seem better designed to repel vertebrates than invertebrates. In contrast *Pseudomyrmex* (in *Acacia*) attacks in a crowd and produces many mildly irritating stings that do not last though the barbed sting sheath is often inserted (Janzen, 1966, p. 264). They are fast and agile with large compound eyes and hunt singly. Captured insects are stung without inserting the barbed sheath and may then be thrown away rather than eaten since the plant hosts provide the ants with oil and protein in Batesian bodies.

These ant-plants use their ants to protect themselves from other plants as well. In the canopy this means killing vine leaders, and pruning the branches of shrubs to reduce the danger of shading. An area of the ground vegetation is also kept clear from herbs and seedlings which reduces the risk of damage by fire. *Tetraponera* keep an area of 2–3 m diameter clear under *Barteria* by repeatedly chewing the foliage. The dolichoderine *Azteca* that lives in *Cecropia* (Moraceae) in the American tropics is not of much use against invertebrates or vertebrates but very effectively arrests the growth of vines, epiphytes and neighbouring plants (Janzen, 1969).

All vespid wasps have stings which they use only for defence. They are very sensitive to sound and smell and have good vision and guards may fly at a strange object nearby. The tendency to do this varies, of course, with the species and with the size of colony; bigger colonies are more aggressive. Wasps with a nest-envelope and an entrance hole have several on guard all the time apparently queueing to do this job. They check each comer and can identify non-members even of the same species; such intruders are pinned down and stung. There is an alarm pheromone which in *Vespula* comes from the poison gland (Maschwitz, 1964, in Jeanne, 1980). The venom is a mixture of amines and proteins. The amines comprise histamine and the catecholamines, dopamin and serotinin which affect the cardiovascular system of mammals. The proteins are mainly peptides such as wasp kinins which reduce blood pressure and there are various amino acids and enzymes. Important amongst the enzymes are: phospholipase, which stops blood coagulation, and hyaluronidase which increases the permeability of connective tissues and spreads the poisons. A strong dose even in one sting, is calculated to hurt, frighten and deter vertebrates and kill invertebrates (Edwards, 1980).

Some wasps give a fair warning before their onslaught. The tropical American *Synoecia surinamus* with a flat comb, starts a rhythmic drumming against the brittle envelope when molested; further disturbance makes the workers come out and raise and lower their wings in time with the drumming. As a last resort they use barbed stings that stay in firmly and allow all the poison to be injected (Evans and West-Eberhard, 1973, p. 180).

As a defence against invertebrates, mainly ants, wasps may rub a repellant on the nest petiole. For founder queens this is effective even when they are away and it is widely used even amongst temperate *Polistes*. It is secreted by the sternal gland of the last segment (Jeanne, 1975; Post and Jeanne, 1981).

Frequently, polygyne swarming wasps leave the nests and brood when attacked and either return later or start a new one. *Polybia sericea* forms a cluster, using visual cues, near the damaged nest. When a new site is found the leaders mark it with a secretion of a ventral gland and also make a trail of spots on prominent objects. The cluster is then activated by erratic runs and the wasps fly along the trail to start a new nest (Jeanne, 1981). A more insidious form of damage is suffered by *Polistes canadensis* in the lower Amazon where it nests on trees in the open; larvae of a moth (Tineidae), present in 56% of colonies, scavenge on the meconia. Unfortunately, they burrow from cell to cell in the process and damage both the comb and the pupae. In response the wasps build several combs near together and when one is infected use another (Jeanne 1979b).

Halictine bees constrict and guard their nest entrance against ants, mutillid wasp parasites, cuckoo bees (*Nomada*) and bees of their own species but of a different family group. An *Agapostemon* in which several females share an entrance hole but build their own cells off the main shaft share guard activity in spells of about an hour, and effectively exclude ants, isopods and others. A cuckoo bee (genus *Nomada*) also tries to get in and lay in the cells where its larva destroys the host larva and uses its provisions. Female *Agapostemon* that live alone are unable to stop this intrusion (Abrams and Eickwort, 1981).

Lasioglossum zephyrum in the laboratory learns the set of odours belonging to residents and to the nest structure itself (Kukuk *et al.*, 1977). This comes more easily if they are related or brought up with them; otherwise continuous social interaction is essential and they soon forget the odours of unrelated bees. Co-operation has obvious gains in cases like this for whilst one guards, others construct and provision. Guards allow relatives in more readily than non-relatives; inbred cousins if 74% related are accepted like sisters (75% related). No doubt the bees learn the smell of kin early in life (Greenberg, 1979). Guard bees of *Bombus* produce repellents in their mandibular glands; the *B. lapidarius* pheromone contains butyric acid and two ketones in the C5 to C13 range (Cederberg, 1977). Less aggressive *Bombus* roll on their back and expose their sting which may carry a repellent; buzzing accompanies this performance. On the other hand, *B. atratus* of tropical America is notoriously savage when its nest is threatened.

In the Apini this method of defence is carried to a degree that is otherwise found only in some Vespidae. The sting of *Apis* has barbs on the lancets that assist penetration and then hold the sting in firmly, steadily pumping venom into the victim (Hermann, 1971); the venom used is like that of wasps and while it lacks serotinin and adrenalin amongst biogenic amines it has two extra peptides called apamin and melittin. Amongst enzymes, acid and alkaline phosphatases are present as well as esterases (Benton, 1967). As bees that sting vertebrates die after 'autotomy' they benefit their society but not themselves; hence only eusocial Hymenoptera can evolve this arrangement (Hermann, 1971). The alarm and recruitment pheromones of *A. mellifera* are generated by glands near the stings themselves which produce a set of at least eight acetates with three alcohols and a series of normal alkanes and alkenes. Isopentyl acetate alone will alert and draw bees to the sting point but it is not as strong as the sting mixture. *A. florea* and *A. dorsata* respond to the pentyl ester for longer than do *A. mellifera* and *A. cerana* (6–15 min compared with 3 min) but another less volatile ester (2-decen-1-yl-acetate) present in the former pair of species extends the duration of alarm. The acetates can be blown out on the air by lifting the abdomen and fanning with the sting out. In addition to the sting localization and alarm pheromones there is another set from the mandibular glands which are released by guard bees if persistently harried and they contain, as principle ingredient, 2-heptanone. This causes a sudden mass exodus of workers which throw themselves on the intruder; those that cannot insert, grip tight and buzz frighteningly.

Some bears and many humans are attracted by the honey stored by *Apis mellifera*, and bees are very sensitive to large dark moving forms near their hive especially if they shake it. Vibration stops the medley of buzzing for a moment and *A. cerana* actually hisses at 700 Hz. This in itself reduces their aggressiveness (Koeniger and Fuchs, 1973)! *Apis dorsata* can be persistently aggressive. A disturbed bee goes back to the comb and runs quickly; the normally smooth shape of the comb changes and clusters form near the bottom. Within moments they fall away to near the ground and the bees fly off in groups: 50 to 5000 can start a hunt for the aggressor in response to a single disturbed bee. This soon stops if no quarry is found but if a sting has been inflicted the acetate drifting downwind gives the searching bees a clue and they quickly 'home' into the area or on to stings stuck around by an experimenter. This upsets the whole colony for several days. Koeniger points out that these are jungle bees; those that live in temples are less aggressive; evidently they are either habituated to humans (Koeniger, 1975), or else are a peaceful strain evolved as inquilines in human society.

Honey-bees defend their hive using their stings against many insects, in particular against species of *Vespa* and *Vespula*, but the wasps' agility and persistence can enable them to destroy weak colonies late in the season. In extreme adversity such as wetness or under regular and repeated interference by wax moth or human beings the bees abscond. In *A. mellifera* a preparation is made that stops brood rearing and fills each bee up with stores; the West African races and *A. m. adansonii* are especially prone to abscond, so too is *A. florea*, which moves to and fro transferring stores (Free and Williams, 1979). The supreme example is *A. dorsata* which undertakes the regular seasonal migrations already described.

8.2 Toxic smears and repellants

Higher ants have lost the use of their stings for hostility as well as for predation. They either secrete a toxic repellent from the vestigial sting or use especially strengthened jaws. The latter arrangement arises from the anterior bias in growth (allometry) which gives large jaws with strong muscles to work them. In extreme cases these large workers can no longer forage or nurse and they devote themselves to situations that need strength including hostility: in the Dorylinae, for example, the big workers line the trails where they stand with outward facing jaws. In genera such as *Labidus* and *Eciton* and in epigaeic *Aenictus*, stings are still used in hunting although they are not strong enough to inflict a sense of pain on man (Hermann and Blum, 1967, Kugler 1979a). Other genera, more generally hypogaeic such as *Anomma* (Fig. 8.2) and *Dorylus* and most *Aenictus* have weaker stings but attack in greater numbers (Schneirla, 1971). *Neivamyrmex nigrescens* and probably other army ants have the antibiotic skatole (methylindole) in their mandibular glands which kills some bacteria and fungi (Brown *et al.*, 1979).

A few examples from the Myrmicinae will help to show the full range from those that hunt with stings, such as the Dacetonini, to those which do not use them except as a trail-laying pen or as a dispenser for repellent (Kugler, 1979a). *Myrmica* represents a primitive condition. Their sting is unbarbed and weakly sclerotized; the venom contains a mixture of proteins and amines; there is also a trail pheromone which is volatile and polar and of a different chemical group (Cammaerts-Tricot *et al.*, 1977). Chemicals from Dufour's gland are used to augment the trail effect and mark occupied areas in and near the nest and around aphid clusters; in *M. rubra*, they include a number of hydrocarbons and the terpenoid farnesene. The hydrocarbons are mainly heptadecene (C17:1), nonadecene (C19:1), pentadecane (C15) and heptadecadiene (Cammaerts *et al.*, 1981b). The mandibular glands in *M. rubra* are primarily an aqueous solution of protein but they also release 3-octanone, 3-octanol, acetaldehyde and butanone. These are 'alarm' pheromones that create activity (3-octanone especially) and direct movement (especially 3-octanol) and so enable help to be summoned to a place where invasion threatens. Ethylketones such as 3-octanone and 4-methyl-3-heptanone are especially characteristic of myrmicine mandibular glands (Blum and Brand, 1972). Other species of *Myrmica* have the same set of compounds in different proportions; these results are summarized in a paper by Cammaerts *et al.*, (1981a). Earlier it was mentioned that *Leptothorax*, when attacked by slave-makers, vary their defence. If *Leptothorax duloticus* attacks, they bite and sting in reply but if *Harpagoxenus americanus* attacks there is panic; it is presumed by analogy with *Formica* that a repellent is discharged in the latter case into the *Leptothorax* nest under attack (Alloway, 1979).

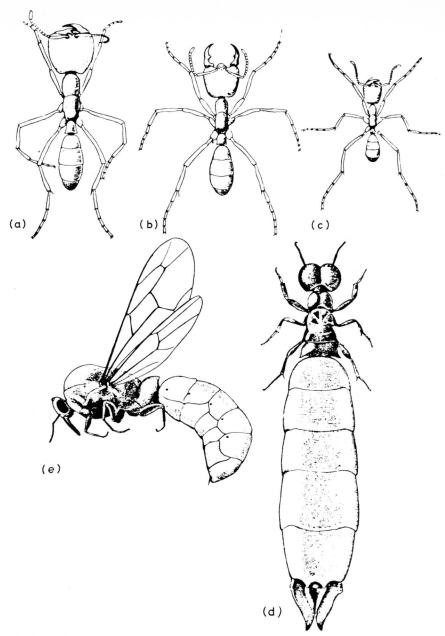

Fig. 8.2 The castes of *Anomma nigricans*. (a), (b), (c) Workers and, (d), gyne are all wingless. The male, (e), is winged. (From *Grassé*, 1951, after Forel.)

Solenopsis invicta well-known as the 'fire-ant' because of its propensity to attack people, holds on with its jaws and forces its sting in deeply whilst injecting a venom mixture of dialkyl piperidene alkaloids (Blum and Brand, 1972). These unusual compounds are all lipophylic and spread well on cuticle when other arthropods are attacked. Their Dufour's gland releases attractant hydrocarbons including *n*-heptadecane. *Pogonomyrmex* are also famous for their sting; in fact, they are the only ant genus to have autotomous stings (Hermann, 1971). The venom is the most toxic insect venom and is more lethal to humans than hornet or honey-bee venom (Schmidt and Blum, 1978). It contains haemolytic factors such as hyaluronidase which promotes internal spreading, lytic factors such as phospholipases which break down blood and mast cells, pain factors such as histamine, serotonin and other enzymes such as lipase, acid phosphatase and esterase. This 'algogenic' venom may have been evolved to deal with vertebrate predators such as lizards, for it has little effect on invertebrates (Hermann, 1971).

The sting of the Attini is plain, unbarbed and weak and is used as a pen for trail laying (Hermann, 1971). Most of this group show no hostility when their nest is disturbed but start salvaging brood and garden material (Weber, 1972). *Atta*, however, 'boil' out of their nest openings dramatically and bite hard enough to draw blood. Their defence relies on their mandibles which can inflict deep wounds in vertebrates by cutting their way in as do the stylets of *Apis*. The mandibular glands of *Atta texana* and *A. cephalotes* are well developed and contain 4-methyl-3-heptanone as part of an alarm and recruitment pheromone. In contrast *A. sexdens* has two isomers of citral which are toxins, not alarm releasers (Blum, 1974, p. 239). It is interesting that this species is much more sensitive to citral than to the corresponding 2-heptanone (Moser *et al.*, 1968, quoted Blum and Brand, 1972).

Pheidole is a large, variable and highly polymorphic genus (Fig. 8.3). *P. biconstricta* produces a repellent in its pygidial gland which alerts nest mates and after application to an aggressor leaves a sticky, toxic residue (Kugler, 1979b). *P. fallax* majors have unusually large poison glands which contain skatole (methylindole), a repellent applied with the sting tip (Kugler, 1979a). The majors of *P. pallidula* and *P. teneriffana* are defensive and difficult to recruit experimentally to food; in the nest they crush seeds (Szlep-Fessel, 1970). *Pheidole dentata* has a defence recruitment system. If in danger, small foragers come back, penning a trail and display to the large ones, which constitute 10–20% of the total, who then go out along the trail and with strong jaws but feeble stings, snap the foe to pieces. Wilson (1976a) found that if *Solenopsis geminata*, a fire-ant native to North America, were numerous they pressed on to find the *Pheidole* nest and could break through by releasing a volatile liquid from their sting held high in the air. Then suddenly whilst the major *Pheidole* continue to defend by biting, the minor evacuate the nest of brood and scatter with the queen. Wilson suggests that

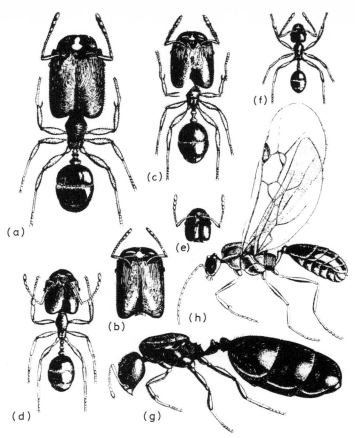

Fig. 8.3 Polymorphism of the female in *Pheidole*. (After Wheeler, 1910.) (a)–(f) Workers; (g) dealate queen; (h), male.

this pair of species are habitual, even hereditary enemies, with a special sensitivity for each other. *P. militicida* majors stay within 5 m of the nest on trails and defend sexuals that emerge for nuptials. If trails of two neighbouring nests cross, though the workers show no hostility, soldiers back at the nest decapitate workers which have taken the wrong trail (Hölldobler and Möglich, 1980).

Plate 15 *Opposite*
Above: A large worker or soldier of *Eciton burchelli* (Dorylinae) in Trinidad. Large workers of doryline ants may form a defensive 'skin' round columns of workers as well as carry large prey. This varies from species to species.
Below: A colony of *Eciton burchelli* resting under a rock, by day. They form a cluster by hanging from hooks on their tarsi and can hold a temperature a few degrees above ambient. (Photographs courtesy of D.J. Stradling.)

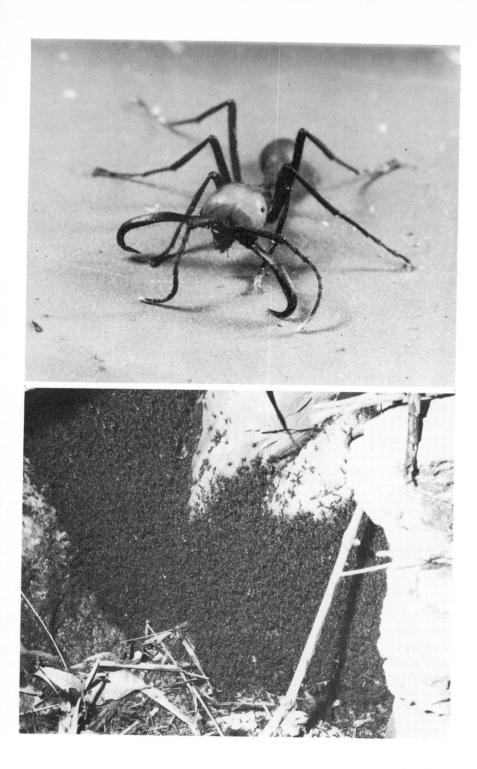

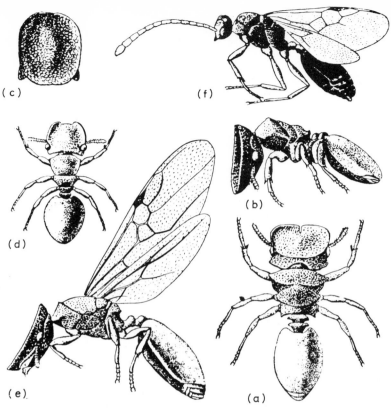

Fig. 8.4 *Zacryoptoceros varians* castes of female: (a), (b) soldier; (c) head of soldier; (d) worker; (e) gyne; (f) male. (From Wheeler, 1910.)

Finally, in this subfamily some bizarre forms of defence exist. *Zacrypto-cerus varians* lives in hollow twigs in mangrove swamps. Major workers defend their nests against other ants by plugging the entrance holes which are carefully tailored to fit their heads (Wilson, 1974a, 1976b) (Fig. 8.4). They have no functional stings but Dufour's gland is huge and their mandibular gland contains 4-heptanone and 4-heptanol; these cause foragers to stop moving and flatten their bodies against the twigs. Curiously, they can run as fast backwards as forwards (Olubajo *et al.*, 1980). *Crematogaster* have spatulate stings from which to disperse repellents and toxins (Fig. 8.5) that drive off much bigger ants (but which are useless against vertebrates). Enemies that will not take the hint get a dose in the mouth where it sets into a sticky glue. One species has Dufour's gland enlarged to contain a lipophyllic repellent which is ejected from the base of the reduced sting while fighting. The metathoracic gland usually secretes organic acids that kill micro-

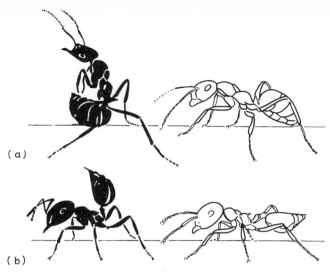

(a)

(b)

Fig. 8.5 Alarm-defence behaviour in (a) *Formica polyctena* and (b) *Crematogaster ashmeadi*. The ant on the right is in a normal posture, that on the left is alarmed. (After Hölldobler, 1977.)

organisms when spread over the body, but in some species swollen glands repel and gum up adversaries with a tar-like exudate; this alarms nest-mates too (Maschwitz, 1974, 1975b). The use of the metathoracic glands, peculiar to ants, has been studied by Maschwitz *et al.* (1970); in the Myrmicinae the secretion is normally acid and odourless except in *Atta* where it smells of honey and contains phenylacetic acid, indolylacetic acid and hydroxydecanoic acid; it is thought to inhibit weed moulds in the fungus garden.

The Dolichoderinae are famous for their toxic secretions. They retain small stings in some species, e.g. *Iridomyrmex humilis* but they are not used for the injection of poison (Pavan and Ronchetti, 1955), which originates in the well-developed pygidial gland whose reservoir opens near the anus (Hölldobler and Engel, 1978). A worker smears this material on an adversary and quickly retreats; it contains alarm pheromones, is toxic and dries into a viscous, sticky material; chemicals are of the cyclopentanoid monoterpene class and include iridomyrmecin (two isomers), and dolichodial both related to the basic iridodial (Cavill and Robertson, 1965). They are toxic to insects and other organisms and their secretion by the large pygidial gland may be one reason for the success of the subfamily (Wheeler *et al.*, 1975). *Azteca* often lives in carton nests or as a myrmecophyte in the tropics, and has workers that swarm out over a human intruder and bite painfully; they release three cyclopentyl ketones which cause alarm, concentrate attack and repel other ants. An interesting structural device for combating an invader is shown by the Australian *Iridomyrmex purpureus* whose nests have many

openings into chambers and galleries that are blind. Ettershank (1968) suggests that this reduces penetration and spread of *Camponotus consobrinus*
which after invasion plasters the galleries with a black resin.

Whilst other subfamilies have been replacing the injection of venoms with
the evaporation of repellents from toxic sticky smears, the Formicinae have
perfected a squirt of the simplest of all organic acids, with only one carbon
atom. This formic acid whilst highly volatile is moderated by short chain
alkanes from Dufour's gland, principally undecane. One or other of these
aided by a range of some 40 alcohols, acetates and sesquiterpenes creates
alarm (Bergstrom and Lofquist, 1972; Blum, 1974). Formic acid is effective
against both arthropods and mammals, either directly or after the skin is
bitten.

The slave makers, *Formica pergandei* and *Formica subintegra* have large
Dufour's glands which contain the acetals of C10, 12 and 14 alcohols. They
spray this secretion into the nest of a slave species and whilst the slaves are
confused take their pupae (Regnier and Wilson, 1971). Big slave workers may
go down into the nest and block the galleries with their bodies to prevent the
attackers reaching the brood (Marikovsky, 1963, cited Brian, 1965). This
recalls the myrmicine *Zacrocryptoceros*. Similarly, *Camponotus (Colobopsis)*
which also lives in twigs in mangrove swamps uses its head to block nest
holes. The *Colobopsis* majors are less specialized than those of *Zacrocryptoceros* as they serve as repletes for storing liquids and so *Colobopsis* can
afford to have one major for every three minor workers instead of 1 to 10.
Their work repertoire has been compared in detail by Wilson (1974c, 1976b)
and by Cole (1980). Two other *Colobopsis*, in Malaya, have given up poison
glands; they use mandibular glands in defence and smear an adversary with a
yellow, odourless, oily fluid that becomes sticky and gums up. These glands
fill much of the body including the gaster and the fluid is released through a
muscular contraction which bursts the whole body (Maschwitz, 1975a, b).

However, these bizarre suicidal defence methods are unusual: the common
genus *Lasius* uses formic acid and methyl ketones from Dufour's gland to
create alarm. Unlike *Formica*, their mandibular glands have a rich variety of
terpenoids and furans (cyclic terpenes) such as citronellal and citral.
Dendrolasin, one of the furans is toxic and useful in underground defence; it
is found in *Lasius fuliginosus* which has big mandibular glands (Bernardi *et
al.*, 1967). Its use indicates a convergence towards dolichoderine defensive
techniques (Wilson and Regnier, 1971). *Oecophylla* show a graded response
to interference; they alert and attract nest-mates with a mandibular gland
pheromone (Fig. 8.6) which is composed (in majors) of the very volatile
hexanal with a kinetic action, the less volatile 1-hexanol which attracts from
10 cm but repels closer up, and a group of at least 30 even less volatile
compounds. These perhaps only have a moderating and spreading function
though they may help to localize the target and induce biting; they include

Fig. 8.6 *Oecophylla longinoda* worker in threat posture. (After Hölldobler and Wilson, 1977a.)

2-butyl-2-octenal and 3-undecanone. Later if hostilities persist, formic acid from the poison gland and a mixture of hydrocarbons, mainly *n*-undecane from Dufour's glands, are released. Straight-chain ketones, alcohols and acetates are absent (Bradshaw *et al.*, 1979a, b, c). Parry & Morgan (1979) summarize pheromone chemistry and ant behaviour.

Meliponine bees are, like the higher ants, stingless, and have evolved similar weapons to replace the sting, concentrating more on the mandibular glands than the poison and Dufour's glands. They rush out very often in a buzzing, clinging mass and whilst biting release from their mandibular glands toxic, irritating and painful poisons. Simultaneously, an alarm pheromone excites workers and directs their attack. In *Trigona (Scaptotrigona)* this contains 2-heptanone and 2-nonanone; in *Trigona (Geotrigona) subterranea* it is citral, also found as a trail pheromone (Blum and Brand, 1972). Much of meliponine defence relies on their well-hidden nests approached through long, and often winding tubes, whose entrances are camouflaged, constricted and often surrounded with sticky repellent materials. The robber bee *Lestrimelitta cubiceps*, gets most of its honey, pollen and cerumen through groups of bees entering nests of *Trigona (Hypotrigona) braunsi*. After finding their narrow, hidden entrance they have to pass a barrier of sticky honey. This delays and even satisfies the robbers temporarily (Portugal-Araújo, 1958). The American *Lestrimellitta limao* releases citral from its mandibular glands (Blum, 1974) giving a lemony smell which apparently disrupts *Trigona* defensive behaviour.

Termites never had an evolutionary stage that included an ovipositor/sting like the wasp group. Their weapons come at the front end, and, like ants, they have evolved a dependent caste, the soldiers with strong jaws (some even lock shut!) and mouth part glands which contain repellents, toxins, viscous adhesives and alarm pheromones (Noirot, 1974). Some have a snout above their mouth through which a jet from a frontal gland is shot; intersegmental glands produce defensive materials in others.

The soldiers of *Mastotermes darwiniensis*, have big heads full of muscle and strong jaws, which are effective against arthropods of about their own size. A liquid is ejected from the mouth which darkens and sets into a rubbery structure and binds up the foe. The setting is like 'tanning' and seems to result from combining *p*-benzoquinone with protein to form a sclerotin. Probably the quinone is from the cephalic gland which opens internally on the frons or nasus and the protein originates in the labial gland (Moore, 1968). The quinone may well release alarm and cause convergence of nest-mates.

In *Zootermopsis nevadensis* (Termopsinae) alarm spreads by contact in two ways: first, workers move up and down and make a noise on the woody substratum and second, they move from side to side silently. There is no evidence that the sound is received or a pheromone emitted but workers collect and mend damage with faeces and debris guarded meanwhile by soldiers who block the hole with their heads (Stuart, 1969).

Two groups, the Rhinotermitidae and the Nasutitermitinae (Termitidae) have evolved the famous termite snouts. The jaws have atrophied and their muscles are used to squeeze the frontal gland reservoir and eject toxins and agglutinators. These highly specialized soldiers are dependent on workers for food. They can follow trails but cannot lay them as they have no sternal glands. In *Schedorhinotermes* (Rhinotermitidae) the snout is not well developed and soldiers retain jaws though they rarely use them: instead they rub their labrum, covered in juices, against the skin of the foe; the juice has five ketones which include two vinyl ketones both toxic to insects. In another species of this genus three ketones have been found (Prestwich, 1975, p. 149). In the same family is *Coptotermes lacteus* which has a snout that exudes a milky liquid from the frontal glands. This is composed of *n*-alkanes of the order of C24 emulsified in an aqueous mucopolysaccharide composed mainly of glucosamin units. In air this exudate sets into a resilient film. In *Nasutitermes* (Nasutitermitinae) the frontal gland produces sticky, rubbery compounds together with stimulating, attracting pheromones. Six species of this genus contain mainly α-pinene but some have β-pinene and some limonene too. *Trinervitermes* (same subfamily) produce the five ketones found in *Schedorhinotermes* including the vinyl ones and α-pinene as well; but the jets of glue are of unknown composition (Quennedy, 1973).

Three species of *Amitermes* (Amitermitinae) vary according to whether terpinolene, limonene or α-phellandrene are the principal compounds. The compound most often encountered, α-pinene, induces a biting frenzy. Alarm and recruitment chemicals used by *Amitermes* have not yet been identified, but in six species of Macrotermitinae soldiers release a labial gland secretion when fighting which contains quinones (Maschwitz and Tho, 1974). *Macrotermes carbonarius* soldiers in forage columns, stab and lock an enemy in their sickle-shaped jaws. A brown, watery secretion from the labial gland which contains mainly toluquinone but also some benzoquinone is then pumped over the foe (Maschwitz *et al.*, 1972). Thus in this group the salivary

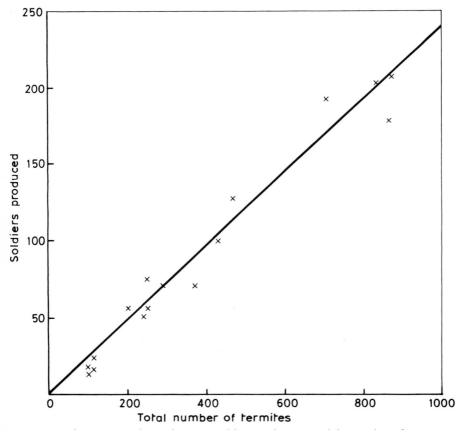

Fig. 8.7   The positive relation between soldier production and the number of termites over 12 weeks in groups of 'workers' of *Coptotermes formosanus*. (After Havarty, 1979.)

gland has become a defensive gland and the soldiers are dependent on workers for food.

In *M. natalensis* small soldiers which mature from early larval instars defend damaged places in nests whilst the workers repair them but outside the nest in foraging columns bigger soldiers take their place. Termites that build covers over their tracks, such as *Nasutitermes corniger*, use soldiers, standing head outwards to line them during construction. *Hospitalitermes monoceros* (Nasutitermitinae) makes tracks 1 cm wide up trees in search of lichens. These tracks are flanked by soldiers who react quickly even though blind; they probably secrete terpenoids from their frontal gland as an alarm signal (Maschwitz and Muhlenberg, 1972).

The proportion of soldiers in a termite society is probably kept as low as possible since they are totally unproductive. They comprise 1 in 6 sterile individuals in *Trinervitermes geminatus*, 1 in 40 in *Macrotermes natalensis*, 1 in 143 in *Apicotermes desneuxi*, and 1 in 300 for large soldiers in *Acanthotermes acanthothorax* (Bouillon, 1970, p. 197). In *Macrotermes bellicosus* soldiers comprise about 4% of the adult neuters (Collins, 1981a). There is experimental evidence that this is self-regulatory (Fig. 8.7).

Ants and termites have converged to the point where *Globitermes sulphureus* soldiers burst and spread a sticky, yellow fluid in defence as do *Colobopsis* species (Maschwitz, 1975a, b).

So the basic defence practised by all social insects is to make an envelope or a shaft with a constricted entrance hole guarded by a sequence, or a single individual, who either is alert for a hostile threat and attack or has an armoured head that just fits the hole. Repellents on pedicels or the emission of repellents from head glands (usually the mandibular) come in this class of guard action, and can have the actual dual function of warning an enemy and summoning help. This applies particularly where soldiers are included in the caste range as in the case in some advanced ants and all termites but no wasps or bees. The soldiers of ants are less specialized than those of termites; they can feed themselves usually and they can do heavy work that is not strictly defensive.

Social insects have to face two main classes of enemy: the vertebrates and the insects, including other social insects. They use the same method: stings, homologous with ovipositors, are naturally confined to female wasps, bees and ants, and the poison has become highly painful, effectively deterrent and in most cases lethal. Even so, advanced bees and ants have given it up and evolved toxic squirts either from the anus or the mandibular glands. In the latter case the jaws are usually strengthened to bite a wound first. In this they have converged on many termites except that the soldier termite has gone further in its dependence on workers for food so that its salivary glands can be used to synthesize toxic gums. The supreme exponents of jet offence in termites are those that have evolved a frontal gland and snout. Apart from salivary glands and poison glands, intersegmental glands have been developed to secrete poison. Soldiers not only defend their nest but the foraging galleries too.

CHAPTER 9

Food processing

In the nest, food is shared out or stored according to its nature. The energy supplying components, lipids and carbohydrates, go mainly to the workers that represent the engine that drives the society; the growth components, proteins, go to the female reproductives that supply eggs and to the larvae that hatch from these eggs; young adults, especially female ants, bees and wasps, take a lot of protein as well as energy foods (Figs. 9.1 and 9.2). The conversion of food into social insects bodies must proceed at a greater rate than they die and decompose (in all developmental stages) though loss is reduced by recycling eggs and juveniles. Naturally as the number of individuals increases more space must be taken over and more metabolic waste either used structurally or eliminated. It is with the organization of these conversions that this and the next chapter deal. First, the refinement and concentration of food, next the distribution of this by nest-workers, then colony initiation by queens or by reproductive pairs, and, finally, the growth spurt that follows colony establishment.

An important principle is to make the young work as much and as soon as possible. Several light inside jobs are available for pale, unsclerotized, physically weak and small individuals: grooming, cleansing, sorting, distributing food, extricating moulters from their cast skins and cocoons. Food malaxation can be tackled later when the teeth are well sclerotized. For light, intricate inside jobs small size can actually be an advantage and in general nest workers are younger and smaller than outside ones.

This is achieved in lower termites by letting the juvenile stages work (they are then called pseudergates, i.e. 'false workers'), and in ants and bees by a slower maturation rate for the smaller workers, which may be carried to such a degree that very small ones never go out at all. Larvae of the wasp group (Hymenoptera) are so different from their adults that they can be used for special purposes such as the manufacture of silk, storage of water, sugar and oil and as a source of proteolytic enzymes.

This chapter could just as well be called 'the evolution of a nursing class' for purification and concentration of foods progresses together with the establishment of a longer and longer adult interval in the nest before going out to forage. Primitively, one presumes, foragers bring food in and dump it so that

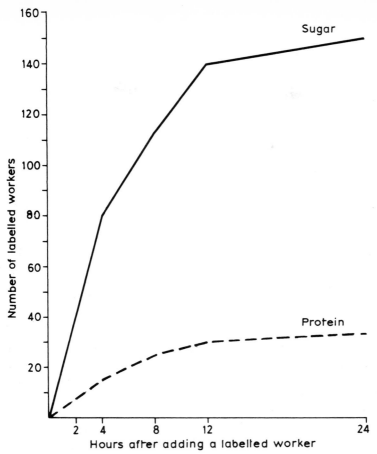

Fig. 9.1 To show that sugar is quickly dispersed through the worker population but protein less so. Workers were starved for 4 days before the experiment. (After Markin, 1970.)

the reproductives and juveniles can help themselves. In the next stage of evolution they chew it, then they mix in digestive juices of their own, then they collect larval juices (in Hymenoptera) and put them on and finally they suck up the digest and regurgitate it to larvae. In a later stage of evolution they themselves swallow the juices and concentrate the main product into a glandular material that is secreted to larvae, whilst the water is excreted as usual. This means more effort by adults and less by juveniles but it must speed growth and maturation and give the adults control over these important processes in polymorphic species. This is particularly vital in Hymenoptera since the larval gut has a septum between the mid and hind portions that filters out any colloidal particles that are able to get down the extremely

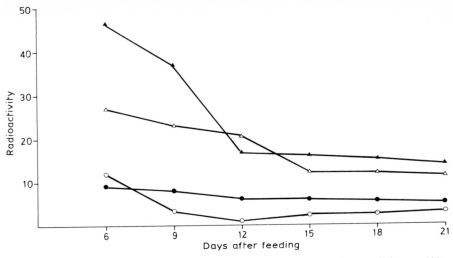

Fig. 9.2 The spread of radioisotope-labelled food in queens (▲), small larvae (Δ), large larvae (●) and workers (o) of *Iridomyrmex humulis*. (After Markin, 1970.)

narrow oesophagus; these are only ejected in their 'peritrophic' membrane after feeding has stopped.

9.1 Mastication, extraction and regurgitation

The labial glands of *Vespula* queens during their nest founding stage secrete a protease for the digestion of prey (Landolt and Akre, 1979). Although the queen starts as a general labourer she gives up foraging and stays in the nest either when she acquires subordinate associates, as frequently happens, in the subfamily Polistinae, or when the first workers start foraging. She then takes loads of prey or pulp off them and after helping herself, feeds larvae and develops the nest. The castes in wasps do not differ much; even in the Vespinae the differences are mainly of size and reproductive behaviour rather than of shape and equipment. In line with this there is only a slight difference between young and old workers, for the cuticle of the newly emerged worker quickly hardens and darkens and it goes out to forage. In *Dolichovespula sylvestris* newly emerged adults after a few hours will malaxate flies brought in by others and put the paste on the mouth of larvae, giving bigger pieces to bigger larvae. They go out to collect prey (flies) on their first or second day. The extent of pre-treatment varies: in *D. sylvestris* it takes only a minute to prepare a fly but the larva rejects the chitinous debris as a pellet which drops out of the nest whereas in *Vespula germanica* the same species of fly is treated by workers for several minutes and reduced to a finely grained paste which the larva swallows whole. The improved extraction efficiency that this implies is obtained at the cost of a greater residue inside the larva unless the adult filters crude particles out herself which is uncertain (Brian and Brian, 1952); which then is the better system? In this species young wasps take prey off foragers and both prepare and deliver it (Edwards, 1980); so for a few days they constitute a nursing class.

In *Vespula vulgaris* newly emerged workers spend up to 20% of their time in the first 5 days resting in cells but after this they attend to larvae, and distribute prey brought in by older workers (Roland, 1976). In general there is probably an increase in co-operation between wasp workers from the tribe Polistini to the tribe Polybiini and so on to the Vespinae; workers go out to forage at a slightly later age and whilst maturing will apply or process materials brought in by older ones. An old problem in wasp nutrition concerns the exchange of liquids between larvae and adults: after adults have delivered food, larvae will give up saliva though not necessarily at once. In *Vespula* this contains proteases and amylases (Maschwitz, 1966) which the adult uses to digest prey, its own mouth-part glands being deficient in these enzymes; though it has a rich supply in its midgut these come too far down the gut for the digests to be regurgitated (Edwards, 1980; Grogan and Hunt, 1977). The need of workers for larval enzymes is pronounced in the hornet, *Vespa orientalis*; here, too, the adults do the mechanical work and the larvae supply the enzymes, though sometimes unprepared meat is put direct on larvae (Ishay and Ikan, 1968). Big larvae are visited nearly a hundred times an hour and if fed on ^{14}C protein they salivate a peak of radioactively-labelled proteases $\frac{1}{2}$ h later and go on for a further 2 h. There is, thus, a useful mutualism in food preparation between adults and larvae: one provides the motor and the other the chemical necessities.

Yet there is more to it than this, for the larval saliva contains 9% sugars, four times their concentration in haemolymph and mainly trehalose and glucose (Maschwitz, 1966), whereas the concentration of amino-acids and protein is only one-fifth that in the haemolymph. However, the sugar production from ^{14}C labelled food comes after the proteases; it starts some 3 h after the meal and goes on for 20 h (Ishay and Ikan, 1968); there is thus quite a lot of metabolism taking place and though it is difficult to believe that wasp larvae de-aminate protein, which they must need desperately, for the workers, they could reasonably digest lipids, of which there are many in prey, and resynthesize sugars. If so, in the early morning larvae might be a rich source of energy for wasps waiting to start work. There are three other features that need to be considered. First, do the adults 'cream off' oil into their pharyngeal glands as they ingest malaxated prey (like some ants); if so, then the larvae have no need to supply sugar to them. Second, the adult wasps collect sugar in nectar and honey-dew and give at least some to their larvae (Brian and Brian, 1952); why then do larvae later pass it back? Perhaps they act as stores and are unable to satisfy the adult need for sugar, though Maschwitz says one feed of saliva would last a worker half a day. Third, wasps cannot regulate water content through their excretory tubules since nest-wetting would be bad for the fabric; it has been suggested therefore that water content is regulated by the salivary glands instead. So these larvae may have three subsidiary functions in addition to their main one of converting food into wasp material: enzyme production, sugar secretion and water excretion.

Only the most primitive ants go out to work young. Novices of *Amblyopone pallipes* escape from their cocoon unaided and immediately tend sister pupae. After only 5 days their cuticle hardens and darkens and they go out to hunt. In fact, in eight different activities, Traniello (1978) could find no age trend. Food is paralysed but never prepared and larvae are just put on it. Chance success in any activity that arises tends to reinforce itself but some individuals respond to social needs more than others and may be called 'leaders'.

9.2 Yolk food supplements

Doryline ants cannot have a specialized nursing class; their nomadism necessitates that the young workers mature, harden and darken quickly, even if nocturnal, and in *Neivamyrmex nigrescens* quite pale workers, 3–7 days old, can take part in raids (Topoff and Mirenda, 1978). The main class distinction is based on size: small workers feed larvae inside the resting cluster surrounded by a layered skin of big workers (Schneirla, 1971). It is uncertain whether trophic eggs are produced. Schneirla describes how *Eciton hamatum* egg shells are left after the contents have been eaten, but these were presumed to be normal 'thick-skinned' eggs.

In *Myrmecia* the pattern of higher ants is present and both size and age affect activity. Some very small workers of *M. gulosa* never forage (Gray, 1973, cited Brian, 1979a) but other species have no such variation even when size varies by a factor of two (Haskins and Haskins, 1980). Yet young workers of *M. forceps* stay a month in the nest and have special adaptations for nursing, the chief being the ability to produce purified food in their ovaries as encapsulated yolk or 'trophic' eggs (Freeland, 1958). These, unlike true eggs, have a thin, flexible chorion and are soft, spheroidal and easily pierced and sucked; they are sterile both in a developmental and in a microbial sense. All age and social categories feed on them and they can, apparently, be solicited by palpation of the genital aperture. What better source of nourishment? Although the yolk is not quite the same as that in normal eggs it may be specially composed for larval and adult requirements; this is unknown.

In the Myrmicinae the period that young workers spend in the nest depends on their size; if they are small this can last their whole life, if large they stay only a few weeks but they still have time to develop and use special nursing features such as the ovarial yolk glands, the pharyngeal glands and their sharply toothed jaws. In *Myrmica*, prey are brought in and dumped by foragers; these are the larger, older, browner and more tooth-worn workers. Nurses puncture prey with their still sharp teeth, add digestive juices and suck out the fluids. Pulped pieces are put on the venters of larger larvae as they lie on their backs; they reach over with their mouths, and add their own saliva (Abbott, 1978). Only a pellet of chitinous debris remains. Some protease comes from the mandibular glands of workers; but most comes in larval saliva, as with wasps (Ohly-Wüst, 1977). As this external digest is sucked into the worker crop it undergoes two refinements. First, suspended particles are sedimented out and collected into a pocket in the tongue (the infra-buccal pocket) where they are pressed into a lemon-shaped pellet that is, at a suitable moment, dumped in the rubbish area. Second, oil droplets are 'creamed off' in the pharynx and added to the oily contents of the post pharyngeal caeca from which they are absorbed directly into the haemocoele. Finally, the aqueous solution of amino-acids and proteins travels back along the thin oesophagous through the mesosoma into the crop which lies at the front of the gaster. Here it remains until it is either forced forwards and regurgitated to small larvae or passed through into the digestive mid gut for final degradation and absorption. A valve regulates this passage and prevents back flow during regurgitation. Presumably a worker that cannot find larvae to feed relaxes the valve and absorbs the food which later forms yolk. The bulk of the oil in the pharyngeal caeca is absorbed and may well be an important source of energy for workers. It is uncertain whether it can be passed from nurse to forager directly nor can one rule out the possibility that these pharyngeal glands (or caeca) are sometimes secretory and synthesize food for larvae. Thus the *Myrmica* worker has a very refined mechanism for digestion, filtration and fractionation of its prey.

Workers lay trophic eggs in response to an egg cluster, whether or not this is composed of their own reproductive eggs (laid when queenless) or of the eggs of queens. Indeed, a laying queen is the best stimulus of all; the more she lays the more trophic eggs she induces and so the more concentrated her own supply of food (Fig. 9.3). Moreover, trophic eggs remain for months in the egg cluster and are easily pierced and sucked by newly hatching larvae (unlike normal eggs), which do not, as a result, need worker attention until the next stage. Young workers respond to Dufour's gland secretion but not to the alarm pheromones of the mandibular gland and they have very little venom in their poison gland (Cammaerts-Tricot, 1975, cited Brian, 1979a).

The production of yolk food by definite nurses seems to be quite common in the lower Myrmicinae, e.g. queens and larvae of *Temnothorax recedens* (Leptothoracini) can induce it (Dejean, 1974; Dejean and Passera 1974). In *Messor capitatus* workers lay eggs of normal shape and chorion but they are infertile and have a disorganized yolk; they too are laid under queen stimulation (Delage, 1968) and appear to be intermediate between normal and

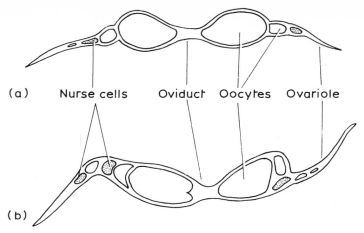

Fig. 9.3 *Myrmica* worker-laid eggs in the ovarioles. (a) Quickly produced trophic eggs; (b) reproductive eggs forming more slowly after fusion (from Brian and Rigby, 1978).

trophic eggs. This genus of ants has an age-determined nurse class and workers are at least a month old before they forage (Goetsch and Eisner, 1930). Attini may also produce trophic eggs. In *Atta sexdens* they are twice as big as reproductive eggs and are given to the queen by the medium-sized workers who attend her; fungal pellets are given too (Wilson, 1980a). This advanced fungus-ant has evolved a variable worker size that comprises some five 'castes'. The smallest (modal head width 1.0 mm) cares for the garden and brood but also rides on forage leaves as they are carried back. Slightly bigger workers (head 1.4 mm) are more general gardeners and nurses; they pulp leaves and separate oils into their pharyngeal glands (Peregrine *et al.*, 1973). Somewhat bigger ones that later forage, nurse when young. Workers with heads around 1.4–1.6 mm assist with ecdysis, those with heads around 1.8–2.2 mm carry substrate inside the nest but the very big ones (head >2.2 mm) do nothing when callow. Later they cut and carry leaves, excavate or defend the nest.

Messor capitatus extracts oil from seeds and with *Pogonomyrmex*, is unusual in not collecting or distributing sugary liquids in the society. Moreover, *Messor* do not have a mandibular gland and so the workers certainly have no supply of protease anterior to the mid gut unless it comes from their labial glands (Delage, 1968). These oddities are at the moment inexplicable.

It has been suggested (Ohly-Wüst, 1973) that a proctodeal food from larvae is circulated in *Myrmica, Monomorium pharaonis* and other ants. This is a watery liquid that collects in the larval hindgut and is derived from the excretory tubules and perhaps by filtration through the mid gut septum. In *Myrmica*, at least, it is frequently turbid with granules of crystalline uric acid from the Malphigian tubules and the workers collect it and either drink it or carry it away in their open hydrofuge mandibles, and dump it with waste. However, amino-acids and proteins with protease activity have been shown to occur in *M. pharaonis*; these could clearly be leaking from the mid gut and might well be worth collecting and recirculating.

Bees, like ants, hardly prepare food at all in the most primitive species. The Halictidae mix pollen with honey and roll it into balls but such parental care is a feature of non-social bees too and is certainly not a social prerequisite. Workers of *Bombus* show age and size bias in their work: *B. pascuorum* workers, if big, go out at 5 days old but if small at 15 days or thereabouts. The pocket-making group of Bombini store pollen in the centre of a group of larvae which feed themselves directly; pollen loads are just pushed into the pocket. In *B. atratus*, a tropical species in which pockets are made for pollen soon after the larvae hatch and are later extended as they grow, the larvae begin to spin silk partitions as they feed and come to lie horizontally. Later workers, in spite of the pocket, cut holes in the wax and push an antenna in to sample larval condition before injecting a liquid food. Finally, the larvae orientate vertically and pupate in a silk cocoon (Sakagami and Zucchi, 1965, Sakagami *et al.*, 1967a). In higher species of *Bombus* regurgitation is the rule and this enables workers to keep in touch and exercise control over larval development. The fluid given is a mixture of pollen and honey, the honey being largely hexoses but with some amino-acids. Most is used almost as soon as it is gathered, but some is preserved by glucose oxidase action and is concentrated to around 20% water and stored in wax-capped cells (Alford, 1975). The part played by the adult digestive glands, in particular the hypopharyngeal, in this food preparation is uncertain. They contain carbohydrases and other proteins (Delage-Darchen *et al.*, 1979). Both pollen and nectar are put in special separate storage cells directly by foragers as they come back since food is rarely exchanged between adults in *Bombus* even though stores are communal.

From some stage like this the Meliponini stem. For, in their standard worker cell, they put a mix of food, mainly pollen and honey, with a dose of 'bee-milk' from the hypo- or propharyngeal glands; on top of this dose of food the queen lays an egg and the workers seal the cell and leave the larva to grow and develop until it is ready to spin a vertically orientated cocoon. The bees do not have to check the size and stage of development of the larva though they must inquire if it is still alive and if it has spun a cocoon, for when it has they take the surface covering of wax away. This seems to be an ideal arrangement and one wonders why higher Bombini and all Apini devote so much time to feeding larvae. Besides injecting food into the cell, workers of Meliponinae lay eggs during the routine of cell provisioning and sealing (Michener, 1974). The queen usually eats these eggs, indeed even 'demands' that they be laid but in some species they evidently get left inside the cell so that the larva has them instead (Beig, 1972).

Although the queen can be said to induce or even to solicit trophic eggs, internal factors are involved too. In *Melipona quadrifasciata* the right ovary is bigger than the left and, unlike most insects, activity of the corpora allata reduces rather than increases egg size. In *Scaptotrigona postica* the trophic eggs laid in the proximity of queens are large, flabby, anucleate and therefore genetically sterile. Although both castes have four egg tubes in the ovary only one on each side develops and whether these yield nutritive or reproductive eggs depends on the age of the workers. Reproductive ones are more common in the left ovaries of young workers and trophic ones in the right ovaries of old workers! In these two species egg yolk protein appears in the haemolymph after 5 days and increases until 14 days when eggs are visibly growing. Yet by 4 weeks they break up and by 7 weeks the egg tubes are empty (Engels and Engels, 1977). Perhaps a worker can lay either sort of egg according to circumstance? Reproductive eggs yield males and there is no evidence that queens inhibit this (da Cunha, 1979). Yet the ceremony of cell provisioning and egg laying first by the queen then by the workers followed by egg eating recalls that of *Myrmica*, in which a laying queen apparently speeds up egg formation in workers, and so causes them to lay trophic eggs. Is there any evidence that Meliponini have switched from yolk food to glandular food as ants have? No bees yet examined have lost their ovaries entirely like some ants so that a switch, if it exists, is not obligatory. The hypo- or propharyngeal glands of *Apotrigona nebulata* are very poor in enzymes, especially carbohydrases and the head section of the labial gland contains a colourless oil that is also poor in enzymes and, as in *Apis*, of unknown use; both could therefore be secreting food (Delage-Darchen *et al.*, 1979). The thoracic labial gland and the mid gut are especially rich in sugar-hydrolysing enzymes and the authors suggest this might be related to the fruity diet of *Apotrigona*.

9.3 Head food glands

The production of yolk food is by no means universal in Myrmicinae and workers of the most advanced genera such as *Tetramorium*, *Monomorium* and *Solenopsis* no longer have ovaries at all. The alternative to yolk food may be a brood and queen food elaborated in the postpharyngeal gland but there is no convincing evidence for this in ants yet (Abbott, 1978). In *Solenopsis invicta* as in *Myrmica* these glands are certainly used to separate oil from food digests (Glancey *et al.*, 1973); this oil can be stored in the crops of the major workers and of gynes until it is needed during colony foundation. It contains hydrocarbons, sterols, glycerides and free fatty acids. All workers, including nurses, share a sugary substance and absorb the sugars from their crop (*sic*); soluble proteinaceous compounds are left for regurgitation to larvae

(Howard and Tschinkel, 1980). However, the pharyngeal glands, though they actively attract ^{32}P in phosphate solution, do not attract radioactive triglycerides or fatty acids injected into the haemolymph so that oil is an unlikely product (Philips and Vinson, 1980; Vinson *et al.*, 1980). Stages in the abandonment of yolk as a food can be detected in the Dolichoderinae (Torossian, 1979). *Dolichoderus quadripunctatus* workers (with queens) produce droplets of yolk that are passed out without a tough chorion; workers suck the yolk up and feed it to larvae. Queenless workers are able to lay reproductive eggs giving males but only 4% are fully formed; they can, however, induce yolk emission. In *Tapinoma erraticum* the droplets are small and infrequent whilst in *Iridomyrmex humilis* they are rare. The last uses its pharyngeal glands to prepare a similar food instead. Markin (1970) found that ^{32}P phosphate travelled to these glands in workers whereas it travelled to the ovaries in queens and he suggests that since this ant never feeds solid food to its larvae the pharyngeals are used to concentrate amino-acids collected in honey-dew. The contents move into the crop and are regurgitated to small larvae and fecund queens though the queens certainly also help themselves to food directly.

Young workers of *Tapinoma erraticum* are fed inside the nest for 2–3 days where they usually remain motionless on the brood near the queen. Later, as nurses, they lick, sort and pile brood though they may not feed them; this attention prevents the growth of surface microbes. About two-thirds of young workers nurse and are at their best aged 10–15 days when they receive malaxate and suck prey and regurgitate juices to larvae. The rest (one-third) are intermittent foragers, some even going out when only 4–5 days old. Young foragers are reluctant to give up their catches to nurses but age reduces this unsociability and also increases foraging tendency. This age effect overlies a bias towards either nursing or foraging that is noticeable from an early age (Lenoir, 1979b).

In Formicinae food preparation takes many forms. In *Plagiolepis pygmaea*, workers with queens present lay small, sterile eggs with weak shells that are used as food generally and are obviously 'trophic' (Passera, 1969); they arise from an early termination of nurse cell activity and could perhaps be nurse cells (trophocytes). In *Camponotus vagus*, workers only lay male-giving reproductive eggs and then only in the absence of both queens and larvae; the ovary is not used as a food gland, and all food is passed by mouth. Work bias is related to age and size since small and medium workers tend brood and collect honey-dew whilst bigger and older workers dig and forage but there is much individual variation (Benois, 1969). Workers of *Lasius niger* lay unfertilized eggs that are diploid as well as haploid and which develop into females and males respectively (Brian, 1980). The period in the nest is prolonged, for even after 40 days only half the workers can be regarded as foragers and some of the smallest ones never go out (Lenoir, 1979a). Big workers develop more quickly than small ones and soon become foragers, a change that is accelerated by nursing larvae but is not easily reversible for once a worker has foraged it finds it difficult to revert to nursing. In *Oecophylla*, workers lay regularly when queenless and produce, like *Lasius*, both the usual large male-yielding haploid eggs and smaller ones which are not trophic but are diploid and develop parthenogenetically into females. In *Cataglyphis cursor* the queen influences egg formation, oocytes are absorbed and again yolk food is not produced but queenless workers can make both diploid and haploid reproductive eggs (Suzzoni and Cagniant, 1975, in Brian, 1980). Worker egg production is thus well established in this subfamily but is deployed more towards reproduction than towards nutrition. The subtle queen-implemented control over egg formation in workers, that enables the ovary to be used as a yolk gland without risk of producing males, appears to be rarely, if ever, achieved; further research may throw more light on the situation. Schmidt (1974), in fact, says that *Formica polyctena* lays trophic eggs late in the season and that the ovaries of workers grow in spring from stores in the fat body; do they then lay reproductive eggs? Evidence points toward the use of oral food glands in spring for the postpharyngeal glands actually secrete oil; in fact, radio-labelled triolein accumulates in them after injection, and later turns up in queens and larvae. A store of oil may be useful for metabolism during metamorphosis (Peakin and Josens, 1978). *F. polyctena* also has a propharyngeal gland which is homologous with the hypopharyngeal gland of *Apis* (Emmert, 1968) but its function is unknown or uncertain at present (Abbot, 1978). In the *Formica rufa* group, task bias in workers is again related to both age and size, for the first few days a worker stands still, grooms and is ready to be picked up; after 3 days it accepts food and then starts to nurse; finally, after a period of transition it goes out to forage. Many investigators have confirmed this, e.g. Rosengren (1977). Workers, if 'born' in late June, may hibernate before foraging as they spend at least a month in the nest.

The early stages in adult life involve a lot of behaviour development (instinctive or species characteristic). In formicine ants workers have to help larvae spin cocoons and pupae extricate themselves. During this event novices learn the smell of their own species. Workers of *Formica polyctena* exposed to the cocoons of *F. sanguinea* or *F. pratensis* or even *Camponotus vagus* or *Lasius niger* for 2 weeks, later tend these rather than their own. The presence of queens during this period increases the certainty of the reaction (Jaisson, 1975). The most sensitive stage is 2 days after emergence (Fresneau, 1979; Jaisson and Fresneau, 1978). This result has been confirmed for *F. rufa* and *F. lugubris* by Le Moli and Passetti (1978). However, naïve workers can adopt the correct behaviour at the age when it is needed, by following older workers (Le Moli, 1978); a case of apprenticeship? In *Myrmica*, Carr (1962) showed that workers which emerge into a queenless colony (or experimental fragment of a colony), though they respond to queens in later life, do not respond so much as those that experienced queens when young. This has been confirmed (Brian and Evesham, 1982) and there is growing evidence that they pass through a sensitive period for the reception of queen stimuli when 3–4-weeks old (Evesham, 1982).

Another aspect of nursing behaviour concerns the role of chance in their development. This has already been brought out by *Amblyopone pallipes* research (Traniello, 1978). In honey-pot ants (*Myrmecocystus*, Formicinae) the crops of some workers accumulate honey to a degree that distends their gaster and prevents them moving: they hang from the ceiling. Wallis (1964) thinks chance events when young determine which individuals adopt this role but the general rule in other ants is that the larger individuals, especially major workers and soldiers, store liquid in their crops. These sugar solutions are preserved by pre-treatment with glucose oxidase which releases hydrogen peroxide and kills microbes as in bee honey (Burgett, 1974).

The pollen and honey mixture, called bee-bread, that *Apis* use, ferments during storage but the lactic acid formed is an appetizer. Their head glands are highly developed for food production and no yolk food is produced at all, though reproductive eggs are often eaten. Workers do have ovaries and do lay eggs but they are normally entirely inhibited by the queen and though at one time it was thought that their mandibular gland pheromone, 9–ODA, acted against their corpora allata this is now doubtful. The food glands are the hypo(pro)pharyngeal and the mandibular gland which lies near it (Beetsma, 1979; Emmert, 1968). *Apis* uses the former for making food which is called worker jelly (WJ) for larvae in ordinary cells and in male cells and it uses a mix of this with mandibular gland secretion for gyne rearing, called gyne jelly (GJ). WJ is rich in proteins but has 74% water, 4% sugars and some lipids. Larvae grow quickly on it but cannot metamorphose; for this sugar (honey) is needed and is given after 3 days of strong WJ. Pollen is present only as a contaminant (Bailey, 1952; Simpson, 1955). The hypopharyngeal glands also produce all the glucosidase and glucose oxidase and so function as honey-making glands, for nectar is subject to a long ripening process involving hydrolysis of sucrose, liberation of hydrogen peroxide and, finally, concentration by evaporation often from the tongue of a worker bee. Eventually the honey is sealed in a peripheral cell with a wax cap. The hypopharyngeal glands may be useful in pollen digestion too (Arnold and Delage-Darchen, 1978; Simpson, 1960). The long-chain lipids that occur in pollen are not digested but absorbed directly; apart from their obvious use as a dense energy store they carry vitamins including cholesterol (Herbert *et al.*, 1980). Acid phosphatase and phosphoamidase are also present and concerned in protein synthesis. A study of the protein production in relation to worker age by Halberstadt (1980) shows that the proportions differ in many respects from those in GJ, of which the secretion forms a large part. Whereas the gland secretion shows a definite enzyme content at all ages even in young bees there is only a trace in GJ. Bees of 30-days old, and also swarm bees, have a high saccharase content which, of course, they use to release carbohydrate energy in flight. Mature bees also initiate honey making, though on return from a trip they give their collected foods to nurses.

Work bias is predominantly age-related rather than size-related though foragers are, on average, slightly bigger than nurses (Kerr and Hebling, 1964). The period spent inside the nest can range from 5–41 days and average 20 days (Michener, 1974). Young bees are adapted for house work and undergo a series of developmental changes that are not rigidly programmed but sensitive to the current situation in the society and to chance contacts which new workers make. However, once the activity of a glandular system has been started it goes on for several days. Nurses use their mouth part glands in feeding and their ventral intersegmental glands for supplying wax for building; both systems can be active from the age of 2 days. Thus a young bee, after emerging from its cell, cleans and lines other cells with an unknown material that induces the queen to lay in them. It may then build comb if this is in demand (as it is in a developing swarm) or it may deliver food to larvae in a fast-growing colony. In feeding it shows no bias either towards small or large larvae as was, at one time, thought. Nurses prepare cells for pollen storage, receive nectar and eventually make orientation flights prior to foraging (Free, 1965). Thus *Apis* has a storage system, an adult digestion system that supplies all the necessary enzymes without recourse to larvae, head gland food but not yolk food, continuous contact with larvae and a largely age-related, but size modified, work bias that is sensitive during the long in-hive phase to the needs of the society. They can even, in emergency, reverse the ontogenetic trend from nurse to forager (see Brian, 1979a). A lot remains to be found out about how these adjustments are made, what stimuli induce them and how much individual behaviour is affected by chance contacts.

In Termites work is done by juveniles who are merely smaller, weaker and paler individuals than the adults but quite well equipped for work in the nest. In *Zootermopsis nevadensis* (Termopsinae) sterile larvae and nymphs dig and build whilst young larvae distribute food (Howse, 1968, 1970). In the nests of *Hodotermes* (Hodotermitidae) the younger, paler individuals process and distribute the grass collected by darker older workers (Hewitt *et al.*, 1969a, b, cited Brian, 1979a; Nel, 1968); size affects the task performed in termites as in ants and bees by speeding development and pigmentation. Thus, in *H. mossambicus*, workers derived from 5th stage larvae are bigger than those from 4th stage ones and do heavier more defensive work.

In the Termitidae, even inside activity is done by pigmented sclerotized and active workers derived in a single moult from larvae, though these may be in different stages of development at the time and so become workers at different sizes. Again, age biases the task; in *Nasutitermes lujae* young workers of stage 1 have better developed salivary glands and function as nurses whilst older ones have better developed sternal glands and collect food (Pasteels, 1965). Sex influences work type through its effect on size and strength, for in the Macrotermitinae the smaller males nurse whilst the larger females forage (Noirot, 1969, 1974). Apparently, there is no general rule about which sex is the larger.

Digestion of vegetable material by termites is internal. The mandibles play a big part in cutting it into small pieces and in humivores, for example, they show clear signs of wear as a result. After ingestion it is further ground up by a gizzard and passed into a hind gut with one or more enlarged pouches containing the symbionts. These are prevented from passing forwards by an enteric valve. In Termitidae it is thought that the gut bacteria produce many of the enzymes and may even fix nitrogen. In the lower termites the protozoan symbionts can hydrolyse cellulose and lignin, though, for the latter process, oxygen is needed and is not usually available in deep gut interiors (La Fage and Nutting, 1978). The diet of termites is poor in lipids, and wood, in particular, usually has less than 1%; yet the fungus combs of Macrotermitinae contain as much as 17% lipid. Lipids also accumulate in nymphs and imagos (the reproductive line) and probably form a vital store for respiration during colony initiation, for it may take years to break down the flight muscles in termites (Noirot, 1969, cited La Fage and Nutting, 1978, p. 208). Vitamin requirements are unknown.

Saliva is a clear opalescent fluid stored in reservoirs before emission. Initially, of course, it is a digestive fluid and so in *Calotermes flavicollis* it secretes an amylase but in general it is used as a food gland or for cement or even as a fungistatic. In *Odontotermes* weed fungi are controlled with it (Batra and Batra, 1966). As a food gland it is used to sustain juveniles, reproductives and soldiers, all of which are poor in gut bacteria (Noirot and Noirot-Timothée, 1969). It contains various lipids and is probably rich in proteins. In *Nasutitermes lujae* it is secreted by the first stage workers which stay in the nest and tend dependent forms (Pasteels, 1965). Saliva is the food transferred by oral exchange, not regurgitated crop material which consists of a paste or pulp of humus and may be given to soldiers in some species. It is digested in the early stretches of the gut.

Anal food is not the same as faeces (Noirot and Noirot-Timothée, 1969). It is a viscous hind gut fluid containing a suspension of fine particles of wood and symbiotic protozoa. Only the non-termitid families circulate it and undoubtedly it is concerned largely with the transfer of flagellates to freshly moulted individuals. However, it must contain nutrients other than these protozoa and can be regarded in *Calotermes flavicollis* as a normal part of the diet (Alibert-Berthot, 1969). Although it has not been analysed it is likely to contain some products of digestion.

Thus in all the main groups of social insects there has been an increase in the time and energy given by the adults to preparation of food for young. This had been anticipated presocially and is a form of parental care. Yet in social insects it can lead to enzyme sharing in which the adult masticates with digestive juices from larvae. The extract is then purified by filtration or sedimentation and concentrated as a glandular secretion which includes yolk from ovaries which function as both food glands and ovum producers. Although this system has been adopted by ants and stingless bees it is not universal, and it appears that some ants and honey-bees use digestive (salivary) glands as food glands though the glands used by each group are not homologous.

This trend towards great food care has been necessarily accompanied by a trend towards greater differentiation of a nursing caste, that is, a group of workers either younger or smaller than average (or both) whose glands are able to develop and secrete food and who can effectively contribute to the society before their pigment and exoskeleton are sufficiently mature for them to forage outside. The ultimate in this trend comes with species such as *Pogonomyrmex owyheei* in which less than 10% of workers forage and their longevity once outside is only 14 days (Porter and Jorgensen, 1981). Before they go out they are tooth-worn, lack tissue reserves and perhaps normal sensory equipment too; their disposability is programmed!

The number of castes which a social insect evolves is never very great. In *Atta sexdens*, four are based on size and shape (allometry) and the three smallest of these groups have age trends within them giving at most seven castes (Wilson, 1980a). The arguments against excessive structural specialization are given by Oster and Wilson (1978) and Wilson (1979); they all depend on the fact that increasing differentiation of function necessitates a larger number of individuals to complete an operative set. This brings increases in cost of maintenance and distribution. In this sense distribution is not just the spreading out of the food appropriately, it is finding the situation in which a worker functions most efficiently: in fact finding a suitable job.

CHAPTER 10
Early population growth

In ants, bees and wasps most of the protein fraction in food is distributed to reproductives and larvae, whilst the oils and carbohydrates are shared between workers. This distribution is not just an anarchic scramble, certainly not in the Hymenoptera where juveniles at least are immobile as in ants or fixed as in bees and wasps. With termites, though eggs are carried, juveniles can run about, and less is known about the rules of food distribution or crowd control. In ants, bees and wasps those that hatch first, get the best treatment and establish an accruing advantage. Queens can take food away from larvae and eat it themselves and they also deprive subordinates by force in polygynic associations. However, the queen of *Melipona favosa*, though she grips workers' heads in her forelegs and antennae, rarely gets food. She takes some of the food provided for larvae just before laying and eats any trophic eggs that workers deposit (Sommeijer *et al.*, 1982). Also, though reports of rapid food transmission exist in the honey-bee, this is not necessarily true in all circumstances. Many reciprocal oral contacts between workers are brief and involve no food at all; there tend to be 'askers' and 'donors' whose states are determined by more profound physiological conditions than their crop contents. In fact, an 'asker' may beg when it is standing by food (Korst and Velthuis, 1982). Queens are only fed by workers about six times an hour (Seeley, 1979) and it seems much more likely that pheromones are transmitted by oral contact.

10.1 Food distribution

In two 'pollen storing' *Bombus* species (*B. terrestris* and *B. terricola*) larvae are grouped into communal cells (chambers) until the final instar. Workers cut a hole in the wax to inspect, then close it again. In the last instar it is left open all the time. The interval between inspections (average 11 min) is about half that between feeds (average 20 min). Feeding is irregular because workers often have to queue for pollen at the store; 5 min can be spent in this way. Is this a shortage of storage facility or of pollen itself? When Pendrel and Plowright (1981) removed half the workers, the others compensated by raising the feeding rate/individual by 28%. If they can keep this effort up it suggests that normally they are by no means overworked in the nest.

Sharing a pollen mass like this is uncommon; even pre-social bees split their brood up into cells provisioned with balls of pollen. These may be made in two different sizes; some *Evylaeus*, both social and otherwise, lay unfertilized eggs on small balls and fertilized eggs on large balls and thus, in effect, have smaller balls for males than females (Knerer, 1977). In *Lasioglossum zephyrum* of this group, once a queen has been differentiated from a group of equals she takes up a position in the nest from which she can influence worker activity, e.g. at the end of a queue of bees near the entrance or at tunnel junctions. From such places she encounters and 'nudges' workers. She also makes brief inspections of cells and Buckle (1982) suggests that she may be gathering information on the state of development of cells and using this to guide workers to places where they are needed. In the Ceratonini with chambered nests, pollen is not specially prepared and stored as in other bees; after mixing with a regurgitated liquid, probably nectar, it is usually put near the larvae which feed communally. Other larvae have special pieces of food put on their bodies within reach of their mouths. In *Exoneura* the larvae appear to signal a need for food by moving their jaws and heads; workers then hold them between their jaws and regurgitate clear fluid (Sakagami, 1960). Larvae are often quite methodically arranged in these linear hollow-stem nests since eggs are laid at one end far from the entrance and larvae move up as they grow eventually pupating near the open end (Michener, 1974). This arrangement develops gradually; it is not implemented by workers as an immediate response to the existence of brood classes. In a stingless bee (*Melipona favosa*) though many builder-caste bees start a cell, only a few (some five to eight workers) both finish and provision it; between 10 and 18 food discharges are put into each cell. The queen, normally on the floor of the hive or in the stores zone, inspects new cells from time to time and, if she finds one with a smooth collar, drums on top of it. She rarely gets food in this way and it may be an oviposition signal. She helps herself from larval food just before laying an egg and later eats any trophic eggs that the workers associated with cell finishing and provisioning may lay (Sommeijer *et al.*, 1982). In the honey-bee, cells are not in short supply in spring owing to the broodless winter state, but the queen is forced to restrict her egg laying to cleaned and lined cells in the zone where workers can hold the temperature up. This, of course, is a function of the volume of workers. The queen is fed on saliva made by workers out of their body reserves and out of stored pollen; during the summer she is licked and nibbled a lot. A nurse with actively secreting hypo(pro)pharyngeal glands examines each cell minutely before putting a liquid droplet either on the cell wall or on the larva. She does not seem to remove either saliva or urine from the larva. With such a minute supply unit it is no surprise that the number of food visits needed is extraordinarily high and one individual bee can only rear two to three offspring in her lifetime.

This should not be thought of as concentrated on a group of cells, for any individual nurse will probably supply many hundreds, if not thousands. An important regulatory feature, however, is that when food is short it is concentrated on the occupants of a few adjacent cells (Gontarski, 1949) and distribution is entirely worker controlled as larvae cannot compete for food when they are separated by a cell wall from their neighbours.

Although larvae do not 'feed' workers as happens in wasps, workers do exchange food between themselves occasionally. A solicitor pushes its tongue into the oral region of a potential donor who extrudes a droplet of nectar when it would not otherwise do (Free, 1959, p. 194). No doubt a donor with an empty crop takes the first opportunity to go out and forage. Pollen eaten by nurse bees is brought in by foragers and put in prepared cells for temporary storage; foragers that cannot relieve themselves of loads in this way are reluctant to collect more and if pollen is given to the colony, less is collected. Brood stimulates feeding by the young bees mainly through a tactile, but partly through an olfactory, channel. Queens stimulate collection of pollen even if there is no brood (Free, 1967; Free and Williams, 1971).

The population of the ant *Amblyopone pallipes* consists only of a queen and 9–16 workers. Eggs are laid in August and this brood is then reared to maturity yielding sexuals and workers the following August. Thus, at any time, one brood stage is predominant and a highly periodic system exists which is perennial, and linked to season (Traniello, 1978). In the Myrmicinae small larvae get regurgitated fluids and big ones solid pieces of food, either prey or seed (Le Masne, 1953). In the first instar trophic eggs are eaten if available, in the second regurgitated juices are added to the eggs and in the third pieces of prey. The third is a long instar and includes two stages of dormancy which arrest growth in summer in preparation for hibernation. The first blocks the growth of those on a worker path of development, the second those on a gyne path. For simplicity it would be attractive to think of two actual types of larvae; worker and gyne forming, but the latter are, in reality, castle-labile and can switch back to the worker path.

In winter the worker/larva ratio varies about an average of 0.88 and shows a correlation between the numbers of larvae and workers and an even closer one between their weights. This is because the worker population determines the number of eggs laid by the queen, the survival of these eggs and of the young larvae from them, as well as the growth of the larvae; this is especially so in the third instar when their weight at least trebles (Brian, 1965, 1979a). This assumes that enough food is available; if it is restricted a limited number of the bigger larvae are fed and the rest are destroyed, usually as they moult. Probably the combination of 'hungry' workers with a strong predatory urge and feeble larvae with weak recognition cues is enough to finish them off.

To select the bigger larvae and spare pupae also makes sense as they are nearest maturity and can soonest aid the workers domestically and release

older ones to hunt for food. Third instar, dormant larvae can hold out
without food a long time even at growth temperatures, but if they hold out
too long they run into competition with newly hatching spring larvae which
are favoured by workers. In natural breeding chambers larvae in the third
instar are spaced and spread out, each one on its back and each accessible
from the ceiling or the walls and they are fed fairly evenly; in experiments on
flat plates without walls they are piled up (this is probably so that they can be
surrounded by defensive workers) and even though the pile is searched and
turned the upper ones get more and so grow more. Thus, the evenness of food
distribution, even where food is ample, depends on the chamber structure
giving sloping walls and ramps and providing a physical retaining structure in
which workers can spread out larvae without risk of interference.

Apart from nest structure queens also encourage the even distribution of
food, perhaps because of the general stimulus they give to food delivery. This
is not because the queens themselves use much food (for their dead body is
almost as good a stimulus;) it is just part of the catalytic effect which queens
have in their societies and which has increased their evolutionary fitness.
Larvae (Fig. 10.1) are not very demonstrative in species of this genus and,
unless they are very deprived they wait to be offered food, which they
sometimes accept and sometimes reject, apparently an inefficient procedure
for if they are persistently refractory the workers move them away into an

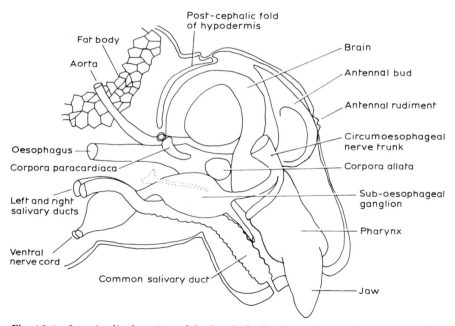

Fig. 10.1 Longitudinal section of the head of a final stage larva of *Myrmica*. (After
Weir, 1959.)

inactive group. This comprises diapausing larvae in their third instar and larvae that have stopped feeding prior to metamorphosis (pharate pupae). The former category can have their 'appetite' stimulated by worker attention. Although their head forms a focus for worker attraction and extending and waving it about a signal that they will accept food, it is their ability to take it off nurses, absorb it and then take more that makes the nurses go in their turn to foragers and beg (Brian and Abbott, 1977).

A combination of the fact that larvae can be induced to take more and that queens stimulate food dispersal, affects the 'decision' made inside larvae about whether to grow straight into an adult or to diapause and hibernate in the last larval stage. At temperatures below 22°C, workers favour larvae which have a marked tendency to metamorphose but many less prone succeed in getting enough food to pass the first physiological barrier to growth or primary diapause, and they complete development into workers before winter sets in. The size of the residue, as we have seen, is related to the number and weight of workers. Hibernating with larvae instead of just adults is a way of fitting a second brood cycle into the year so that there is a summer cycle of eggs that reach the adult stage and a winter cycle in which larvae are active in autumn and spring but dormant during mid winter when all larvae are in the third instar. The summer cycle produces workers; the winter cycle produces sexuals and workers.

A new batch of eggs can be started in the next year before the previous batch has become adult and although this creates overlapping batches or brood streams which compete for worker resources, it has the advantage of enabling workers to extend their active season into winter for as long as larvae are able to absorb food which they can do down to temperatures near 10°C (Brian, 1965).

In constructing a model of a *Myrmica* society (Brian *et al.*, 1981), competition between queens and larvae for food was difficult to formulate without causing instability. The system adopted assumes that the food available is divided between the two categories in the ratio (by weight) of an egg to a pupa so that for every pupa produced just one replacing egg is laid. This keeps a steady stream of growing units through the feeding chamber, spreads the load over the worker population and maintains the continuous presence of larvae by inflow of eggs and outflow of pupae at the same rate. Food storage is another device for diminishing the population pulse. In *Myrmica*, trophic eggs accumulate in the queen's egg cluster after most larvae have metamorphosed. Foragers can thus continue to collect food and nurses to process and store it as yolk during periods of larval scarcity (Brian, 1977).

10.2 Colony foundation

New colonies arise from single queens, pairs of termites, groups of queens or swarms of workers and queens. How these are prepared is the theme of a later chapter. Here the early stages of population growth of a few types are described.

In all wasps and bees the queens forage until workers are produced; this is a very risky period when outdoor activities are best minimized. One way is for the queens to aggregate and establish a power gradient or hierarchy that amplifies small initial differences at the time of joining, in size, activity and, perhaps, in intelligence. This enables one female, the top one or queen to obtain first pick of any food that is brought in and to augment it with any eggs that others may lay. This, plus rest, enables her to actively intervene when necessary and so to maintain her position. The discomfort and privation in the nest to which subordinates are subject probably causes them to go out and seek food. That they return instead of making a new nest is, at first, surprising, but in dense populations new sites are difficult to find and the subordinate queens may have outlived their 'instinct' to explore these even if the nest experience has not changed their attitude. It is usually thought that starvation or just lack of rest suppresses the activity of the corpora allata and so of the ovaries and enables them to forego aggressive competition for oviposition sites. In general, ovary activity and corpora allata activity are associated with dominance in the social system to form a reproductive syndrome (for examples see Brian, 1979a).

In the tropical wasp, *Parischnogaster mellyi* (Stenogastrinae), though most nests are founded by single queens, sometimes as many as four co-operate (Hansell, 1982). The senior female who is sometimes, but not always, the mother of the others, receives food from them but never gives any away, except sometimes to larvae; the associate females exchange food, feed larvae and even males who, though they are threatened by butting, never ripost. Butting between females usually takes place head to head and is followed by a motionless crouch or no apparent response, rarely by pellet sharing or regurgitation. The senior female is thus able to rest on the nest near the top, develop ovaries and lay all the eggs but she co-operates in building cells and feeding larvae. In fact, since daughters that stay gain technical expertise and may even take over eventually, it probably pays them to help especially if the mother is elderly.

Most interesting is the fact that in some polygyne Polybiini, e.g. *Metapolybia*, in which workers and queens start new colonies together, workers recognize and 'salute' all young queens at first but later come to distinguish those that grovel and regurgitate from those that respond viciously. The low status ones they harass and penalize until they either leave for good or become workers in all senses of the word, even resting with them in inferior nest zones. The workers thus combine to elect one female, the least easily intimidated, as a queen during this phase of the society. Should this one die or be removed experimentally several associates take over and the society again becomes polygyne for a while (West-Eberhard, 1973, 1977). Thus the workers are actively selecting one queen by withholding and even withdrawing food from all the other contestants, some of the food so reclaimed no doubt goes to the queen or to her daughter larvae. Association benefits the dominant female by reducing her dangerous outside activities and her nett energy expenditure.

In *P. fuscatus* the chances of survival of the nest are enhanced by pair-formation, but production is not (Gibo, 1974, cited Brian, 1979a). In *P. metricus* there is a danger of usurpation when the female is away if she has no associates to act as guards and at high population densities single queens do poorly (Gamboa, 1978; Gamboa *et al.*, 1978). Nest that are started earlier attract joiners and do better (Gamboa, 1980). Often the conflict between queens occupies too high a proportion of their time and energy and interferes with the smooth development of a nest, e.g. *P. canadensis*, in which queens can start colonies and copulate later (West-Eberhard, 1969), and *P. annularis* (Hermann and Dirks, 1975). The ability to settle down to a dominance order may be easier if the females are related, e.g. sisters or half-sisters (Klahn, 1979; Metcalf and Whitt, 1977a; West-Eberhard, 1969). In another polistine, *Mischocyttarus mexicanus*, studied in Florida, co-operative tendency varies seasonally: high in autumn low in spring, when a strict hierarchy and effective monogyny develops. Litte (1977) suggests that the greater abundance of food in spring enables queens to collect it quickly and so to reduce the time that the nest is left unguarded. However, single queen nests, though less vulnerable to birds than multiple queen nests, are unable to keep ants out. The ecological situation is not simple. Nevertheless, there is a good positive relation between cells made and the number of initiating females in the colony if these do not exceed six. Survival is better too and damaged nests are more likely to be re-started.

In *Polistes* a few females often found nests jointly and there is good evidence that this is advantageous. Alone, a queen will construct 20–30 cells, depending upon the species, lay eggs in them and stop to feed larvae on fresh food (with some eggs included) when the first hatches. Fewer than 10 pupate and mature into workers and whilst the queen waits for them to emerge she resumes laying in empty cells (Brian, 1965, gives more details of this). Nevertheless, many single queen nests do not reach the worker phase, possibly because queens cannot incubate well enough or get killed whilst foraging.

In *Dolichovespula sylvestris* (Vespinae) queens start nests alone and kill would-be joiners or usurpers. They make cells, surround these with up to four envelopes, lay in the cells and feed larvae as they hatch, relaxing egg laying meanwhile. Larvae are clearly given priority over egg production by the queen herself, presumably because of their movement, smell and perhaps salivary exudates. Their size is estimated by adults since bigger ones make bigger pellets and are obviously given bigger pieces of food. In several Vespinae large larvae scrape the cell walls and the sound recorded and played back attracts wasps with food to donate (Ishay and Schwartz, 1973; Jeanne, 1980). These larvae readily accept food and this sound may well improve distribution and reduce the advantage that larvae in cells near the entrance to a nest have over those inside. On average, *D. sylvestris* queens produce 40 cells, 38 with brood; of these 7 contain cocooned pupae, 19 larvae and 12 eggs (Archer, 1981a). It can be argued that this species lays too many eggs, since many of the larvae the queen starts off are later starved. They yield, after the workers have taken over, undersized adults that are even smaller than the first workers the queen rears entirely alone (Brian and Brian, 1952). However, it is difficult to decide whether 'few large' are better than 'many small' in an undefined ecological situation. Queens vary a lot in their level of activity and only 10% produce a large worker force; the biggest queen nest usually has four envelopes. Archer (1981b) noticed that the paper made by queens is paler than later additions made by workers and was able to show that the size of the original queen contribution correlated with that of the final nest. Presumably once a good place is found success depends on the energy and sense of priority that the queen brings to the work: not too many cells so that envelopes are neglected and not too many eggs so that larvae are neglected. Envelope making is wholly beneficial in that it does not lead to future commitments; it is, in fact, an ideal time-filler.

Bombus are, like *Vespula*, extremely intolerant in spring of other queens. Even so, they are subject to social parasitism by species with suitable entry tactics, e.g. the thick skinned *Psithyrus*. Colony growth is probably limited by the capacity to make wax; this may save them from over production of eggs, some 12 of which all share a small wax cell in the first place and are guarded and incubated close to the body of the queen in her fibre-lagged nest. She has to forage for pollen and nectar to keep the larvae alive and use the sugars to produce wax and heat for incubation. Larvae spin cell walls between themselves when they grow large and later they make cocoons; the wax scraped from these starts a new egg cell and brood cycle. In pocket-making *Bombus*, several larvae feed on the same pollen mass (Fig. 10.2) and it is obvious from their difference in size as pupae that they have conflicted, those nearest the centre growing more than those at the side. There is also an influence of the worker/larva ratio on adult size (Plowright and Jay, 1968). This indicates the importance of worker and larva number being adjusted in the growth of the society especially as worker size affects their work bias. Only three weeks is needed to produce workers and then only a few days more before the bigger ones are able to forage and relieve the queen.

Like *Amblyopone pallipes*, primitive ants take a long time to start a colony. *Myrmecia forficata* lays some eight eggs which take 4 months to yield about three workers; *M. regularis* takes up to 8 months to get workers (Haskins and

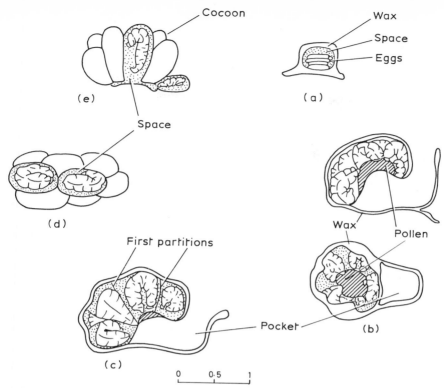

Fig. 10.2 Five stages in the development of a cluster of eggs into pupae in *Bombus atratus* (a) eggs in wax cell; (b) larvae eating communally from a pollen mass on the top of a wax pocket; (c) larvae making silk partitions, leading through (d) full grown larvae, to (e) cocoons replacing the wax which is scraped off to make an egg cell. (After Sakagami *et al.*, 1967a.)

Haskins, 1950). *Myrmica* found colonies with single or grouped queens as well as by colony division. The queens forage a little and convert their fat body stores and wing muscles into eggs. One can lay 25 eggs and produce, in the first season, three new small workers and five dormant larvae (Brian, 1951). This takes 9–10 weeks and is much slower than the rate for annual bees and wasps, partly due to the fact that the ant queen cannot incubate and relies entirely on the microclimate she can find locally. Specially small workers are produced at first and the worker-biased eggs that give rise to these are a feature of young queens. The establishment of a small group of workers before winter can be very important for some species. The northward extension of *Tetramorium caespitum* seems to depend on the ability to have nuptials early enough in the summer to accomplish a full-scale development before autumn. The queens store an orange-coloured oil in the crop and pharyngeal glands and start to lay as soon as the soil chamber is ready. They may produce a dozen or so pupae giving some 10 workers within 2 months.

In *Solenopsis invicta* single queens establish colonies without foraging and in about 20 days at 30°C small workers appear (Markin *et al.*, 1972). The queen lays within 2–3 days of completing her chamber and accumulates 20–50 eggs; after these start to hatch on day 5 most of the other eggs, which are perhaps sterile, are eaten. The queen extracts larvae and feeds them on regurgitated liquids derived from autolysed wing muscles, and in rearing a brood she loses half her weight in this way. Under natural conditions the mode is to have single queens but as many as five (rarely more) may group together. In laboratory culture where they are free from disturbance and the effect of predation is excluded the advantage of additional queens is slight. Thus, in rising from one to four the egg yield increases from 35 to 86 (2.5 times) and workers increase from 14 to 22. However, the percentage survival with one, two and three queens rises from 62 to 73 and 80% but with four queens drops to 45% on account of a parasitic fungus. The laboratory culture data shows that three queens together gives a much lower yield of workers (18) than three separately (42) and it is clear that association carries many disadvantages.

In *Messor aciculatum*, up to four founder queens occur together in dense populations (Taki, 1976). Queen number does not affect the time taken to produce the first worker but total production increases to a limit of around 50 with eight experimental queens. This limit is apparently caused by increasing larval mortality and interference between queens, since their weight loss in big groups, is less than in small, and weight loss correlates with the number of workers produced. So the production per queen is probably most with single individuals and total production most with eight individuals leaving a compromise optimum of around four queens.

Both *Lasius niger* and *L. flavus* queens show co-operation with only brief periods of tension as eggs are laid (Waloff, 1957). Although more eggs are laid with more queens, there are fewer per queen and fewer eaten and it seems likely that some queens drop out; yet workers appear after 4 months with two to three queens instead of 7 months with only one queen. With pupation, the queens disperse either peacefully (*L. flavus*) carrying pupae with them or agonistically (*L. niger*). In the latter species monogyny is the rule, but *L. flavus* is known to be less strict in this. In view of the diminishing return for each additional queen the optimum would be less than five. This takes no account of the possible use of supernumary queens in defence or of the risk of disease at high densities.

Oecophylla longinoda queens start nests singly and obtain larvae within 10 days at 30°C. These are used to silk over the crevice where they live, and after 18 days workers appear (cf. *Solenopsis*). Thus, for *Oecophylla* and *Solenopsis* living at 30°C, 3-weeks delay before a worker appears is short compared with, say, *Tetramorium* (4–5 weeks) or *Messor* (10 weeks) and certainly with *Myrmica* (12 weeks). Of course, the optimum temperature for the last genus is 20°C not 30°C. Two special ant cases can be briefly mentioned, both in the subfamily Myrmicinae. *Carebara* queens, which live in termite mounds and feed on the inhabitants, have such small workers that they can cling to their legs during nuptial flights and be ready to help (Wheeler, 1936), thus cutting out much risk and time. *Acromyrmex octospinosus* queens have plenty of reserves for making small workers, but the queen forages for leafy matter to start a garden which she inoculates with hyphae from the parent nest carried in their infrabuccal pocket. The first workers take 40 days to appear after which there is still enough store left over for 2 weeks (Jutsum and Fisher, 1979).

With termites a pair, male and female, start the new colony. In *Calotermes flavicollis*, which eats dry wood, the 10–20 first young that hatch after nearly 8 weeks are fed on a digestive gland secretion and later on wood paste. After a year, 50–60 individuals are present, 65% of which are soldiers; this figure drops to 3% as the colony grows (Lüscher, 1961, cited Brian, 1965). *Neotermes tectonae* of the same family starts in dead wood in trees in the tropics and may have only 30–60 workers (including 3–7 soldiers) after 2 years (Kalshoven, 1959, cited Brian, 1965). In the laboratory, two colonies started close together fused and one reproductive pair was eliminated. Their rates are thus much slower than ants in comparable climates and one can suggest that this is due to poor protein supply in the food which holds up growth once the sexuals' stores are finished. *Reticulitermes hesperus* (Rhinotermitidae), also a wood-eater, starts even more slowly. It is a month before any eggs are laid and incubation lasts 2 months; after 5 months there are no more than a dozen individuals in the nest. First and second stages last only 17 and 20 days, respectively, but the third varies from 18 days to 6 months and depends on the intensity of competition. Again, soldiers are more common in the small groups and in *R. lucifugus* the young are fed orally by the male (Buchli, 1950).

In Termitidae the pair digs in quickly and makes young from their reserves. *Tenuirostritermes tenuirostris* (Nasutitermitinae) starts at five eggs a day, 3 days after pairing but has only 60 at the end of a month (Light and Weesner, 1955, cited Brian, 1965). Young larvae are fed on saliva and grow so quickly they can forage for dead grass 2 months from pair formation; nasute soldiers comprise 25–35% of the first brood. With *Ancistrotermes guineensis* (Macrotermitinae) egg laying is again rapid since the adults have internal stores: five a day for 10 days and by day 60 there are small and large workers and soldiers but it is not until day 75 that they break out and start to form fungus combs of faecal pellets (Sands, 1960). This seems a very casual process compared with *Acromyrmex*.

A species of *Microtermes* (Macrotermitinae) that make small fungus combs underground and connect them with galleries, has pairs that dig a shaft and chamber, copulate and begin to produce eggs within 3–4 days (Johnson, 1981). The male actually grooms the female and helps her lay eggs as well as copulating from time to time. She starts at four eggs/day and drops to one/day after 11 days but stops when the first larvae hatch after 28 days. After 13 weeks there may be 20 big workers (male), 15 small ones (female) and 4 soldiers (female). This sex ratio is compatible with unity. After a week, 10 small workers break out and move up towards the soil surface and a few days later they start a fungus comb as a cluster of faecal pellets that form a pillar. The fungus that grows is apparently derived from spores in the gut of the pair. Soft wood is collected and more faeces added to the comb but it is usually 5 weeks before it is used as food.

The foundation rates of five species of *Trinervitermes* have been compared by Sands (1965); all start egg laying within a day of copulating but the rate declines progressively. Like *Tenuirostritermes* they have some 30–60 eggs after 12 days though *T. suspensus* has fewer than the other species. Larval hatching rate is variable and smooths out a peak of egg production. Sclerotized workers have two white instars beforehand and are all female. The males have four instars including the pre-soldier which precedes the minor soldier. As male stages are shorter, soldiers manage to mature at about the same time as workers. Again, the latter stages take longer, probably through food shortage. The percentage of workers at the time the first escape duct is made (about 3 weeks after first appearance) is for *T. ebenerianus* 70.5%, for *T. carbonarius* 55.5%, for *T. oeconomus* 60.0%, for *T. auritermes* 55.9%, and for *T. suspensus* 66.5%; quite variable.

Since male and female already form a co-operative team with the male not only helping build and defend but supplying salivary food, co-operation between several pairs provides hardly any additional advantage and is rarely found. Records indicate that pairs are normally hostile to each other and even if the infertile castes pool to form one colony the reproductives fight (Nutting, 1969). Termites probably use eggs as food during nest initiation, but the biggest difference between termites and ants is that the former make soldiers much earlier.

10.3 The growth spurt

This comes soon after the initiation of swarms. With a worker force, and soldiers if necessary and food sources located and protected, a colony can build up a worker population quickly. Although it is intuitively clear that more workers mean more work done, this is easily checked experimentally and a numerical relationship formulated. In *Myrmica rubra*, if a single queen is given an increasing number of workers she lays more eggs up to a limit which is reached with only 20 workers (Brian, 1969). Thus as the average

Plate 16
Chamber and reproductives of *Macrotermes michaelseni*.
Above: A vertical section to show the horizontal floor and arched ceiling. (Photograph courtesy of J.P.E.C. Darlington.)
Below: A chamber with the lid removed showing two queens on a smooth floor with small workers, a few large soldiers and a male (on the left).

number of workers/queen in this polygyne species is 76 (Elmes, 1973) there are many surplus workers to look after larvae and there is no reason to suppose that a queen cannot maintain maximum output of eggs whilst most workers (say, two-thirds) are looking after larvae. The more queens per worker, the lower the average individual output of eggs (Brian and Rigby, 1978); some probably suffer and even drop out of the competition for food. The best number, which maximizes egg yield and minimizes mutual interference clearly increases with the number of workers, but the right number of queens for a given worker force will certainly not correspond with the right number of larvae for the workers to feed. These problems await experimental analysis and numerical modelling.

The importance of worker number can be demonstrated in relation to egg survival too. Fifty eggs given to *Myrmica ruginodis* workers in groups ranging between 3 and 40 yielded larvae on a diminishing return curve with an upper limit of around 12 (Brian, 1953); only 20 nurses were actually needed to obtain 11 larvae. A compromise between group output and individual efficiency gives 11 workers as the most economical number able to develop seven or eight larvae. The same can be worked out for the growth of larvae, thus using 50 third-instar larvae just after hibernation and giving them a range of workers from 5 to 320 it is possible to show that 80 workers suffice for maximum growth and maximum pupal yield with negligible mortality. Since again each worker contributes less as the group 'grows' an economical value of 42 workers, which gives about two-thirds maximum yield, can be calculated. This gives an optimum ratio workers/larvae of 0.84, virtually the same as that found in wild colonies collected in only one season: 0.88. This certainly suggests that wild nests are functioning near their optimum as regards brood rearing. This means, of course, that economical functioning is favoured by natural selection, not that the workers are always able to work efficiently.

In a much bigger sample of *Myrmica* colonies of various sizes and species, the ratio workers/larvae averaged 0.85 with 2000 workers (Elmes and Wardlaw, 1981). The seven species all showed the same regression between workers and larvae but not proportionally: with 1100 workers the ratio was unity, with 1000 workers it was 1.14 and with a hundred workers it was as high as 3. So that, instead of rising, or at least staying steady as number of workers increases, as one would expect if density reduced efficiency, it actually falls slightly. This may mean that larger colonies are more efficient at brood rearing than smaller ones. Inefficiency arises from overlapping activities, redundancy, interference with communication, collision or just avoidance of impact. In heavily crowded conditions, such as when many workers try to get at a few larvae, the larval message just may not be received by the outlying individuals who, as a result, wait quietly on the group fringe. This is good; it reduces the size of the jostling crowd of workers trying to service larvae.

Given that workers produce workers and so create a positive feedback growth, i.e. geometric or exponential growth rates, it is still difficult to realize how dynamic a population of social insects really is. Some examples may help. The queen *Evylaeus marginatus* lays only six eggs in her first year and then after help from the subsequent workers she produces in a lifetime of 5 years, over a thousand eggs; workers live at most 1 year (Plateaux-Quénu, 1962). *Bombus pascuorum* begins slowly but accelerates to a maximum of 12 eggs a day; their neat method of matching egg input to adult input by laying eggs on cocoons gives an average of about three eggs/pupa (A.D. Brian, 1951, 1952). Figures for survival in an indoor colony of *B. pascuorum* (and one of *B. humilis*) with free access to the outside, show that 71% of eggs hatch, 73% of these pupate and 90% emerge as adults; this gives a hollow curve of survival fitted by a negative exponential function with a death-rate constant at 4% day, a mean life of 25 days and a half-life of 17.5 days. Thus a simple model with twelve eggs/day input gives four new adults/day and could build up a stationary population of 100 bees balanced by wastage of death at a rate of 4%/day (Brian, 1965). For *B. morio* in the tropics, with both larger and older foragers, the mean length of life for nurses is 73 days and for foragers only 36 days (Fig. 10.3) (Garofalo, 1978). The nurses' survival curve is half-normal in shape and that for the foragers is hollow but both are less prone to accident than the European bees. *B. atratus* near the tropics in South

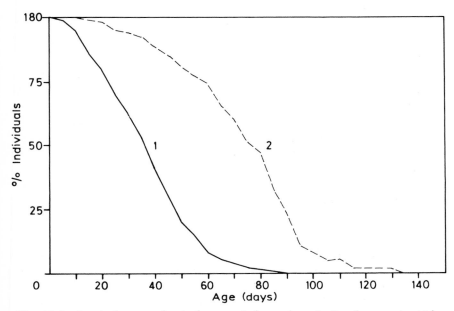

Fig. 10.3 Survival curves for 1, forager, 2, house-bees in *Bombus morio*. (After Garofalo, 1978.)

America has a similar brood production pattern but adult survival is not given (Sakagami *et al.*, 1967b).

With *Polistes* the necessity to construct cell foundations limits egg laying to two a day; at least this is so whilst the worker/larva ratio is high as it is after the first batch of workers has emerged. There is no doubt that feeding larvae is given priority so that at low worker/larva ratios cell making and egg laying are reduced in spite of the fact that recently vacated cells are re-used. In *Dolichovespula sylvestris* (*Ds*) and *Vespula vulgaris* (*Vv*) data collected from wild colonies have been used to build a model simulating the life history of an average colony (Archer, 1981a). For a substantial period around 45–50 days every small cell is full of brood. In *Ds* cell making goes on steadily for 40 days to reach 300 or so before the first cells become available for re-use and in *Vv* it goes on for 80 days and then stops at a total of 7000 cells. For both species this means that one worker makes slightly over one cell a day. *Vv* merely goes on for longer and with a worker population growing geometrically, creates a much bigger system. In both species the queen is able to keep the cells full of brood, though in *Ds* the queen only lays 13 eggs a day compared to 34/day rising to 348/day in *Vv*. This only takes into account the pre-maturity phase; in a mature colony of *Ds* with big gyne cells the queen can reach 65 eggs/day. Presumably she could not achieve the output of the *Vv* queen even in a much bigger nest, though, of course, this is not known.

The maximum worker populations are 175 for *Ds* and 2200 for *Vv*, but 330 workers are produced altogether in the first and 10 293 in the second species. In fact, the mean age of workers at death is only 10.3 days for *Vv* (excluding juvenile stages) and about 14 days for *Ds*, but data on the latter species is sparse; in hornets (*Vespa*) workers live about 5 weeks. The death/age curve of *Vv* is asymmetrical with a mode at 9 days and a median between 8 and 9 days which suggests that much mortality is due to accidents whilst foraging. The protected in-nest period of nursing is short or non-existent.

The stationary population of adult workers that is possible in a model of *Vv* with 10% mortality/day depends on the production capacity (number of small cells) and the period from egg to adult and the brood survival rate if food is plentiful. Assuming 5000 effective worker cells and a development period of 25 days this implies that 200 workers are added each day and if they survive on average 10 days there will be 2000 at a stationary state. For the *Vv* model this exists for only 3 weeks between the 74th and 95th days. In reproduction by swarm a mortality of 10% is high but with the aid of a vast, productive capacity and an extraordinary intake of food, high populations can be maintained.

In *Apis mellifera* a queen can lay about 2000 eggs/day depending on her race; if there is no room on the comb for them they are dropped and eaten by workers. This high value declines with age from a peak in the first and second year and with season from a peak in spring. The queen can rarely have the right conditions for full activity: in winter with ample comb space she is limited by the comb area that the bees can keep warm and in summer in a swarm with no comb she is limited at first by the rate of making cells and later by the space occupied by sealed brood from earlier eggs. Thus, basically, as in wasps, the queen is limited by the number of workers, either as a source of heat, as a source of wax, or as a source of food, and the rate of growth of the colony will depend primarily on these, as long as the worker/larva ratio is not allowed to fall too much. The queen may lay a proportion of unfertilized eggs in 'worker' cells; these have to be destroyed along with the non-viable but fertilized eggs, and egg infertility can rise to 50%, though it is normally less than a third. In larval and pupal stages as in wasps, mortality is slight as long as diseases and predators (e.g. mites) are absent and in a good hive sheets of sealed brood cells have few gaps. A hundred eggs can give rise to 94 larvae, these to 86 pupae and in time to 85 workers (Fukuda and Sakagami, 1968).

The workers live substantially longer than wasp workers and have the typical 'convex' or physiological survival curve of cloistered conditions in which death rate increases gradually with age to about 3% in foragers. It also changes with season: they often live 5 weeks in March, 4 in June and then longer again with the appearance of the soporific hibernating bees which can last all winter, especially if produced late (Free and Spencer-Booth, 1959). Like most insects in autumn they have a lower metabolic rate and a tendency to store materials in their fat body (Maurizio, 1946). If they can be induced to rear brood their survival is somehow reduced but they usually stop in good time so that over winter there are only adults. Avitabile (1978) has suggested that preparations for winter start soon after the summer solstice through a worker sensitivity to shortening days, and the winter solstice may re-start brood rearing as a response to lengthening days. The social and physiological mechanism is unknown (Kefuss, 1978). An African race of honey-bee, *Apis mellifera adansonii*, cultured in Brazil with summer temperatures of 24°C and winter temperatures of 18°C does not hibernate and does not create bees with a hibernation physiology so that the average longevity is 26.3 days in summer and 24.3 days in winter (Terada *et al.*, 1975).

In a stationary population of about 60 000 bees (a strong colony in June) with a total mortality of 3% a day (or 1800) the queen needs to lay (Assuming 70% viability) 2600 eggs a day which is just about possible in good conditions. A model of this bee system made by McLellan *et al.* (1980) incorporates an egg laying schedule that rises quickly to a peak in May/June and then falls off gradually to zero in October (a curve from the family of equation: $y = xe^{-x}$). The data they use includes: 1–1.2 thousand eggs/day, an average of 20 days to an adult, 10% juvenile mortality and an adult life of either 30 days in summer or 200 days in winter. This simulation agrees well with the known life history of a honey-bee society and can readily be applied to other perennial social insects, for the shape of the production curve is quite general (cf. *Myrmica* in Brian, 1965).

A. *m. adansonii* in South America has a swarm cycle lasting about 2 months (Winston, 1979). New workers appear 20 days after swarming and arrest the decline of the population which 20 days later still begins to rocket and a swarm emerges. Winston says that colonies swarm during a period of rapid population growth and in his summary comments '. . . swarms occur before high growth rates peak . . .'. Sixty per cent of the workers go with the old queen, especially the young ones and in the 20 days prior to swarming new workers comprise 40–60% of the adult population. From none to four after-swarms may come away with gynes, the number being proportional to the volume of sealed brood. After a swarm has left, brood survival is reduced to 50% but improves as young bees appear again though a colony still has to wait 20 days for a new laying queen. This 2-month cycle may be the cause of the August secondary peak in swarm frequency in temperate climates to be mentioned soon.

Meliponini with their clear in-nest nursing phase have the same age-increasing mortality as the honey-bee. *Nannotrigona postica* and *Plebeia droryana* have been studied by Garofalo (1978) who found that the mean age of the latter species is 42 days longer than the mean age of *Apis mellifera* (Fig. 10.4). No stingless bee so far as is known makes special long-life winter bees.

Termites start with small fecundities, in the order of 2–300 eggs a year in Calotermitidae and Termopsidae but rise to astronomical figures with *Macrotermes*, the queen of which may lay 30–40 thousand eggs a day (Table 10.1 and other citations in Brian, 1965). Very little is known about juvenile survival which is not so easy to determine as the juveniles do not stay in one place and moult frequently (Nielsen and Josens, 1978). Larvae of *M. bellicosus* diminish from 60–64% of the total population to 28% in colonies over 10-years old (Collins, 1981a). The growth of these colonies fits a geometric series in the 5 years after reaching 10 000 individuals, and goes on to over a million. The natural rate of increase is 0.78 units/year of the same order as 1.06 for *Neotermes tectonae* according to Brian (1965).

In ants the same range occurs: *Amblyopone pallipes* with a few eggs to *Eciton burchelli* whose queen is estimated to lay 1.2×10^5 eggs in a batch every 5 weeks, averaging 3500 a day. *Myrmica rubra* queens range from 410 to 826 a year (Fig. 10.5). The record seems to be held by the oriental termite *Odontotermes obesus* with 1 egg a second. Unfortunately, data on survival and longevity are not available yet.

Some simple models of social insect societies growing exponentially can be found in Brian (1965, p. 40). All social insect groups have evolved some large societies: many thousands in wasps and bees, millions in ants and termites. They gain more power and stability but as their food collection power increases the fecundity or number of reproductives must also increase.

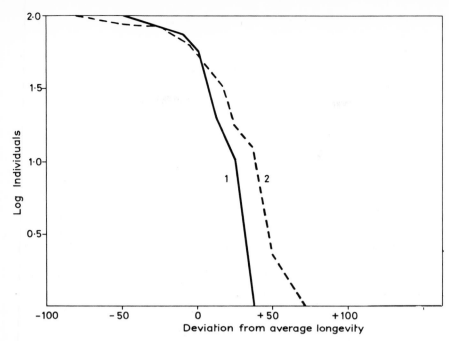

Fig. 10.4 Comparative survival curves for 1, *Nannotrigona postica* and 2, *Apis mellifera*. (After Garofalo, 1978.)

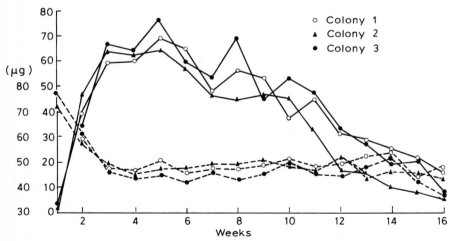

Fig. 10.5 Weekly egg production and egg weight (μg) of queens from three colonies of *Myrmica rubra*. Each curve is the average output of four queens. The worker-biased eggs start at the peak of production in weeks 2 to 3 and fade out slowly. (From Brian and Hibble, 1964.)

Table 10.1 Production of eggs for some queens of different ant and termite species. (From Brian, 1978.)

Species	Number of eggs produced per queen	Author*
Ants		
Monomorium pharaonis	400–500 per year	Peacock, 1950
Myrmica rubra	400–800 per year	Brian and Hibble, 1964
Formica polyctena	2.1×10^3 per year	Gösswald, 1951*a, b*
Formica rufa	6.3×10^4 per year	Gösswald, 1951*a, b*
Eciton burchelli	2×10^5 per month	Schneirla, 1971
Dorylus nigricans	$3–4 \times 10^6$ per month	Raignier and van Boven, 1955
Termites		
Some Calotermitidae	200–300 per year	Grassé, 1949
Cryptotermes havilandi	8 per day	Wilkinson, 1962
Apicotermes gurgulifex	400–500 per day	Bouillon, 1964
Apicotermes desneuxi		
Trinervitermes geminatus	500–1000 per day	Josens, unpublished
Microcerotermes arboreus	1690 per day	Emerson, 1938
Nasutitermes surinamensis	3917 per day	Emerson, 1938
Macrotermes subhyalinus	36 000 per day	Grassé, 1949
Odontotermes obesus	26 208 per day	Arora and Gilotra, 1960
O. obesus	86 400 per day (= 1 per second)	Roonwal, 1960
Cubitermes severus		
Immature society (< 10 000 individuals)	50–200 per day	⎫
Mature society (10 000–40 000 individuals)	up to 600 per day	⎬ Bodot, 1966
Senile society (> 40 000 individuals)	80 per day	⎭

*See Brian (1978) for full references.

The increased fecundity is achieved by feeding purified, concentrated food to the queen, by multiplication and lengthening of egg tubules in each ovary (and expansion of the gaster to hold it) and growth of the endocrine system, especially the corpora allata which supply the ovary with neotonin. In many ants the number of queens can be increased to hundreds. This, rather than increasing the longevity of the workers, is the rule. In fact the longevity of workers usually decreases, e.g. from 2 years in *Myrmecia* (Haskins and Haskins, 1980) to about 7 weeks in *Monomorium pharaonis*. The latter may have over 100 queens, each laying 1.5 eggs/day and if only one in three survives to adulthood as in *Myrmica*, this balances a loss of 2%/day and each queen can keep a stationary population of 25 workers (Peacock and Baxter,

1950, cited Brian, 1965). With a death rate of 2%/day the average length of life, which is the reciprocal of the death-rate, is of course 50 days. This is not because it is a tropical species, for *Lasius alienus* has an average life of a few weeks (Nielsen, 1972). Yet as domestic in-nest functions are extended it becomes possible also to increase the survival of young workers whilst decreasing that of the older foragers. What advantage does a high turnover have? Chiefly the ability to specialize a worker on a brief, ephemeral food source. In bees, on a single flower crop; in wasps, on a single prey type; in ants, perhaps on a single seed crop and in termites, on a particular harvest of litter or grass. They can be bred, learn the required skill, do their bit, and then be disposed of. They do not need to have the capacity to master several skills sequentially like their simpler evolutionary prototypes that live a longer, more flexible and varied life and learn and relearn continually. This presumably makes it possible to standardize and simplify worker structure, but so far little exact information on this exists.

In this chapter the stages in the transition from food to ant have been examined. After the workers have taken the energy foods the growth foods are given to larvae and reproductives (queens in Hymenoptera). Young larvae and egg-laying adults get the most purified and concentrated product, older larvae a mix or just raw food. Probably large size helps attract service from nurses for food is not distributed at random. In the next chapter the way in which some larvae are rationed to produce small workers, under the influence of the reproductives will be described. The importance of competition between queens and larvae, and its bearing on social stability through ensuring a match between embryo supply and food supply, has been stressed, introducing three alternative solutions to total fecundity: bigger ovaries in one queen, more queens, and workers to lay male eggs when female production is in a lull.

Colonies may be founded by fission, by reproductive pairs (termites) or by inseminated queens, either singly or in small groups. Egg production is not a problem in small new colonies but both disturbance and predation whilst foraging are, and it is in defensive functions that groups have an advantage. Queens which sequester are conspicuous and run a greater risk before starting a colony than afterwards. Several cloistered queens can lay into a common egg pool and use their eggs to feed larvae as they hatch; small groups gain but if too large they are liable to disease. Once a colony is established its numbers multiply but so do its organizational problems.

CHAPTER 11

Maturation

The spurt in growth caused by positive feedback from the increasing worker population eventually wanes and gives way to a stable state which may be either stationary or oscillatory. One obvious restraint is the fecundity of the queen or queens just discussed. Less obvious restraints are spatial; more room is needed for brood rearing, for traffic in and out of the brood chambers, for storage of food and of inactive foragers. Additional facilities may present insuperable architectural problems concerned with air conditioning and mechanical support. On top of all this is the greater energy cost of food since the more workers there are the more food is needed to replace and service them and this must be collected further from the nest. So travel is greater and traffic is thicker and more congested in spite of the trackways. In fact this one word 'congestion' summarizes the state which threatens to engulf a society with spurting growth. The theme of this chapter is that societies of insects are designed or programmed to anticipate this eventuality. They avoid it by reproducing and dispersing whilst still at their greatest growth rate. In so doing they maximize their contribution to later generations, i.e. fitness in its evolutionary sense.

11.1   Simple models of reproduction

The simplest models involve only two adult categories: worker and sexual; egg and food supply are assumed adequate. In one, the number of workers present in the next time interval is an exponential function of the number present in this one (with the exponent less than 1) and gyne formation is proportional to the numbers of workers too (Brian, 1965). So many sexuals may be formed that they destroy, or at least undermine, their worker production base or, at the other extreme, so few may be formed that the society is under-productive, if this is measured as gynes/worker. Between these extremes of instability and immobility is a condition that approaches equilibrium by damped oscillations. The risk here, of course, is that chance deviations may cause extinction, that the safe, perennial society will underproduce rather than overproduce. On the other hand, the acceptance of annual extinction with winter survival by specially prepared, fertile queens, e.g.

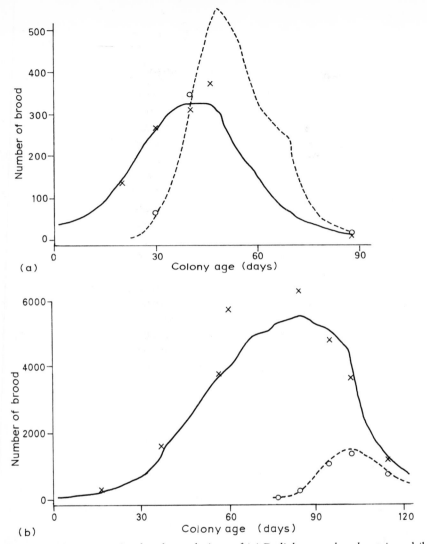

Fig. 11.1 Computer simulated populations of (a) *Dolichovespula sylvestris* and (b) *Vespula vulgaris*. The number of small cell (———) and large cell brood (- - - - -) are given in relation to colony age. (From Archer, 1981a.)

Vespula or *Bombus*, enables them to aim at high rates of sexual production and high yields per worker during summer. This can be achieved by switching over completely from worker to sexual production but for this the timing is vital since waiting too long may not allow time to copulate and prepare for hibernation, whereas starting too early will waste productive summer days; a question again of fitting a population cycle into an annual cycle.

Lövgren (1958) has modelled this for wasps. Now, Archer (1981a) who reviews various models, whilst simulating annual populations of two European Vespidae has found differences in the way their sexual production is linked with season (Fig. 11.1). In *Dolichovespula sylvestris (Ds)* the first big cells for gynes and males are made when the rate of building of small worker cells is highest, and the colony is only 30-days old. This thus fits the theory about maximizing sexual output. By contrast, *Vespula vulgaris (Vv)* starts at a time when the small cell construction rate has started to drop, 7–8 weeks later than for *Ds* and thus builds up a much bigger worker population. Whereas *Ds* switches from workers to sexuals entirely, *Vv* continues to produce small cells for some time after it has started large cells. This is not as strange as it seems for males are largely made in small cells in this species and the continuing production of these cells can be thought of as part of the sexual production programme. Since it goes on longer *Vv* produces more gynes per queen in the end but it uses its workers much less efficiently, that is to say in terms of gynes/worker, values are 0.18 for *Vv* compared with 1.66 for *Ds*. The bigger nest of *Vv* enables them to keep temperature up in late summer and makes possible the quicker development time which again reduces the cost of producing a sexual. So by using a longer season and producing sexuals in late September instead of late July *Vv* achieves the bigger output of gynes, in spite of using its workers less efficiently. Perhaps it could exist in warmer climates without pressure of hibernation? Spradbery (1973b) found that the sibspecies, *V. germanica*, introduced into Tasmania did do this. *Ds*, on the other hand, could certainly do well in a shorter summer and Archer lists 20 differences that would seem to design this species for short summers and higher latitudes and altitudes than *Vv*.

A model which includes perennial societies that collect food and have a food-related variable egg input has been developed for any species, e.g. *Myrmica* by Brian *et al*. (1981a) and even more generally by Frogner (1980). The former model considers a colony that starts from a queen, a few workers and brood generated out of her stored muscles and fat body. The new workers have a fixed longevity and capacity to collect and convert food but as their numbers increase they become less efficient due to congestion and the food collected increases much less per worker the bigger the total population. Allowing a respiratory need for food that is proportional to the worker number produces an actual maximum food collection before the energy cost of a large worker force begins to tell. This is still well above the amount of food needed to replace the workers that die and this surplus is available for sexual production (Fig. 11.2). If sexual production is excluded from the model it grows until the queen's egg output stops it or until the extra workers choke food collection and only just allow enough to replace the workers that die. This equilibrium is preceded by a phase of damped oscillations, due to the time lapse between an egg being laid and a worker being formed, i.e. to the

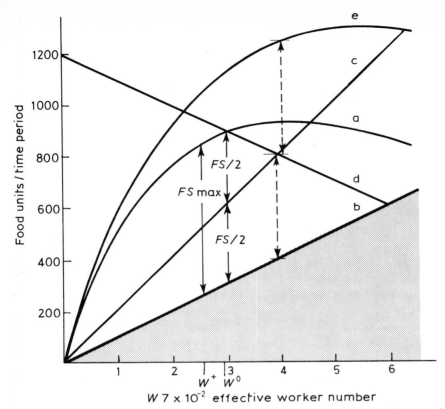

Fig. 11.2 In a numerical model of a *Myrmica* colony with a single queen, the rate of collection of food is related to the size of the worker population (line a). Line b represents the amount of food to keep the worker population steady and hence a − b is that available for sexual production. This reaches a maximum at $W=W+$. However, this is not the amount that gives the genetically optimum sex investment (see Chapter 2); the smallest colony that divides the surplus equally between the two sexes is $W°$ and $W° > W+$. If the colony is in its best habitat with a food supply of suitable density it will collect e food/worker population and the lines c and d will partition the surplus into males and gynes, respectively. Where these intersect equal sexual investment will occur; for this $W = 400$, still greater than $W°$ which is based on a state of under-nourishment (From Brian, *et al.*, 1981a.)

period of development (Maynard Smith, 1974), so that the worker population that determines the egg input at any instant is not the same as that which receives the resulting population increment at a development time onwards. In a later chapter an adjustment is made for differential sex investment, and a model of *Myrmica rimanica* production described (Uchmański and Petal, 1982).

11.2 Social control over caste

In ants, wasps and bees, only the females are caste differentiated but in termites both sexes are. In this section the power of the reproductives over this is considered. In general there are two aspects: a time in the life history or development of the colony and a time of year appropriate to the production and success of the sexual forms. In simple societies caste is settled by the dominance of the queens over their peers, and their failure, through injury, age or diminished fecundity will open the way for successors. For example, in the neotropical polistine *Mischocyttarus drewseni* as the colony grows, the ageing queen can no longer force all the females that emerge into worker roles and gynes appear (with males simultaneously) (Jeanne, 1972). Again, in another wasp, *Metapolybia aztecoides*, a young female is likely to become a worker if queens are present when she emerges, otherwise she will copulate and become a queen (West-Eberhard, 1978). In the halictine bee, *Evylaeus calceatus*, the first progeny contains some females that are bigger than others and remain in the nest. If the founder female fails they are able to copulate and start laying at once (Plateaux-Quénu, 1978, 1979). In other *Evylaeus*, e.g. *E. marginatus*, males play a part in caste determination since, if they are not available, the females cannot copulate and do not develop the usual reproductive behaviour of young queens. In such cases, therefore, male production initiates queen production (Brian, 1980).

Quite another problem is the one which asks: what is the state that changes a society from worker to gyne production? One state might be a high ratio of workers/larvae. This will only begin to fall when the fecundity of the queen starts to decline or is near its limit, which makes it a poor index of the best state for reproduction. Nevertheless, in some simple social insects this index appears to be effective, e.g. pocket-making *Bombus*. Quite early in social development if the worker/larva ratio is raised experimentally by adding workers or removing larvae, gynes are produced. In this case they are just extra large females that intergrade with workers but acquire a reproductive diapause and a tendency to copulate (Brian, 1980). However, the evidence of Cumber (1949, cited Brian, 1965) is that queen oviposition does not decline if sexuals are being produced. In *B. pascuorum* moreover, also a pocket-making species, colonies vary in size from 30–150 before sexuals are produced and so, with one queen, the worker/queen ratio can hardly be critical. In fact, experimental work has shown that in this group of *Bombus*, queens do not affect the brood-rearing behaviour of workers. Thus, for the moment, one must conclude that the supply of food outside the nest and the worker/larva ratio inside are the main influences on gyne formation.

In temperate *Polistes* the caste of females is determined on emergence as an adult by the season acting through temperature and day length (in Brian, 1980). It is possible, though uncertain, that the dominance of the queen is also influential, since experimentally high worker/larva ratios early in the season produce only big workers, not gynes. In the tropical *P. canadensis* social state is more important than the very slight seasonal changes and the gynes gradually increase in proportion as the colony grows (West-Eberhard, 1969). Correlated with this drift are the worker/queen ratio (it is a polygyne species), the worker/larva ratio and, of course, the average age of the queens. Thus in the simpler bees and wasps, the queen that builds up a good population of workers is the one that keeps them busy by repeated demands for food and regular high rates of egg laying.

A collection of inseminated sisters of *Lasioglossum zephyrum* can develop a queen by interaction in a few days. Such a queen then behaves distinctively: she shows a greater tendency to copulate again, develop and lay eggs, nudge workers which tend to avoid her, sit near the nest entrance at the end of the queue of workers that forms there, and also rest in other places that give her a command over worker movement and activity. She also makes many brief visits to cells during which she learns the state of construction and provisioning; this can later be used to co-ordinate or direct worker activities (Buckle, 1982).

In *Bombus* that regurgitate food mixes to their larvae all the time, queens undoubtedly control the caste of female offspring. The evolution of this new control factor overrides the influence of worker/larva ratio since less food just means fewer gynes not more workers. The continual examination of larvae (Pendrell and Plowright, 1981) enables the workers to recognize gyne-prone ones; they then either develop larval potentiality fully by ample and continuous feeding, if the queen is weak or absent, or they oppose it by corrective behaviour if the queen is strong. This means that the workers either dilute or stop food entirely so that the larva cannot fulfil its complete development programme and becomes a worker by default. Which of the two styles of worker behaviour is adopted towards female larvae depends on the frequency of contact with their queen; they seem to need actual physical contact at least once a day (Röseler, 1977; Röseler and Röseler, 1977, 1978). The ability of a queen to control workers in this group will thus depend on the number of workers (strictly the worker/queen ratio) and on the queen's activity and pheromone production which decline with age and probably depend on the daylength and temperature. It is also likely that the age and size of the workers will affect their responsiveness to queen contact. The point of the queen as a regulator seems to be that she can sum up these factors and switch behaviour sharply so that production effort swings towards gynes and away from workers, totally and irreversibly, at the appropriate time of year.

The Meliponini do not use a system of continuous examination and feeding like higher *Bombus*. They simply make big cells with more stores for gynes but there must be a system, so far undiscovered, that regulates the number of these gyne cells. Young gynes can be given protective custody in their cells and fed through holes, only being let out when the workers 'decide' that a contest with the queen is needed. In this contest, described by Imperatriz-Fonseca (1977) for *Paratrigona subnuda* workers appear to take sides at first but later accept the outcome. It is likely that the workers are sensitive to the queen's potency at all times, perhaps through the aggressiveness she shows during the egg-laying ceremony, or the frequency of this event or, perhaps, through the strength of her pheromone output. Some Meliponini of the genus *Melipona* have worker-sized gynes (microgynes is a general term for these females). Gyne-prone embryos are thought to be genetically different but still to need a little more food; only if this is available, can they reach the maximum of 25% in the population. The evidence in support of this idea is summarized by Brian (1980).

In vespine wasps gyne production starts when large cells are first made. This follows the emission of a new signal from the queen not the waning of an old one. Although at all times she shows uncontested dominance and activates and socializes workers through a contact pheromone (Landolt *et al.*, 1977), at some stage in her life she starts to secrete a new substance. This is the reverse of what happens in *Bombus*, and is undoubtedly part of the programme of changes governed basically by age but surely also influenced by daylength, food availability and social condition. In *Vespa orientalis* the active component is a 16C lactone which probably comes from the mandibular gland. It has been shown to influence cell building even when presented to queenless workers on an inert, structureless material like paper (Ishay, 1975).

In *Apis mellifera* maturity is signalled by gyne cell making; these cells are started if bee density is high following the spring push (Lensky and Slabezki, 1981). The queen is normally able to stop the construction of 'emergency' cells in the brood area by several mechanisms; most important, undoubtedly, is the secretion of a low volatility, unsaturated fatty acid 9–ODA (9–oxodecanoic acid) from the mandibular glands. This is distributed through the society by workers which form an ellipse or 'court' facing her whenever she stops walking. They touch her with their antennae and mouth parts and then rush away in a hyperactive stage and pass the material message on to other workers (Ferguson and Free, 1980; Seeley, 1979). In *A. cerana*, workers touch the queen less often and interworker communication is correspondingly reduced. The 9–ODA is intrinsically attractive to workers though they do not form a court round a piece of cork or paper as they do round a live queen. Although this substance is undoubtedly effective for most of the year, in May and June the main season of reproduction, very large, artificial doses are needed to prevent gyne cells being made. Other mechanisms may then be needed to reinforce the 9–ODA such as the footprint pheromone produced in a tarsal gland and left on the rims of combs where the cells are made as the queen walks. Laying queens make an oily colourless trail in much greater

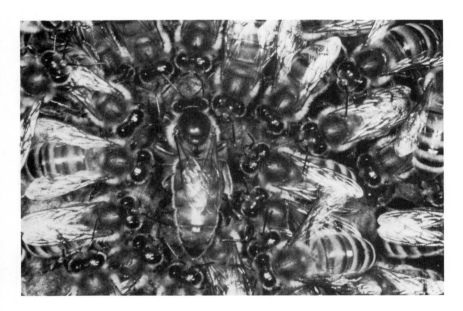

Plate 17

Above: Food transfer between workers of *Vespula* sp. on a comb, with pupal cells sealed with silk and carton. Notice the larvae situated peripherally.

Below: Workers of the honeybee forming a court round their queen. They all face her with antennae extended. (Photographs courtesy of J.B. Free.)

Maturation   189

amounts than can be made by a worker (Lensky and Slabezki, 1981). By itself it is not an effective inhibitor even of gyne cell cups but with the mandibular gland secretion it is very effective. There is also an abdominal pheromone which appears to be water soluble but about which little is, so far, known (Brian, 1980). Again, in *A. cerana*, swarm clusters communicate queen presence up but not down the chain of bees holding each other's legs and this communication, whatever it is, can pass a 2 mm gap, but not a 4 mm gap! Crushed queens are effective, so the signal must be chemical rather than behavioural, though this does not necessitate continued chemical transmission up the worker chain (Rajashekharappa, 1979). Even young gynes still in their cells can inhibit the formation of other cells (Boch, 1979).

In a special study of factors which precede gyne rearing in *Apis mellifera* (the Africanized race which lives in South America), at 5°N, Winston and Taylor (1980) found that a set of stimuli add up to a threshold trigger value. The main contributions come from colony size, comb area and number of adults, all of which need to be large and increasing though the bees do not have to fill the nest cavity. The authors found that they stopped at 20–25 litres when they covered 8–10 thousand cm² of comb surface. Also necessary was a high percentage of young bees, a high congestion on the brood zone where the bees tend to push inwards and a local food glut (Fig. 11.3). These both dilute the queen pheromones and reduce, or abolish, their secretion

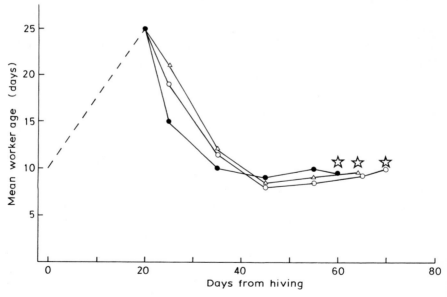

Fig. 11.3 The reproductive cycle of the africanized honey-bee. Swarming occurs 20–30 days (stars) after the average age of workers falls to 10 days. (*FS* is first swarm.) (After Winston, 1979).

temporarily. At 5°N with little variation in daylength and no winter the bees show a periodicity with a duration of about 50 days. Once cells are made and the queen has laid in them she cannot stop the larvae being fed. With pupation, however, and the sealing of the cell she can attack cells directly and once they are holed the workers finish the destruction off. In large, congested hives especially, the workers actively protect these cells against queen attack and attack by the first gynes to emerge; multiple swarming follows (Simpson, 1974). The workers have a form of vibratory dance that is thought to pacify these aggressive gynes (Fletcher, 1978a).

In ants, queens also play an important part in regulating reproduction, both as regards gyne and male formation. In a ponerine, *Odontomachus haematoides*, workers with queens ignore the demand for food by large gyne-prone larvae that can be reared to gynes by the experimenter. Thus, in contrast to the honey-bee the queen can control feeding. Undoubtedly the quantity of food late in larval life, not its quality, is crucial (Colombel, 1978, cited Brian, 1980). In *Eciton* removal of the queen causes heavy mortality of small, worker-biased larvae probably from neglect and the maturation of a few gynes, presumably from ordinary female larvae that get the benefit of special attention and re-cycled food. In dry seasons to which, no doubt, the whole colony are sensitive, the queen is isolated in a broodless part of the resting society and in the other part, sometimes separated by some physical feature, such as a bank or a log, a large brood of male and a few female larvae are reared. The females become gynes. The boundaries of these two zones are often the site of conflict between workers. It is presumed that the female eggs are laid by the queen shortly before segregation but it is unknown whether the males are laid by the queen or by the effectively queenless workers (Schneirla, 1971). In *Anomma* a similar process takes place (Raignier, 1972). The whole arrangement has the appearance of being engineered by the workers in response to a seasonal change, but the first small changes may take place in the queen and involve loss of effectiveness as a pheromone generator or egg layer or more likely both.

In the unspecialized genus *Myrmica*, colonies reach a reproductive state through dilution of queen influence on worker behaviour both directly and through the formation of unresponsive workers. Colonies of M. *ruginodis* mature at populations varying between 900 and 3000 workers. It is not thought that they stop growing at these figures, indeed they may continue to grow after emitting sexuals in the way that many organisms do. As the species is polygyne a lot of this variation in size at maturity is explained by the number of queens. In fact the crucial figure is in the order of 1000 workers/queen (Brian, 1965, p. 57, Brian *et al*., 1981b). In M. *rubra*, an even more polygyne species, it was originally thought that the queens cluster together and form effectively a single group, but though this is true in spring when they come up to warmth it is in general incorrect: they are only slightly aggregated. In fact, queens tend to be hostile to each other and disperse into territories where they are more attended and in communication with workers than if they are moving about as in the *Apis* 'court' (Evesham, 1982).

The basis of this dilution effect is that the queens have to touch each worker from time to time if they are to retain control over their brood-rearing behaviour. A dead queen is effective but lipid extraction destroys potency entirely. There is no evidence that communication of the queen status of the colony can take place from worker to worker, though it has not been ruled out altogether (Brian, 1973c) and could be propagated by oral contact as in honey-bees or some other ceremonial behaviour, that cannot be passed across a wire gauze too fine for a worker to squeeze through.

Control consists, as in *Odontomachus*, of under-nourishment of gyne-prone larvae, mainly, it seems, by feeding them on a watery regurgitate instead of directly on lightly malaxated prey (Table 11.1). There is another

Table 11.1 The effect of queens on *Myrmica* worker behaviour towards larvae and egg laying.

Behavioural type	Queens present	Queens absent
1. Worker biassed larvae	Nurses feed these actively in preference to other larvae and they metamorphose into workers	These larvae are neglected and they stop growing half way through the third instar
2. Labile female larvae	Workers (mainly foragers) attack these larvae during the stage in spring when they are signalling an ability to become gynes by means of a cuticular secretion from their venters	Both workers and foragers feed these larvae on solid protein-rich foods, focusing on a few at a time
3. Male larvae	Neglected but not attacked	Fed in preference to worker-biased larvae but not labile female larvae that are forming gynes
4. Egg formation and laying	As long as the queen is laying or there is a cluster of reproductive eggs of either sex, young nurse workers lay imperfect eggs that are useful only as food (trophic) Very young workers (less than 3-weeks old) will lay reproductive eggs that form before they become sensitive to queens	Young workers will lay large perfect reproductive eggs that are unfertilized but capable of developing into males partheno-genetically

action (not recorded for the ponerine): that of biting consistently and fiercely on the larval venter, which brings on metamorphosis and produces a worker. This aggressive activity, for so it appears, against gyne larvae is more easily aroused in workers and travels further from the queen in space and time, than the feeding inhibition. Viewed as a message it penetrates more deeply into the society than the message to ignore gyne larvae. The last only works well close to the queen, but is more reliable as a caste control than attack which varies a lot in intensity. Yet the feeding inhibition is more easily attenuated through social movement and as the society grows is the first queen control to vanish. Queens, by acting as work stimulants, also cause female larvae to omit an important growth phase and, again, to metamorphose into workers. This is discussed in a later section on caste morphogenesis.

Workers which have experienced queens soon after emergence respond better to them later on when they come to rear larvae (Brian and Evesham, 1982; Carr, 1962). They treat caste-labile larvae with more decision: starve them if there is a queen and overfeed them otherwise. The first 3–4 weeks seem to be crucial for this indoctrination though whether it is simply that workers have to learn the individuality of their queen or need their help to learn how to regurgitate to larvae or, more subtly, whether they have to be socialized by queen experience, is not certain. Smeeton (1981) has shown that they are 'borne' in a solitary physiological state and will lay reproductive eggs unless they come under the influence of laying queens. So societies that have a deficit of queens (i.e. a high worker/queen ratio) when the new workers emerge will be more likely to be unresponsive to queens later on.

In *Leptothorax nylanderi*, queen influences also override food supply and worker/larva ratio and queens vary in their power to suppress gynes; like *Myrmica* their power increases with age and peaks early in every season (Plateaux, 1970, 1971). In addition to the variation of queen potency with season and age, workers themselves vary in their response, partly, no doubt, for the reasons outlined previously. Larger and older workers, mainly foragers, are more responsive to queens than the smaller and younger ones which are mainly nurses. As colonies grow, the average size of workers tends, at first, to increase. This is well authenticated in *Myrmica ruginodis* by Brian and Brian (1951), in *M. sabuleti* by Elmes and Wardlaw (1982a), in *Lasius niger* by Boomsma *et al.* (1982) and others. In *Lasius* the fraction of sexual investment caused by gynes increases with worker size to a limit near 0.8, then stops, presumably due to the interference of gyne-making with worker replacement. Support for the queen thus increases as the colony grows for the big old workers are the ones that attack larvae when queens are present and come back into the nest to help feed them when queens are absent. This may be due to their having been queen-sensitized in early life in a smaller colony with a higher queen contact frequency. However, as the colony grows with increasing speed young workers become increasingly common and for this reason alone will be under-sensitized by low contact frequency with the queen(s). Such a population structure is a feature of a stable, geometrically increasing population rather than a stationary one and is exactly the condition in which maximal reproduction should occur (Brian and Jones, 1980). Maturation is thus caused by a large increment of workers being insufficiently sensitized to their queen so that they do not control the feeding of gyne-potential larvae adequately.

For *Myrmica* and *Leptothorax* there is another, quite different change, this time with queens as they age. The eggs of first-year queens are more liable to metamorphose into workers than those of older queens which tend to develop in a gyne direction though they do not necessarily go the whole way. Information, at present, does not support the idea that there are two distinct egg types (see above) but rather a continuous range from one extreme to the other: from those in which early worker-forming metamorphosis is in-built even at temperatures as low as 20°C, to those that metamorphose into workers only if they get the active attention of nurse workers in spring in the presence and under the stimulation of queens. If they do not get this attention then except at high temperature (e.g. over 25°C) they go into diapause. Diapause larvae hibernate and slowly establish (at low temperature) their capacity to grow into gynes; but they are caste-labile, not gyne-determined. So, as queens mature they tend to lay fewer worker-biased and more caste-labile eggs, that are potential gynes and only await the right social conditions to mature. However, even queens more than a year old have a brief spell in spring when they can lay some worker-biased eggs. An important feature is probably speed of laying rather than small egg size, a fact which implies that worker bias is a deficiency which interferes with embryogenesis and larval growth potentiality but does not entirely disrupt development as in trophic eggs. Maternally-induced caste bias in eggs is unknown in bees and wasps but common in ants; it clearly provides a maturation process built into the queen and transmitted to the society through her eggs (Brian, 1980).

In *Monomorium pharaonis* this aspect is very pronounced. Young queens lay eggs that mostly give small workers (though they are big eggs); as the queens mature they lay eggs that are capable of growing into big workers and, finally, eggs that yield gyne larvae, near the end of a 120 day cycle. These gyne larvae are different in appearance: rotund, hairless and golden-gutted. Some of the eggs laid by queens are unfertilized and give males. After sexual production and copulation in the nest the cycle restarts (Petersen-Braun, 1977, cited Brian, 1980). However, an important point is that the cycle has safeguards against error, for only large, mid-term workers, together with ageing queens, will actually rear gyne larvae, otherwise they are destroyed and recycled. Queens in addition to their maternal effect on eggs, thus have a strong, inhibitory effect on sexual production through worker behaviour; one that is not subtle enough to deflect the gyne larvae into a worker path of development as in *Myrmica* but eliminates them altogether (Edwards, 1982). Queens are recognized by the workers through a non-volatile surface lipid as in *Myrmica* (Berndt, 1977; Berndt and Nitschmann, 1979; Brian, 1980) but they also emit neocembrene from a posterior exocrine gland and this is a signal for workers to attack gyne larvae (Edwards, 1982). Thus all the features of social development in the temperate *Myrmica* are present in *Monomorium*, a tropical, cyclic species; *Myrmica* has two cycles each year, one with sexuals and one without whereas *Monomorium* has three complete cycles with sexuals each year. It is not known how *Monomorium* workers or queens recognize a sexual larva, though to humans they are conspicuously different both in shape and surface texture. No doubt they have a special surface chemistry too as in *Myrmica rubra*, where an oil melting at 20–25°C is exuded by dermal glands (Brian, 1980). Different species of *Myrmica* will behave correctly in relation to each other's labile larvae but only rear their own worker-biased ones. Presumably the sign of sex potentiality is common to all species in the genus (Brian, 1975; Elmes and Wardlaw, 1982c). In *Solenopsis invicta* sexual larvae may secrete triolein. This oil, put on a filter paper and cut into pieces of larval size, is collected by workers and stored with their pharate sexual pupae; it is not eaten even though this species is very fond of oily foods (Bigley and Vinson, 1975).

In *Pheidole pallidula*, another myrmicine, mature queens also lay more labile eggs, mostly in early spring. Larvae from these have a characteristic pear shape and are only nourished if there are about 1400 workers, for there is never more than one queen (Passera, 1980a). The maternal influence on these eggs may well arise from juvenile hormone (JH) in the haemolymph for queens treated with this lay more labile eggs (Passera and Suzzoni, 1979).

Species of *Myrmica* vary in the amount of growth that these labile larvae make prior to winter. In *M. scabrinodis* they are indistinguishable in size and appearance from other larvae but in *M. rubra* and *M. sulcinodis*, especially the latter, they can be much larger than worker larvae during winter (Elmes and Wardlaw, 1981). Nonetheless, they are still able to move 'sideways' and become workers but with *Tetramorium caespitum* the size of gyne-forming winter larvae vastly exceeds that of worker larvae and it is doubtful whether their caste can be switched (see Brian, 1974a). Perhaps this advanced myrmicine has become, like *Pheidole* and *Monomorium*, a 'recycler' of caste-determined brood rather than a 'transformer' of caste-labile brood. Worker determination from the egg stage could well be at the root of their loss of ovaries.

Plagiolepis pygmaea (Formicinae) controls caste through its queen who, as long as mated and active, can stimulate early metamorphosis into workers as in the myrmicines discussed; any larva which goes too far is recycled. There is a pheromone, since dipping the queen in acetone removes activity (Passera, 1974, 1980b). As labile larvae hibernate in a third stage (more advanced than worker-biased larvae or males) it is likely that they are laid early the previous summer and need a period of cool winter weather before they can develop fully, as with *Myrmica*. This, of course, co-ordinates the life cycle with the climatic cycle (Brian, 1977).

Before discussing wood ants we must consider the special case of social parasites. *Plagiolepis grassei*, a parasite of *P. pygmaea*, is able to prevent gyne formation by its host. *Harpagoxenus sublaevis* (Leptothoracini, Myrmicinae) has queens that can stop queens of one of its hosts (*Leptothorax muscorum*) laying even though they are inseminated, but not queens of its other host, (*L. acervorum*). The parasite thus appears to control its host's sexual production. In this parasite the queens are often wingless, though with full ovaries and spermathecae, and Büschinger (1978) has evidence that the wingless state is due to a single dominant gene. So, too, it is likely that *Strongylognathus testaceus*, a parasite of *Tetramorium caespitum*, has queens which inhibit gyne production by the latter. Presumably, all these parasites produce more, or a stronger, inhibitory pheromone than the host queen does.

Formica rufa and its allies are known through the work of Gösswald and Bier (1953) and Bier (1953) to produce two sorts of eggs with few, if any, intermediates. The first eggs each season are labile but only form gynes in nature because the queens, after laying them in the warm surface chambers, migrate deep into the nest and leave the workers to bring them up alone. This, of course, becomes easier the larger the colony and can be thought of as one of the processes by which queen power is diluted in this species. After a rest deep down the queens lay small eggs that develop into workers whether near queens or not, they are in effect worker-determined but Gösswald and Bier showed that if cultured by another less polygyne species, *Formica pratensis*, they are able to grow into gynes; thus they are only 'determined' within the natural context of their own species. The difference between these two egg types is due to the temperature and rate of manufacture, the smaller ones lacking RNA amongst other things. The way in which queens influence worker rearing is not understood but it is likely that queens cause workers to recycle gyne-prone larvae since these are easily recognizable from the second stage onwards as having less hair and bigger heads (Schmidt, 1974).

In termites similar controls exist: if the reproductive pair are removed substitutes form. In lower termites these come from larvae and nymphs whose sexual organs mature so precociously that their wings and eyes do not develop properly. In Termitidae this is not possible and winged forms have to be made in the usual way at the correct season as in ants. They then form pairs and take over. Some termites seem to vary in their capacity to use one or other method. If they are in the stage of small, young colonies with many small larvae they make substitutes but if they are in the stage of mature colonies they make winged adults. The seasonal element shows up since hibernated larvae are the most sensitive and most likely to become nymphs and so, after further moults, adults (Brian, 1979a). As in the honey-bee, too many substitutes are formed; they fight each other, and workers finish off the wounded leaving one pair. Whether this is a genuine knock-out contest which selects the fittest will be difficult to test.

Both sexes play a part though a rather different one in the control of replacements. Each sex might control its own replacements but this is not so. In *Calotermes* removal of the male does lead to the formation of new male adults but removal of the female leads to the formation of not only more females but of more males too. Evidently the male cannot suppress its replacements by itself; it needs the help of a female but surprisingly another male will do instead (Lüscher, 1964). Clearly a companion even of the same sex supplies an important part of the pair function: they are not just zygote manufacturers. Quite possibly they communicate better with each other than with other castes and this could mean a stronger bond and more ensuing stimulation of metabolism. In *Neotermes jouteli* again heterosexual pairs are all powerful, yet homosexual pairs of either sex can suppress all females but not quite all males. The process clearly involves co-operation between adults and does not depend on sex specific inhibitors alone (Nagin, 1972). Pheromones distributed in food have, of course, been suggested (Lüscher, 1974, 1976) and contact is essential (Springhetti, 1972). Varnishing all but the mouth does not stop the control function and at present it is thought that a head gland produces inhibitory pheromones that are passed about by mouth. No doubt this is oversimple and some behavioural signal is also involved (Brian, 1979a).

11.3   Males in social Hymenoptera

For termites male production is not a behavioural problem; like the females they are formed from zygotes with sex chromosome determination. This need not imply, of course, that the sex ratio is unity but it does necessitate that investment in the two sexes at the adult stage is equal. Problems only arise when this does not confer the best ratio between sterile castes (workers/soldiers) as in Termitidae and other families where sex influences the likelihood of soldier formation through genetic differences.

Sex determination in Hymenoptera is haplo-diploid and most diploids are female, provided that outbreeding is stringent. Haploids are always males and arise from unfertilized eggs; these may either be laid by queens or by workers. As males never work in or defend the nest in this group of social insects except in a very few species of ant, their production is a waste of energy and resources and must be controlled so that they only mature in societies which can stand the strain, or in societies which can benefit from slowing down their rate of growth. This means that males can, in some circumstances, be regarded as early growth deccelerators and their negative feedback as a stabilizing influence. Of course, at all times they are gene vehicles potentially able to spread the queen's or the worker's genes.

In primitive bees whose societies are still only associations of inseminated females, sexually mixed broods are often produced. The queen is simply bigger and older than the others and she intervenes in their behaviour and stimulates their nursing activity, e.g. *Lasioglossum zephyrus* (Breed and Gamboa, 1977). Often the queen's associates are uninseminated and can only lay male eggs; though they cannot replace her they can damage her chances of success by laying eggs and producing males too soon in the life history of the colony. Queens can prevent copulation between males and her associate females in *Evylaeus calceatus*; only her presence is necessary, she does not need to intervene actively. If suppression does not stop associate females or workers laying, the queens destroy and eat their eggs, e.g. *Bombus*. A more subtle way is to starve them in the larval stage so that they produce small worker-like individuals, e.g. *Evylaeus malachurus*.

Temperate *Polistes* show convergence with bees in all these respects (West-Eberhard, 1969, 1973). In *P. fuscatus* male production is associated with high rates of egg laying and though this may mean that workers are adding a quota, an alternative view is that the queens are laying so fast that some eggs fail to get fertilized, a situation known in parasitoid Hymenoptera and applied to social Hymenoptera by Flanders (1962, 1969). In *Metapolybia* a polybiine with large polygyne colonies that swarm, the queens are aggressive towards newly emerged gynes whose ovaries may develop before they copulate (West-Eberhard, 1977). Precopulatory male production occurs in parasitic Hymenoptera and experimental interference at this stage, may lead to ovary regression (Flanders, 1969). However, in *Protopolybia pumila*, workers both lay and eat their own eggs as well as those of queens though queens guard their own eggs for some time after laying. High worker/queen ratios increase the proportion of male brood which suggests a quantitative queen control over worker oviposition (Naumann, 1970; West-Eberhard, 1977). These swarming polybiini with many gynes, workers and attendant males, appear to be adapted to tropical forest conditions where disturbance by birds and ants is frequent. Given this multiple structure they can quickly fly away and regenerate a colony elsewhere (Jeanne, 1979b).

In *Bombus* there is little more to add about the carder and pocket-making species. In *B. hypnorum* the queen attacks egg-laying workers during the early part of the season, evidently identifying them by smell or behaviour; later the situation seems to become chaotic and she attacks all and sundry (Röseler and Röseler, 1977). This general aggressivity is associated with an inability to fertilize her eggs. In higher *Bombus* there is little show of dominance before the end of the season and it is now fairly certain that the mandibular gland secretes a pheromone that inhibits worker oviposition

(Honk *et al.*, 1980). It acts through sensory and neural pathways to reduce activity of their corpora allata (Röseler *et al.*, 1981). This would support Röseler's (1977) evidence that the queen interferes with the corpora allata of subordinate workers. Honk *et al.* point out that behavioural effects could also occur, for the workers press near to the queen and they lay eggs together and eat each other's eggs. Top workers develop, dominate subordinates and inhibit their egg laying in turn (though they cannot stop them rearing gynes). They are the older workers and they may sting the queen to death. This co-operative egg laying and egg eating suggests that an egg-laying female tends to attract others with similar tendencies which may then eat the fresh eggs and so lay more of their own; it is reminiscent of trophic egg laying in ants and meliponine bees.

In stingless bees, reproductive eggs are laid by queenless workers of many (but not all) species; workers seal the cells at once after laying (Beig, 1977) and males emerge later. In one species, *Scaptotrigona postica*, workers lay reproductive eggs even in normal queen-right colonies. These eggs are larger than those of the queen and are placed on the food after the queen has laid, by an intrusive worker who later seals the cell. Males hatch in 65 h compared to 67 for females and are larger and more active and quite soon perforate and eat the female egg or larva. Males always emerge from cells with two eggs (Beig, 1972). The workers lay these male eggs in normal cells that contain less food than average though why, unless males need less food for satisfactory development, is obscure (Beig *et al.*, 1982). The authors suggest that the egg-laying workers compete for these cells after the queen has gone and that the amount of food is not important, it is the greater space above that is essential; but why?

The success of male larvae in competition with female larvae has been shown experimentally for *Myrmica* (Brian, 1969; Brian and Carr, 1960). In free chamber culture it is difficult to imagine the male larvae actually eating the female ones; some worker discrimination is thus likely. Where workers lay the male eggs they are, of course, their sons and a bias in their favour could be predicted on genetic grounds.

In Vespinae the queen shows uncontested dominance in young colonies, inhibits workers oviposition completely and stimulates social behaviour. It has been suspected that like *Bombus terrestris*, she produces a pheromone controlling worker ovaries but so far this has not been demonstrated (West-Eberhard, 1977). If the queen is taken out the society becomes anarchic, workers fight and neglect larvae then develop a hierarchy of their own and thus restrict egg laying to a few top rank individuals whose eggs yield males. Normally as the queen ages she starts to lay unfertilized eggs, loses dominance and is attacked and evicted by rebel workers who take over the male-producing function (Montagner, 1966). No doubt the workers can lay many eggs between them but what proportion they produce is uncertain (Kugler *et al.*, 1976). No special cells are made for males and many arise in worker cells in some species, e.g. *Vespula vulgaris*, but in large cells in others, e.g. *Dolichovespula sylvestris*. Are laying workers less discriminating than queens?

In the honey-bee, male production necessitates prior construction of the special 'drone' cells which in naturally growing colonies are formed on the peripheral zones of growing combs just before gyne cells are added along the rims. This is only done in substantial societies but once the cells are there, the queen lays unfertilized eggs in them, methodically, many at a time, not singly. This process is obviously under her control for fertilization of these eggs is extremely rare. Any mistakes that the queen makes are corrected by the workers for they have very fine discrimination where larvae are concerned and can easily identify diploid drones or gyne larvae in the wrong cells and destroy them. There is no doubt that in this bee the queen normally lays the male eggs; however, when queenless some workers do lay but in an irregular manner in both small and large cells. In most races worker eggs give rise to males but in one South African (Cape) race, *A. m. capensis*, they can produce diploid eggs that give workers and in fact sexuals are redundant. The inhibition of worker egg laying by the queen is total; there are no half-way trophic eggs and it is thought that the 9–ODA she secretes is antagonistic to the corpora allata but very large doses are needed to stop them laying and topical application of JHA does not arouse their ovaries, it changes them from nurses to foragers which, in fact, have defunct ovaries; this unusual reaction may depend on dosage. It looks as though the corpora allata in *Apis* either do not function as gonadotrophic glands as in most insects, or are unusually sensitive to concentration of JH in the blood. Their excision in the queen does not prevent normal egg production (Rutz *et al.*, 1976, cited Brian, 1980); so what regulates oogenesis?

The construction of drone cells is favoured by colony size and hindered by queen presence. Free and Williams (1975) and Free (1977) have results which suggest that the workers can assess the proportion of drone to worker cells. They suggest this might be done chemically but no chemical difference between the cells has yet been demonstrated. Normally drone cells are made before gyne cells but experimentally, gyne cells containing larvae will induce drone cell making and there is evidently some sort of connection between the construction of the two different cell types. The queen, once male cells exist, seems unable to prevent herself laying unfertilized eggs in them in spring but the workers destroy these untimely eggs. In late summer she will not lay in drone cells even though the workers construct them. So between the two the male arise at the right season! In August, in preparation for winter, males are starved and ejected, stung by workers as long as there is a queen present, but not otherwise. Thus the whole system depends on colony size, season and the existence of a functional queen who opposes not only egg development in workers but male cell construction. Only in large colonies that are well fed will the workers overcome this queen opposition.

In ants there are examples in all sub-families of workers that lay male-producing eggs but whether this is regular and ubiquitous or only done in emergency by a few is still uncertain in many cases. Where egg laying by young workers is well attested it is reasonable to expect that workers generate many, if not all, the males, but it needs checking. It is theoretically likely (Oster and Wilson, 1978). In *Myrmica rubra* the evidence that this is so is strong (Brian, 1969; Pearson, 1982b; Smeeton, 1981).

In *Leptothorax nylanderi* a recessive gene for pale colour (*pallens*) which shows up well in males was used by Plateaux (1981) to show that males are produced by workers. Obviously in species in which the workers have only vestigial ovaries the queen must lay all the unfertilized as well as the fertilized eggs. Where this happens there is likely to be a season when sperm are shut off: the dry season in Dorylinae or the end of the summer in *Tetramorium caespitum*. In this last species the males and females produced each year correlate for each colony so that the queen may lay a mix of fertilized and unfertilized eggs in repeated batches. *Formica rufa* lays male eggs only in the spring but again whether these are laid by workers or deälate gynes is uncertain; it has been known for a long time that workers can lay male-producing eggs but this, of course, does not mean that they normally do so (Schmidt, 1974, 1982). Such eggs only yield males in spring when young rested workers are available as nurses.

The way in which queen ants stop workers laying reproductive eggs at the wrong time of year varies. In *Odontomachus haematodes*, workers can resist queen interference and will form an enclave, if about 50 are present, which ignores and outcasts the queen (Colombel, 1974). Other queens lay copiously themselves, e.g. *Myrmica*. This acts in two ways: first, by producing more female brood which keeps the workers busy and second by inducing worker-laid trophic eggs. In *Myrmica rubra* queen laying peaks in early spring before the ovaries of over-wintered workers are able to secrete yolk; when they do, female larvae start hatching. Young workers formed late in the previous summer can lay reproductive eggs even in the presence of queens, to whose influence they are immune at first. Only after the age of 5 weeks, if temperature allows, do they start to lay trophic eggs. These new workers, at 20°C, can begin laying when only 3-weeks old and it is possible that some development of eggs takes place whilst they are still in the pupal skin (Smeeton, 1980, 1981). Totally queenless workers lay more reproductive eggs in longer day-lengths. At the right daylength and temperature not only can a young queen stop reproductive eggs being made by workers, but also profit from the alternative trophic eggs and sustain her own egg laying longer. To exert full control in this way the queen must be present all the time for even in small nests each extra day a queen is present reduces the number of reproductive

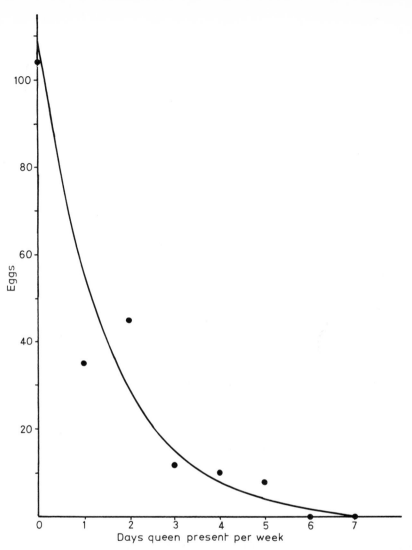

Fig. 11.4 The number of reproductive eggs laid by *Myrmica rubra* workers when their queens are present for different numbers of days in each week. (From Brian, *et al.*, 1981b.)

eggs that workers lay by half (Fig. 11.4) (Brian *et al.*, 1981b). Put another way, if a queen alternates between two small chambers in a nest, spending half her time in one and half in the other, only 10% of worker-laid eggs will be reproductive. An interesting point is that a queen has to be actively laying and produce an egg cluster to stop male production (Brian and Rigby, 1978) but to stop gyne production she need not even be alive! Again, nests with newly adopted queens may be unable to stop workers laying some reproductive eggs, yet they can easily prevent gyne formation (Winterbottom, 1981). Male control relies on a more specific aspect of queen activity and may depend on worker recognition of individual queens.

This conception is supported by the use of the *Myrmica* model in which it is shown that males cannot be suppressed (unlike gynes) in poor habitats which can only support small populations, e.g. *Myrmica ruginodis* in narrow woodland glades. This could be the state of any society that is not well established, e.g. a young society; it will then produce males before it produces gynes and so apply a light touch of the brake to its population growth. Moreover, species whose colonies divide will produce purely male broods to balance the sex investment ratio (see next chapter). This, it is suggested, explains how *M. sabuleti* living in a fairly uniform habitat with strong competition from other species (Brian, 1972) has 'males-only' colonies which average about the same size and are as common as those producing both sexuals. Plateaux (1981) indicates that colonies of *Leptothorax nylanderi* only produce males when there are at least 100 workers.

So far it has been assumed that male control in workers operates before oviposition, i.e. during oogenesis. This is only partly true, for in *Myrmica rubra*, male larvae are underfed just as are gyne larvae when near to queens. This cannot, of course, sidetrack them into the formation of male 'workers' nor does it kill them; they just mature as smaller adults. Moreover, male larvae are not attacked by biting as gyne larvae are and they do not appear to emit a signal from their venter. They thrive in competition with general female larvae as just discussed and workers almost certainly distinguish them and feed them preferentially if queens are absent (Brian, 1981).

In other species there is no doubt that males are killed in the larval stage or even as eggs when the queen influence is strong. Thus in *Monomorium pharaonis* male larvae can be found in egg masses at all stages of the population cycle yet none mature (Peacock *et al.*, 1954, cited Brian, 1980). In *Oecophylla* the workers lay very large reproductive eggs which develop into males, and much smaller ones which are also unfertilized which develop parthenogenetically into females: both are laid in enclaves of the compound nest, apparently secure from the queen's influence (Ledoux, 1950). What determines whether haploid or diploid parthenogenesis takes place is unknown. In *Cataglyphis cursor* queens cause egg absorption in workers but in their absence there is again both haploid and diploid parthenogenesis. In spring, workers can rear both sexes even with a queen present (Cagniant, 1979). In species that can produce females from unfertilized eggs, queens are only needed to activate and influence worker behaviour. Periods of parthenogenetic reproduction save expenditure on males and spread what is, at the time, a genetically good line; sexual reproduction is still possible periodically to generate novelty (Maynard Smith, 1978a).

11.4 Maturation in general

The control of sex in Hymenoptera, made possible by the haplo-diploid mechanism, has thus enabled the social exclusion of males; they carry genes, no more. Indeed, unless they are formed in strong societies at a suitable season they are a severe drain on resources and may interfere with the socialization of females. In contrast, termites do not have this ability to control sex and include males in the society; they come eventually to use this genetic sex difference to enrich caste structure. At first the signs are that the uninseminated females of Hymenoptera (workers) lay eggs that develop into males; this gives them a special interest in producing haploid eggs and favouring male larvae. The whole system only works as long as the inseminated females are allowed to lay many diploid eggs which if underfed or maltreated give more workers.

With the production machinery in full swing the society then switches over from queen domination to queen subordination and males mature, for it seems to be more difficult to stop male production than gyne formation. Workers continue to accrue for some time. A male egg if successfully laid may go undetected and the larva either kill female larvae or outmanoeuvre them in some way. The queen must lay fast to totally inhibit worker-laid reproductive eggs and when these hatch they must compete with females; males have no asexual caste. This compromise between male-producing workers and female-producing queens comes to an end with the queen laying eggs of both sexes. When this happens workers lose their ovaries entirely in ants, and are produced from worker-determined eggs or are pharmacologically neutered and lose the correct oviposition behaviour in bees and wasps. The queen is then able to lay a run of males (sperm valve closed) and a run of females (valve open) and so fix the sex ratio. However, the workers, as brood rearers, are still in a vital position to influence growth and survival, and will only rear males in an appropriate season and an appropriate stage of the colony cycle; selective survival replaces socially sensitive production.

Insect societies pass through a cycle of population changes that are geared to the age changes of the single queen or of the queen population. This cycle affects the ratio of adults to brood, the ratio of workers to reproductives and the composition and ethology of the worker population. Hence there is an interaction which leads to the inversion of dominance (whether overt and physical or subtle and pharmacological) between reproductives and workers. When the colony is young and small (again whether monogyne or polygyne) the queens are able to stop workers laying male eggs and workers do not culture any males the queens may lay. The queens are also able to prevent workers giving excessive food to any female larvae with caste lability and to encourage active aggression against them, but their main gyne inhibition process in ants (not bees and not wasps) is the laying of eggs which have a proclivity to form workers, ending, almost certainly, in worker-determined eggs as distinct from caste-labile eggs. Thus, queens favour worker formation and workers favour sexual production and the outcome of this conflict between the generations shifts as the balance of power shifts from queen to worker.

In termites the pair of reproductives act as a queen, though the female is stronger and two of any one sex are better than one. This shows that control is not wholly related to the capacity to produce zygotes; ability to communicate with another sexual may enhance their physiological potency somehow. In ants a group of queens may have a more potent control than a single one for similar reasons.

The changing characteristics of the worker population are less obvious. Individuals become bigger on average, presumably due to an improved food supply with which the egg supply cannot keep pace. They also come to include more young workers as the age structure nears stable form after exponential growth. There is evidence that young workers are less responsive to queens than older ones. This is probably just a matter of age but it may eventually turn out that if they emerge into a large colony (i.e., one with a high worker/queen ratio) they will fail to meet a queen during a critical period some weeks after 'birth'; this reduces their responses later on. So a combination of an ageing queen population with a growing and more youthful worker population switches power towards sexual production.

CHAPTER 12

Reproduction

Now that the society is mature it faces the need to disperse and this chapter is concerned with the logistics of this: first how the sexuals are differentiated from the workers, then how they disperse and copulate and in some cases disperse again. A little data is also given about the quantity of sexuals produced.

12.1 Caste morphogenesis

12.1.1 TERMITES AND ANTS

In termites other than the Termitidae, workers are little more than larvae toughened and lightly pigmented, more active and more communicative. Development is gradual and proceeds without any change in the structure of wing or gonad rudiments. Moulting allows for growth and development though this does not always take place; and 'stationary' moulting means that the number of stages can vary. However, patterns exist. Thus, in *Mastotermes darwiniensis* the most primitive existing species, there are five larval stages and a worker stage that is darker and forages. It is not developmentally terminal, for it can moult into a nymph with visible wing pads and then after three nymphal stages expand into a fully winged imago with a functional genital system (both sexes). Thus there are 10 stages. Even the nymph is not committed to becoming an imago, for it can take a path of development that ends in a soldier. To do this it must first moult into a transitional stage called a pre-soldier in which re-organization takes place and then moult again into the adult soldier which, in *Mastotermes*, is just a bigger, stronger-jawed, wingless individual with undeveloped gonads. This change, like the change into an imago, involves loss of developmental potentiality and so is totally unlike the metastable worker stage which can remain a worker undergoing 'stationary' moults or become, via the nymphal stages, a reproductive or a soldier so ending its life history. In small colonies narrow-headed soldiers can be formed, possibly by a precocious metamorphosis under the stimulation of the reproductive pair (Watson, 1974). Termites of the Calotermitidae, e.g. *Calotermes*, *Zootermopsis*, have a similar stem-like morphogenesis, work being done by the older larvae whilst they are queuing for a chance to become

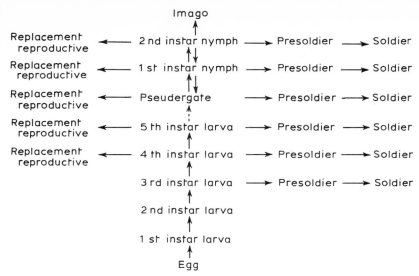

Fig. 12.1 Ontogenetic paths available to *Calotermes flavicollis*. Each arrow is a moult; the broken arrow is a variable number of moults. (After Lüscher, 1976.)

imagos or soldiers through several nymphal or one pre-soldier stage as the case may be (Fig. 12.1). In this family the nymphs can, in emergency, become reproductives that are wingless or stump-winged but fully effective. They can also change into soldiers.

With the Rhinotermitidae, e.g. *Reticulitermes* there are still two larval stages left in common. Thence the single stem of caste-labile individuals is split into two. The new feature is a branch almost entirely for sterile castes, i.e. workers and soldiers. It has five larval stages whose individuals work but can still give rise to pre-soldiers or wingless substitute sexuals but never to nymphs or imagos. They have clearly lost the capacity to develop wings but not quite lost the capacity to develop gonads. The labile stem consists of six nymphal stages which can become substitute sexuals or go to completion and become winged imagos; they are not workers and cannot develop into soldiers. However, this stem is labile in the sense that nymphs can regress into larvae, i.e. lose their wing pads and can then work and later become soldiers through the pre-soldier transition stage. In most Rhinotermitidae, movement into the sterile branch depends on season and on nutrition as well as on colony size but in species of *Schedorhinotermes* entry is impossible for males, and all the work and defence of the colony is done by females.

In the Termitidae wingless or near wingless substitute sexuals do not exist and are probably evolutionarily obsolete; winged imagos appear to make adequate replacements even though they are only produced seasonally as in ants. Workers, too, are no longer toughened larvae; they are made out of

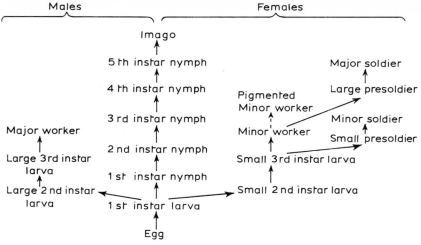

Fig. 12.2 Ontogenetic paths available to *Macrotermes bellicosus*. Each arrow is a moult except the broken one which represents pigmentation, without a moult. (After Lüscher, 1976.)

larval stages by a single moult that is usually terminal, though unlike the imaginal stages they retain prothoracic glands. Although their gonads are not fully developed, the two sexes can be recognized in the larval stage and it is now certain that some sexual bias exists which influences the likelihood of their taking the sterile path. Once this decision has been made their likelihood of changing into soldiers is also sex-biased, e.g. in Nasutitermitinae soldiers are usually male but in Macrotermitinae and Termitinae they are female. Where two caste sizes exist in a species it is more often through a difference in stage of maturation than through a sexual difference (Brian, 1979a; Noirot, 1969, 1974).

The reproductive and sterile stems diverge after only one stage in the Termitidae, possibly even as early as the egg (Noirot, 1969). On the sexual stem five nymphal stages then precede the imago (Fig. 12.2). In *Amitermes evuncifer*, a simple species, there are two larval stages between the egg and the first working larva then three or four stages of worker. Soldiers, formed without sex bias, come mostly from stage 1 workers so there is a rudimentary schism between worker and soldier stems. In *Microcerotermes* (Amitermitinae) workers, after two larval stages show three stages in males and four in females which thus end up bigger than the males. Pre-soldiers again diverge early from stage 1 workers. Other examples can be found in Noirot (1974) or Brian (1979a). In general, one can say of the Termitidae that there is an early divergence, possibly in the egg, of reproductive and sterile lines; that fewer larval stages occur but more nymphal ones; that a sex bias for soldiers but not

Plate 18
The castes of *Neotermes jouteli*.
Above: Winged sexual forms, workers and eggs.
Below: A major soldier with highly developed jaws and jaw muscles. (Photographs courtesy of J.B. Free.)

for workers exists and that size differences due to sex occur, as well as due to the stage of maturation.

The maternal effect in Termitidae has been demonstrated as a difference in JH level in the egg (*Macrotermes michaelseni*) by Lüscher (1976). The female reproductive has a low concentration of JH in her blood during the season when eggs are giving rise to nymphs. Young queens, by contrast, have high JH levels and all their eggs take the sterile path. Lüscher (1976) elaborated this into a general theory of the use of JH in termite caste control, maintaining that a JH-like substance released into the food stream by the reproductive couple could stop the formation of replacements during a sensitive stage in the moulting cycle. Their capacity to synthesize JH is no doubt vast; they need a lot to make eggs but is there enough left over for general circulation? (Fig. 12.3). Soldiers, Lüscher suggested, emit an anti-JH substance which stops the formation of more soldiers; and there is ample experimental evidence that adding JH analogues to termite cultures containing larvae in the sensitive

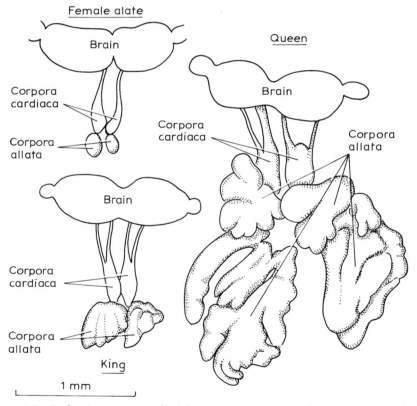

Fig. 12.3 Endocrine systems of adult *Macrotermes subhyalinus*. (After Lüscher, 1976.)

stage (the first 6 days of the third larval instar in M. *michaelseni* according to Okot-Kotber, 1980) boosts pre-soldier formation (Brian 1979a; Nijhout and Wheeler, 1982). One supposes that the JH enables further allometric growth to take place and leads to the typical exaggeration of the head. Nijhout and Wheeler (1982) suggest that timing of JH pulses is more important than the general concentration in the blood.

In ants there is very little information yet about the morphogenesis of caste in primitive species. In the ponerine *Odontomachus* food quantity can differentiate caste very late; there is thus a main labile stem but whether this has an earlier offshoot or even a maternally implanted egg caste bias needs to be explored. In *Myrmica*, all these ponerine controls exist along with several ways in which queens can govern the process. One, possibly two, moults have been replaced by cuticle plasticization and the glandular machinery pressed into service to regulate caste morphogenesis. Two discrete and alternative physiological states can be recognized. These differ in the gearing of the embryo growth rate to the whole larval growth rate ('embryo' refers to the rudimentary adult inside the larva). In gyne formation which represents the normal process of insect development, the embryo proceeds slowly, remains relatively small and most of the food intake by the larva is stored in the fat body. This goes on until the larval brain has moved into the thorax where the adult head forms around it. After this the embryo grows isometrically and develops wings, divided ovaries, segmented antennae and legs; all the necessary parts of the adult gyne start to differentiate visibly at this stage. The ability to do this arises during a cool period (normally winter) which resuscitates the corpora allata. This development is metastable and can be upset by starvation at temperatures over 23°C. The embryo then increases in size and changes in shape; this is easily seen through the larval cuticle as an enlargement of leg buds but not of wing or ovary buds. The total effect is to increase the proportion of the larva occupied by embryonic tissue. This event is irreversible and involves a loss of developmental potential: it is worker determination. The quick transition of the embryo from one shape to the other is a passage between two stable states and prevents intermediates forming naturally except when parasites disturb the nutrition. As the worker embryo grows, the wings remain small and vanish into the exoskeleton during metamorphosis, and the small germarium does not split into tubes as in the gyne but joins its oviduct to form a single tube ovary (Fig. 12.4).

What of the smaller size of workers? Relieved of its wing and ovary and enlarged, the embryo is able to develop and stop the larva eating 2 or 3 days earlier than it does in gyne development; this causes workers to be smaller. Worker determination can occur as a relatively late deviation from the labile stem that leads to the gyne (Fig. 12.5), at a stage after the larva has stored food, diapaused, hibernated and again recovered its growth potential; it is then rather like a nymph of *Schedorhinotermes* regressing into a worker

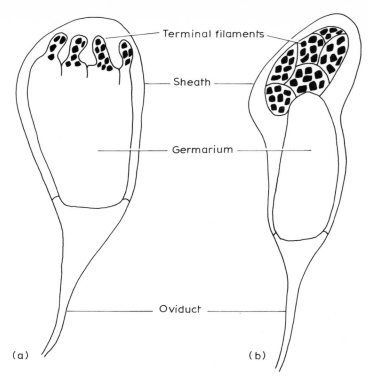

Fig. 12.4 Ovaries of final stage larvae of *Myrmica* developing into (a) gyne and (b) worker ovaries.

though, of course, there is no moult and all the changes take place inside the larval cuticle. Yet many, perhaps most, workers are formed from embryos that diverge from the labile stem well before the pre-diapause storage phase; this includes all those derived from worker-biased eggs. After a typical worker determination they do not rush through differentiation, stop feeding and metamorphose suddenly; the differentiation and expansion of the embryo into a pupa inside the larval skin after feeding is concluded, is a distinct process and may take as long for a worker as it does for a gyne. Nevertheless worker determination can be regarded as a step towards metamorphosis in that it causes growth of the embryo relative to its enclosing larva and reduces the time for growth and so the final adult size of the worker. It is certainly a precocious but unconsummated metamorphic event that distorts the embryo irreversibly (Brian, 1974a). The mechanism of its formation is probably a brain signal that switches off corpora allata activity without triggering the prothoracic (moulting) glands.

In *Myrmica*, as in all ants, only the female sex has two modes of size, shape, behaviour and longevity. Males do not and the male embryo suffers no

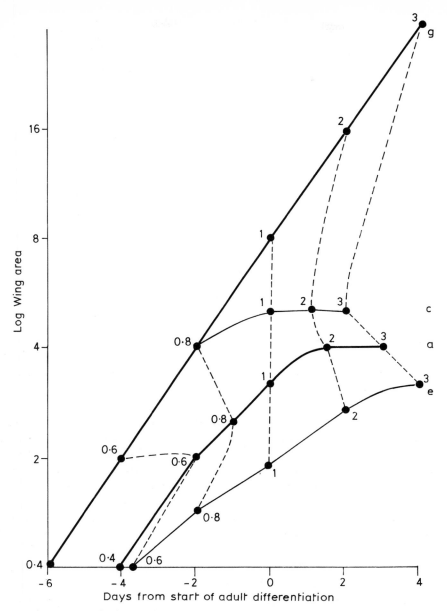

Fig. 12.5 Wing-bud growth in *Myrmica* larvae; log wing area plotted against time measured from the start of leg segmentation. g, gynes; c, large workers; a, average workers; e, small workers. Stages in adult differentiation are marked 0.6, 0.8, 1, 2, 3. (After Brian, 1974a.)

distortion under stress like that of females; all the adult rudiments are developmentally linked and the embryo grows in a resolutely isometric way. The role of juvenile hormone in both sexes is to restrain the growth of the embryo and encourage storage of nutrients in the larval fat-body. Ligating the corpora allata out of a caste-labile female larva leads to worker determination and can only be stopped by applying JH analogues (Brian, 1974b). It is the cell size of the corpora allata not their cell number that relates to their activity in *Plagiolepis pygmaea* (Suzzoni and Grimal, 1980). The worker ant then is derived from an embryo held in a worm-like larva but matured and externalized before it is fully developed. This process of emergence from the larva terminates its further development just as the larvae of advanced termites become fixed workers. Ant and termite workers are thus homologous. Their differences are only due to the termite juvenile being external all the time and the ant juvenile being enclosed in a totally different but temporary vehicle, extrication from which ends the plasticity of its form.

The formation of soldiers in ants is still not clarified. In *Pheidole* the gyne stem is separate from an asexual stem as far back as the egg which is influenced (see above) by the JH concentration of the mother's haemolymph (Passera and Suzzoni, 1979; Wheeler and Nijhout, 1981). The gyne larva is rotund and hairless, the asexual larva elongate and hairy. Soldiers diverge in the last (third) instar when the larva has a golden gut; they are recognizable for their greater size and above all in that they have fore-wing buds but no visible hind-wing buds; worker larvae have no wing buds (or rarely very small

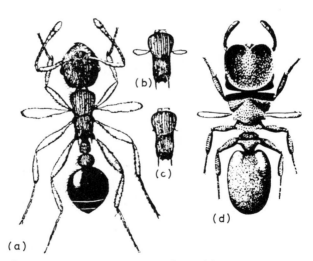

Fig. 12.6 (a), (b), (c) *Myrmica* intercastes with peg-like wing extentions; these arise from labile larvae that have failed to grow into full gynes and are near the caste known as soldiers in, say, *Pheidole*. (d) *Zacryptoceros* soldier with similar wing stumps. (From Wheeler, 1910.)

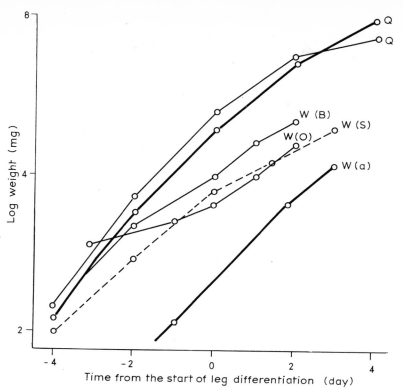

Fig. 12.7 The weight (log₁₀) of growing *Myrmica* larvae plotted against time in days from the onset of leg segmentation. Q, gynes; W(a), normal average size workers; W(O) and W(B), labile larvae cultured before winter when growth potential is poor; W(S), labile larvae cultured after winter recuperation but also just after a period of starvation at higher temperature (25°C). W (O, B and S) are the prototype 'soldiers' (i.e. large workers) of *Myrmica*. (After Brian, 1974a.)

ones). In all these castes the wing buds are small compared to the leg buds but in the soldier pupa it is possible to see a peg-like extrusion where the fore wing would be expected (as in 'large' workers and intercastes of *Myrmica*, Fig. 12.6).

Soldiers and workers are of discrete sizes, and the extra growth of a soldier is obtained if JH is applied topically during a sensitive phase in the last instar. This stops the initiation of metamorphosis and gives the scope for additional growth, which is allometric and so produces the soldier form. It also enlarges the workers but only slightly, and produces a situation like that between two species of *Solenopsis*, *S. invicta* (equivalent to untreated *Pheidole*) and *S. geminata* (equivalent to *Pheidole* after JH treatment).

Under natural conditions soldiers are formed if a queen is present, food is rich, temperature is high (26°C or more) and there are many workers. The presence of adult soldiers inhibits formation of more, though their power to do this declines as the season advances. This is not that soldiers are bad nurses and cannot supply the food their larvae need; it is related to contact frequency and seems to involve an ability to destroy JH (Edwards, 1982; Wheeler, 1982) as in termites. It will be interesting to know when intermediate ontogenies have been charted how the *Myrmica* situation with a labile stem and two main zones of deviation (giving small and large workers) evolves into the *Pheidole* situation with a distinct gyne and worker path right from the egg stage with the worker path producing an extra large and discrete soldier caste (Fig. 12.7) (Brian, 1979a).

12.1.2 WASPS AND BEES

In the wasps and bees, wings are present in both castes and the necessity for wings in workers may have held back the evolution of a substantially different worker caste. Vespine gyne cells allow more and perhaps better food which may contain more salivary protein (Montagner, 1966). Eggs are unbiased and large size differences between larvae only appear after three out of the five instars. Gyne cell larvae regress into workers if put in small cells, i.e. they are caste labile (Spradbery, 1973a), but small cell larvae put into large ones form intermorphs. They are not committed workers and perhaps in normal circumstances caste is determined after the larva has stopped eating. Then its nutrient status may be assessed internally and metamorphosis canalized towards one or other morph giving either a large gyne in reproductive diapause or a small non-diapause worker whose reproduction is blocked by social influences.

In *Bombus*, cells are not regularly distributed but their contents are examined frequently (Pendrel and Plowright, 1981). Eggs are not biased in any way. Some larvae lose caste lability and are determined as workers when 3–4-days old and near the end of the first instar, after roughly a quarter of their larval life. Workers can recognize larvae with gyne potentiality and they give them extra food so that they spend 3 extra days as larvae and 4 extra ones as pupae. In queenless groups, the worker/larva ratio governs how many larvae form gynes but queens can reverse this. The juvenile hormone concentration is higher in the last instar of gyne than of worker larvae and so the corpora allata probably delay metamorphosis as in *Myrmica*. Röseler (1977) suggests that queens inactivate the corpora allata of larvae but since they do not need to touch larvae directly the workers must transmit the inactivator. Perhaps the queen gives out a signal to which workers respond with queen-dependent behaviour as in *Myrmica*.

In Meliponini too, caste differences depend on the amount of food: gynes form in large cells, and they can develop on worker food if enough is given (Darchen and Delage-Darchen, 1974, 1975). Even in *Melipona* food supply can be quite important in bringing out or developing latent genetic differences associated with a two-allele heterozygote. The situation is evidently assessed quite late in larval life just prior to metamorphosis for treatment then with JH gives more gynes and fewer workers. Thus caste potential is assessed internally after feeding has stopped and the hormonal changes necessary for appropriate morphogenesis then made.

The honey-bee has much greater differences between female castes than any other bee. Although the wings are retained in workers the genital system is reduced and special glandular systems and behavioural repertoires developed. Eggs are unbiased and caste differences develop gradually under different feeding regimens. The whole life of the larva is only 5 days for a gyne or 6 days for a worker. Larvae transplanted from worker to gyne cells become gynes if less than 3-days old, but a later switch gives an intermorph, with fewer egg tubes and a small spermatheca. They often die or are recognized as wrong by workers and ejected.

Worker larvae are not starved at all for the first 3 days; in fact they get a protein-rich diet synthesized in the hypopharyngeal glands of young workers and they grow as fast or faster than gyne larvae of the same size (Fig. 12.8). However, the worker jelly (WJ) does not enable larvae to metamorphose and for the last 3 days worker larvae really are stressed by being given a mixture of WJ and honey. As a result worker larvae take longer feeding and metamorphosing than gyne larvae, unlike *Bombus*. Gyne-cell larvae get the same food all the time (GJ, gyne jelly); it is well balanced, has 12% sugars instead of 4% (WJ), is rich in vitamins though these do not appear to be crucial, and its water content rises to about two-thirds whereas that of WJ falls to two-thirds. These and other qualitative differences in food supply (and only a few have been listed here), are important, for females reared in drone cells, where they get ample WJ, grow into big workers not into gynes. GJ deteriorates in store and gives rise to workers: this is probably due to the crystallization of hexoses, for WJ with hexoses added is almost as good a food as GJ. In fact if an analogue of JH is added it *is* as good. Larvae in gyne cells are fed a secretion from the mandibular gland as well as the hypopharyngeal gland 10 times as often as worker larvae and there is no doubt that they get a regular supply of well-balanced food. This is crucial (Beetsma, 1979; Brian, 1979a, 1980).

An interesting comparison with *Myrmica* is possible. Worker larvae 1–3-days old do not store fat or glycogen in their fat body whereas gyne cell larvae do. A similar difference is detectable in last-instar *Myrmica* larvae: it means that gyne morphogenesis is characterized by high food storage and low embryonic growth, whereas in worker morphogenesis the food intake goes straight to the physiologically dominant and active embryo. This, as in *Myrmica*, is likely to arise from a higher concentration of JH in gyne larvae, which is interesting as it has been shown in *Apis* that the concentration of JH in the blood is 10 times higher in gyne larvae 3-days old than in worker larvae

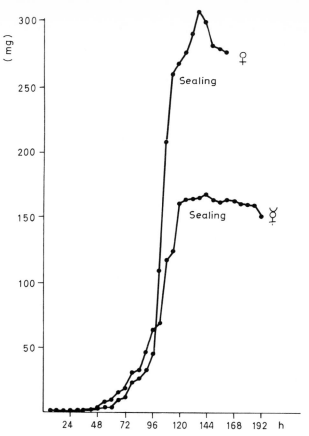

Fig. 12.8 Growth of gyne and worker larvae in the honey-bee. (After Wirtz, 1973.)

(Beetsma, 1979). It is known from experiment that JH switches on all gyne features together. The anti-corpora allata factor precocene which destroys the corpora allata without reducing feeding may prevent gyne formation sometimes. So the high JH food storage gyne-producing syndrome appears in both the ant and the bee.

These differences in JH concentration continue to affect metamorphosis when caste differences first appear in the larval stage. In the gyne, the ovary which is extraordinarily large in honey-bees with about 300 egg tubes, retains all of them but in the worker they are reduced from over 100 at the end of the larval stage to around 12. Thus in worker formation the female that is genetically able to build 300 is epigenetically forced to lose 288! The material released by this degeneration during metamorphosis is no doubt built into wax glands and pollen-collecting equipment and other special worker features. Ovary stabilization in worker larvae can be achieved by 1 or 2 days in a gyne cell where the GJ enlarges the corpora allata and increases the level of JH in the blood about 10 times. One final point of interest is that even in glass cells in the laboratory gyne larvae orientate vertically, head downwards after spinning a cocoon whereas workers lie horizontally. This gyne larva behaviour is in line with bees generally and can be induced in worker larvae by treatment with JH (Beetsma, 1979). The worker larva is obviously adapted to life in vertical combs with near-horizontal cells; some terminal developmental genes never operate.

From all the data, collected over many years by many scientists, a small part of which has been mentioned here, it becomes clear that the different food regimens cause differences in glandular development that affect the plan and process of rebuilding and reorganization in the pupal stage. One question still open is how the organism assesses food differences? Is it that the foods influence the neuro-endocrine system after they are absorbed from the gut, i.e. nutritionally, or is it that the food includes appetisers that stimulate oral sensory organs and so affect the neuroendocrine systems: nutritional or sensory?

Beetsma (1979) and his colleagues take the latter view and have shown receptors sensitive to sugars in the labial region of larvae. They maintain that the sugar content of WJ in the first 3 days is too low to enhance feeding rate. Yet larvae in worker cells grow as fast in the first 3 days as do those in gyne cells and so must be eating as much; it is only later when the WJ is thinned with honey that growth rate drops. Too little is known about the effects of larval nutrition on the epigenetic control of morphogenesis; the current emphasis is on the corpora allata but it is known that the prothoracic glands play a role not only directly but in interaction with the whole system of endocrine glands (Brian, 1980).

Thus caste differences in all social insects are due to feeding differences moulding a divided genotype and so ultimately to differences in worker behaviour. In eusocial species this behaviour is queen-dependent. In higher ants and termites but not bees and wasps it co-exists with a profound effect coming from the internal state of the mother during egg formation. Together these influences enable the embryonic adult, that lives and develops inside the larval envelope, to dominate the food intake. The embryo of course depends on the larval mechanism for collecting and digesting food but as it grows hinders larval feeding, with the result that it eventually cuts off its own supply and never reaches the size that gyne embryos achieve. JH may be the factor in the maternal environment that influences caste bias just as it is in the larval and pupal environment.

There is little evidence that the genes do anything but create the condition of sex-limited dimorphism in females absent in males (of Hymenoptera). In one ant, *Harpagoxenus*, and in one bee *Melipona* genetic factors have been demonstrated. There are also a few cases of isozyme differences between castes in ants: in the cross *Solenopsis geminata* × *S. xyloni* though workers are hybrid, 95% of queens are of the maternal type. This means that the paternal genes are only suppressed when the hybrids develop into gynes (Hung and Vinson, 1977). Although in one *Myrmecia* species, workers also differ from queens in gene frequency this was the only case known when Crozier (1980) reviewed this subject. Polymorphism is genetic but caste morphogenesis depends on the mother's haemolymph and the larva's food regimen.

12.2 Copulation and dispersal

There are two aspects to dispersal: genetic and ecological. First, a need to ensure outbreeding (at least periodically), second, a need to find and occupy new habitats as they become available or else to invade and displace species in already occupied ones. The common sequence is: congregation of the sexuals, pair formation and copulation, dispersal and habitat selection. There are many variants, especially amongst those social insects that bud small, complete new colonies. Copulation between sibs in or on the nest is rare and there are many devices to prevent this. The deleterious effects of inbreeding have been shown in the honey-bee using artificial insemination. Inbred workers survive poorly and store less honey than normal ones and show poor temperature regulation. In view of the theory that haplo-diploid systems are less resistant to inbreeding depression than diploid ones this is noteworthy (Brückner, 1980). The sex determination mechanism can be upset too. In *Apis* there is only one sex locus but several sex alleles and the locus must be heterozygous to give females, and hemi- or homozygous to give males. Inbreeding increases the likelihood of homozygosity and so of diploid males which are inferior sexually to haploid males. Workers recognize the larvae and destroy them when they are only a day old just as they recognize gynes in the wrong cells (Woyke, 1980a, b).

Male *Polistes* try to copulate in the nest but are refused by females and chased off by workers. They gather where females collect before winter, fly at them, fall to the ground and copulate briefly (e.g. *P. fuscatus*, Evans and West-Eberhard, 1973). In tropical *P. canadensis* in contrast, males come to new nests and copulate with females that have already started to build without worker intervention. This is generally true of swarming Polybiini; it probably provides for outbreeding and reduces risk for the gynes. *Vespula* males have routes between areas of vegetation though they leave no perceptible scent (Calam, 1969); copulation is followed by hibernation of the young queens who only disperse and seek nest sites afterwards. *Bombus* males also have routes between sets of posts or tree trunks that are definitely marked with their mandibular gland secretion (Calam, 1969) or with their labial gland secretion (Bergström, 1975; Bergström *et al.*, 1981). These secretions no doubt help to collect females and are species-specific; B. *pratorum* uses a 15C terpenoid alcohol such as farnesol, B. *pascuorum* uses a 16C unsaturated aliphatic alcohol (hexadec-7-en-1-ol), B. *lapidarius* uses palmitoleyl alcohol (hexadec-9-en-1-ol), and B. *lucorum* has two forms, one pale, one dark, both using ethyl esters: ethyl myristoleate in the pale, and ethyl dodecanoate in the dark form. Gynes visit these stations, copulate, go back to the nest and later, hibernate. Besides using different chemical markers species stratify in the vegetation.

The Meliponini swarm all the year round or, as in the African *Hypotrigona*, avoid only the main dry season (Darchen, 1977). They swarm in stages. First dark, old honey collectors prepare and line the cavity, then young ones come and build storage pots and put honey and pollen into them and lastly a gyne and a whole cluster of young workers come and make brood cells. Males hang around new nests and settle nearby ready to copulate when the gyne emerges some 2 days later; she then lays at once. This system must reduce inbreeding as males from all around can attend and it must also reduce the risk of accident to the new queen both during transit and during copulation. Risk of failure in establishment is small as the site is well tested and assessed before the main group moves in. In fact, this system must be nearly perfect in that it provides protection for the gyne all the time except for the brief moment of copulation and yet allows strange males to compete.

In the honey-bee many more risks are taken since the mother queen leaves the nest with a cluster of supporting workers gorged with honey but without a prepared new nest to go to. Several mutually hostile daughters are left and a risky nuptial flight is necessary before restarting the old society (Simpson, 1974). The queen's fitness, in an evolutionary sense, depends on maximizing the number and quality of daughters she produces. To achieve this she, with the aid of her workers, first builds up a vast number of workers as quickly as possible in spring (in temperate sub-species) and then, whilst growth is still maximal, and before congestion or inadequate distribution of food seriously interferes with the efficiency of the colony, produces male cells and gyne cells with appropriate eggs. This is directly geared to long days, warmth and flower abundance. Probably the immediate cause of the construction of male and gyne cells is the dilution of the queen's controlling pheromones by high worker/queen ratios and a general unresponsiveness of the young workers which come to predominate in the population after a period of rapid growth.

One factor in success is choosing the best time of year to reproduce. This must enable the mother queen to found a new colony, prepare the nest, build up a population and establish sufficient stores of honey to survive the winter. Most swarms are in May or June (Fig. 12.9) which just about gives time in a reasonable summer since days are long and most flowers are blooming. Yet a small number of swarms come out in August which can have no hope at all. They probably represent the vestige of a former tropical cycle of about 3

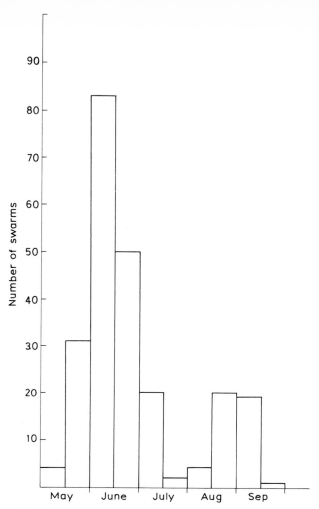

Fig. 12.9 Honey-bee swarm emergence dates summed over 5 years in New York, USA. (After Fell *et al.*, 1977.)

months. Another factor in success must be to produce an optimum number of new gynes. This should be as many as can be given adequate worker support, i.e. the best sized swarm for success. The bigger the swarm the quicker it can make a new nest and establish itself with comb, eggs, brood and stores but to make it unnecessarily big will detract from the size of other swarms and from the stock that stays in the base nest. It seems likely that two or three new colonies is as many as one colony could and should make each year. Of these, only one will be started by an inseminated queen over a year in age, the rest will be started by gynes.

So, how is this arranged? At least 10 days before swarming the workers take up honey in preparation for leaving the hive with the queen. It is said that the workers which go with her are more or less a random sample of the population (Butler, 1940), but in view of this preparative phase this cannot mean that a bee decides to go or not to go spontaneously: some workers for some reason must form a closer allegiance to the queen than others. Can they recognize genetic affinity in some way? One would expect a bias towards young bees to supply wax for comb building and then develop food glands for larval feeding and in *Apis mellifera adansonii* this is the case, for more than four out of five are less than 8-days old (Winston and Otis, 1978).

Consider first what happens in the hive which the mature queen bequeaths to her daughters. The gynes are hostile to each other and the first out of her cell tries to sting the younger ones and if successful heads the new colony. Otherwise workers intervene and perform pacification dances and add wax to gyne cells to stop younger gynes getting out especially in congested colonies (Simpson, 1974); these then mature and harden their cuticle inside the cell with food given from time to time through a hole. Whilst being kept apart in this way the gynes make their famous intermittent 'piping' noise at 650 Hz caused by 'flying' with wings folded and body pressed against the wax substratum. It stimulates swarming and is a feature of large and tightly packed colonies that can generate a few viable after-swarms. When later the gynes are let out, unless an after-swarm emerges they eliminate each other leaving only one, but whether this is the fittest is doubtful. Some sort of elective contest may occur in which gynes compete for worker allegiance as do gyne and queen in *Trigona subnuda* (Imperatriz-Fonseca, 1977).

Sometimes the queen does not leave with a swarm but stays in and is superseded by one daughter. In such cases fewer gyne cells are made and it looks as though the process was planned beforehand as a queen replacement rather than a swarm, perhaps on account of the lateness in the season or because the queen is failing as an egg-layer (Butler, 1957). The least risky way of supersedure is for the daughter to copulate and then eliminate the old queen and for the mother it is certainly better to bequeath the nest to an inseminated daughter than risk total loss through an accident on the nuptial flight (Simpson, 1974).

When conditions are right both externally (long days and fine, warm weather) and internally (sealed gyne cells) some bees more active and sensitive than others start a buzz-dance in which they run in straight lines very fast with their wings partly open and vibrating at 250 Hz. When a bee contacts another it holds it for 5 s and pipes continuously at 500 Hz (not in bursts like a queen). This induces the new bee to buzz-dance only if it is ready: if a buzz-dance is played in a non-swarming hive the bees 'freeze' (Esch, 1967). So the behaviour spreads in a ripe hive until the emigrant bees pour out together much faster than usual and form a dense cloud near the hive.

They do not go straight to a new nest but cluster within 10 m of the hive as long as the queen is present otherwise they go back inside. Queen recognition is complex; Ferguson *et al.* (1979) stabilized queenless swarms with synthetic Nasanoff gland pheromone containing citral, nerol, geraniol and farnesol together with 9–ODA and other components from the queen's mandibular gland. The variety of chemicals is important for Ambrose *et al.* (1979) have shown that swarm bees can distinguish their own queen from other queens and will choose one of a similar age and reproductive state to their own rather than accept a different substitute. In fact, they can identify their own queen with the help of the shellac mark that the experimenters use! (Boch and Morse, 1979).

Once settled the scout bees (some 5%) indicate to their companions the direction and distance of a nest site using the same display as for food. When there are several possibilities it seems that the scouts visit the sites individually and assess them. Only when all agree is the cluster ready to go but it still takes time to get the bees airborne. Seeley *et al.* (1979) found that about an hour before they take off scouts start to collect on the cluster: some are quiet, others do information dances and others do the buzz-run activation dance. The buzz-runs induce the high-pitched piping sound and the chains of hanging bees are disintegrated as other bees take wing and circle round, and eventually move in the direction indicated by the swarm display. They are probably helped by scouts moving to and fro along the line to be taken, whilst others fly on and release Nasanoff gland pheromone at the nest site. The velocity rises to 8 or 10 km h⁻¹ at 3 m height and the swarm flattens as it goes along.

The habitat selected by swarms in a feral state is old, open deciduous woodland which provides nest cavities in old trees and a variety of early food. They rarely choose new sites less than 300 m away but can go as far as 1 km (Fig. 12.10); they prefer sites 1–5 m high, with an entrance 12–75 cm² at the bottom facing south and a volume around 40 litre (Seeley and Morse, 1977, 1978). Scouts spend 40 min assessing the site and may assess the volume in the dark by integrating various walking distances (Seeley, 1977). Avitabile *et al.* (1978) found most feral colonies in broad mature maple (*Acer*) and oak (*Quercus*) trees over 3 m up facing south-west.

Since males do not come to the nests as they do in stingless bees the gynes have to go out to find them and copulate. Males fly and congregate in places where air currents converge and rise about 10 m above the ground (they are invisible without a telescope but audible). With *A. m. adansonii*, in good weather, gynes start to orientate by flying around their nest site for 1–2 min 3–4 days after hatching. They copulate on the fifth day, and some on the next day too; these flights last 13 min and take place mostly at 15.00 h (Fletcher, 1978b). Whilst flying the queen emits a trail of lure substances from her mandibular gland which are detected by special cells on the antennae of the

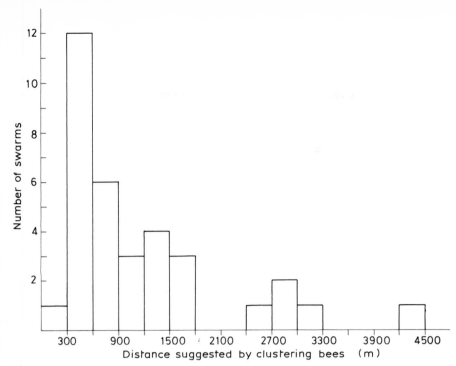

Fig. 12.10   The distance from cluster to new nest site indicated by swarming bees, 1, observed by Seeley and Morse (1977) and 2, by Lindauer, 1955.

males. 9–ODA can be detected when very dilute and enables the male to home in on a gyne from a great distance upwind. The main problem at the moment is to know how the gyne finds male assemblies. After copulation the supply of sperms is around 5–6 million which drops to nearer 2–3 million when 10 000 eggs have been laid, an average of about 30 sperm/egg. As many as 10 may enter each egg cell (Harbo, 1979).

Amongst ants there are many types of reproductive behaviour. In some Ponerinae and all Dorylinae the queens are wingless and remain in the colony all their lives. Males which have wings, seek them out by following foraging trails, e.g. *Megaponera foetens* (Longhurst and Howse, 1979b). In *Eciton* (Dorylinae) sexual broods contain thousands of males but only a few gynes. The cluster divides: a part with the queen and a part with the brood. Between these there is some antipathy. Young gynes whilst still pale are surrounded by worker groups which push their way out and cluster within a metre of the colony; workers from these clusters which grow to 8–10 cm diameter, circulate with the main colony. Some days later males emerge from their cocoons and the excitement probably starts off what Schneirla (1971) called 'the queen-exodus stage' in which the parent queen moves out with workers and goes off on a trail. An hour or so later the first young gynes still escorted by workers, move away and join another trail forming an incipient colony. Others emerge too late and have difficulty attracting workers; in hostile conditions they are 'sealed off' by workers packing round and holding on to their legs. However, some gynes do escape from these clusters and with persistence gain acceptance by a raiding column that then refuses its own queen; such supersedure can occur without colony division if the old queen is very unattractive (cf. *Apis*). In *Eciton*, males normally fly away and try to join a column of their own species. Within the first day they pick up the colony odour, lose their wings or have them pulled off and then mature sexually. Males cannot copulate before they have lost their wings, and once copulation has occurred they die.

In Lamto savannah, Leroux (1979a) found that *Anomma* normally changes its nest every 8 days but when it reproduces, producing many males and a few gynes, it stays 2–3 weeks in the same place but hunts relentlessly. Some colonies divide repeatedly whilst others never do; in 134 days one colony divided three times, but of 46 colonies regularly followed, only 14 were seen to divide. Since Doryline males hunt out colonies of their own species and copulate with the elected gyne only the male runs much risk of death. Male activity is circumscribed; they often come to light traps in the dry season but the period varies from species to species; they fly in humid air after rain (Baldridge *et al.*, 1980). The Doryline gyne, though flightless, is unlikely to inbreed, since her brothers have to fly away before seeking a colony for adoption. She is protected by workers all her life and this makes the risk taken by an *Apis* gyne on her nuptial flight seem enormous since loss to a predator can be the end of that colony as it is normally too late for the workers to return to their parent nest and start again; one wonders whether association with humans has allowed the honey-bee to become careless.

Wingless gynes do not always have to wait for males to come to them. Those of *Rhytidoponera metallica* (Ponerinae), go out and attract them by means of a pheromone from intersegmental glands (Fig. 12.11). *Ponera* normally has winged males and wingless females (LeMasne, 1953) but at least six species have wingless males too. In *P. eduardi* both sorts occur and the wingless ones copulate with wingless gynes soon after 'birth'.

In the Myrmicinae various degrees of dependence on other free-living ants have evolved and this social parasitism is often associated with winglessness in gynes and males. *Formicoxenus nitidulus* has both winged and wingless gynes; the latter go out of the nest, climb on a plant nearby and with gaster raised emit venom from the tip of their extended sting. Wingless males walk towards this from 15 cm at least and, after some 'coyness' on the part of the female, copulate (Büschinger, 1975). The winged ones never seem to fly and break off their wings within 2 days. A lot of inbreeding seems likely but lack

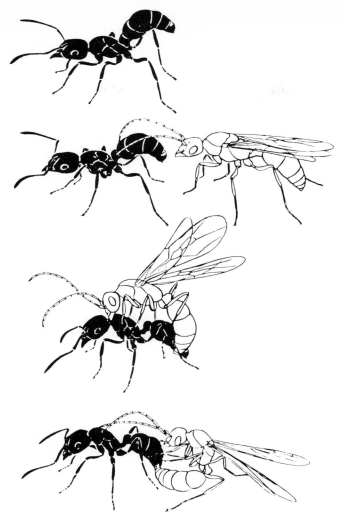

Fig. 12.11 Calling by the gyne of *Rhytidoponera metallica* and then copulation. A pheromone is emitted from the tergal gland near the tip of the gaster. (After Hölldobler and Haskins, 1977).

of wings is a feature of social parasites especially amongst the Leptothoracini. *Harpagoxenus sublaevis* climbs plants near its nests and stays for hours with its sting extruded carrying a droplet of venom; this brings winged males from 3 or 4 m. In this species winged females occur as well but in *H. canadensis* all females are winged (Büschinger, 1974). In *Anergates atratulus* (Büschinger, 1974) males are almost pupoid and copulate in the nest with their sisters soon after emergence. Such extreme inbreeding would appear to be very bad genetically but this is a very specialized workerless social parasite, whose gynes after copulation fly to another nest of *Tetramorium caespitum*. Many other cases occur (Büschinger *et al.*, 1980), and Wilson (1975) has pointed out that these apparent disadvantages do, in fact, save them from losing touch with their host population.

In ant sub-families with predominantly flying gynes it is usual for both sexes to leave the nest and meet outside. The distance travelled varies from only a few centimetres to several kilometres and can give excellent dispersal. In *Myrmica, Pogonomyrmex* and many others, the males go to a conspicuous object such as a rock or a tree and fly in a close, looping cluster on the sheltered side where they fall to the bare soil surface beneath for copulation. Pheromones are emitted that drift downwind and help the gyne to locate the spot. Different species may use the same assembly at different times of day or season, year after year, presumably because it has some special characteristic for it cannot be held by tradition as with bird leks. Wynne-Edwards' suggestion that some sexual selection takes place in ant congregations may be true. To select strongly flying gynes favours those with big wing muscles and hence those which have the best food store to see them through the claustral phase of colony foundation. In *Pogonomyrmex* gynes may copulate with several different males in succession, and then fly off, settle, break off their wings and start off on foot to look for a crevice (Hölldobler, 1976b).

The attractant produced by male *Pogonomyrmex* is related to 4-methyl-3-heptanone or its alcohol (McGurk *et al.*, 1966, cited Hölldobler, 1976b, p. 80). The females emit a poison gland pheromone which combined with a skin substance and their specific shape and behaviour helps the males, of which there are several in succession to connect. Sounds are not used for coupling except when a satiated female cannot escape the males; she then stridulates (Markl *et al.*, 1977). Thus the congregation of the sexes depends first on roughly simultaneous ripening, second on release in the right sort of weather, usually warm, still and humid, at a certain time of day which varies with the species; and third on flight towards a certain type of configuration or landmark whilst emitting a characteristic pheromone. Once a cluster is formed it may attract visually.

Many ants seem not to use a marker object but simply to congregate over ground features that reflect light or generate heat rays. Some just rise to a certain level and stratify. *Solenopsis invicta* workers open holes and the males leave the nests before the females and stratify at 90–150 m about midday for several hours (Markin *et al.*, 1971, in Lofgren *et al.*, 1975). Flights take place several times a year mostly in June and July. Local or widespread, they occur 1–2 days after rain, especially if this is preceded by dry weather, in a light wind at a relative humidity of 80%, and a temperature of 24–32°C but in winter they may fly at 20–26°C (Lofgren *et al.*, 1975). Delayed nuptials can be a serious drain on a colony's supply of food and space and many ants destroy sexuals that have not flown by the end of the season. There is also a danger that though uninseminated, gynes will start to lay eggs. As long as these are edible, gynes are simply helping, but if they are reproductive too many males at the wrong season could result. *S. invicta* appears to have a safeguard: the queen produces a pheromone which stops gynes shedding their wings in the nest before copulation. If the queen is taken out gynes start to do this with the help of workers (if they are young) within a day. After a few days workers begin to kill a lot of these newly deälate gynes leaving only a few with well-developed ovaries. The new 'queen', though uninseminated, is able to stop more gynes removing their wings presumably by developing pheromone production herself (Fletcher and Blum, 1981a, b). The pheromone is on the queen's surface (especially the abdomen) and can be washed off with acetone; it is not volatile and depends either on direct worker contact with her or on being passed round on the mouth parts of workers. Because there is a latent period of 1–2 days after the queen has been removed before deälation, the authors suggest that it is a 'primer' substance that enters the bodies of gynes and changes their behaviour by direct action on the nervous system rather than by stimulating exteroceptors. This inhibitory system is very like that used by *Myrmica* queens to stop sexualization of labile larvae (Brian, 1970); 30 s in ether washes their pheromone off.

Tetramorium caespitum, like *Solenopsis*, appears to be a stratal nuptial-izer. Unlike other heath/grassland ants, they fly as soon as the rising sun touches the nest surface. The male mandibular gland has 4-methyl-3-hexa-none and the corresponding alcohol; this is present in workers but absent from gynes (Pasteels *et al.*, 1980).

Lasius and *Acanthomyops* are also stratal nuptializers and produce ter-penoids and an indole in species-distinctive proportions; males probably signal their emergence by these means. The strata used by different species is unrecorded but it is known that they tend to fly at different seasons, e.g. in England *L. niger* flies in July and August, whilst *L. alienus* flies in September and October, often too late, in fact, for suitable weather in which case they are eaten in the nest by their own workers. *L. flavus* tends to fly at the same season as *L. niger* but later in the day: 17.00–19.00 h instead of 13.00–17.00 h (Boomsma and Leusink, 1981). These authors suggest that a species will fly when the temperature in its nest is about the same as that outside; they thus avoid thermal shock as they launch themselves. The stratum technique with or without ground feature orientation, means that these species can inhabit treeless plains of grass or steppe. These mass flights are well known because they almost always attract generalist predators of many kinds; not only birds that catch them in the air but reptiles and amphibians, mammals and arachnids as well as beetles, bugs and other insects that catch them in the vegetation and on the ground during or after copulation. It is argued that even though many different kinds of predator are attracted by synchronized flights, the mass simultaneous exodus saturates their combined predatory effort and enables a substantial proportion of sexuals and a substantial number of young queens to escape and begin founding new colonies.

Other ants use the wingless gyne's trick of luring males. This is generally true of the *Formica* group; after copulation quite near the nest they return and later join a bud that temporarily retains communication with its stock nest but eventually is either resorbed or released (Mabelis, 1979b; Pamilo and Varvio-Aho, 1979). So too, in *Monomorium pharaonis* the gyne uses a Dufour secretion to attract males (Hölldobler and Wüst, 1973).

To summarize, ants congregate in two main ways: first the gynes fly to males who assemble by a marker or in a zone some distance above the ground; second, the gynes 'call' quite near their own nest and are found by itinerant males. The specialized doryline method in which the male is adopted by the colony whose gyne he subsequently inseminates, is perhaps a variant of this. The first method provides for maximum dispersal of genes and colonies, the second provides adequate gene dispersal but since the colonies divide or bud sometime afterwards, species dispersal is slow though sure. In the first case the fit females are the ones with big flight muscles which give them a chance to found a new group alone. In the second case the fit females are the small ones that return and obtain acceptance by their own colony or are social parasites and enter colonies of another species. The latter do not need long flights or big flight muscles and dispersal into newly created vacant habitats is no advantage either. Their best chance is to stay near their hosts and not be blown away and in the winged males they still have a means of gene dispersal. Where even males are wingless and copulate as they emerge with sisters the advantage of not losing one's host may outweigh the advantage of outbreeding.

The speed with which young queens break off their wings if they have them after copulation suggests that they do not search very diligently for the right habitat: indeed in many cases they are likely to be over it already. The males may play a part in habitat selection, as in bumblebees, but probably only in a very general way (Wynne-Edwards, 1962, in Brian, 1965). *Lasius*, in which the male is so small he can be carried whilst still coupled, may represent a case where the new queen searches by air. There is some evidence that the queens do select, for *Lasius niger* and *L. flavus* in Europe prefer bare garden soil to that covered in *Brassica* crops (Pontin, 1960), and *Lasius neoniger*, in America, a grassland species, avoids woodland whilst still winged and will fly away if it lands there by mistake (Wilson and Hunt, 1966). Queens of *Lasius niger* in Europe settle in moister areas than do those of *L. alienus* (Brian *et al.*, 1966). In grass heaths and scrub, ant queens of *Formica lemani, Myrmica ruginodis* and *M. scabrinodis* will choose open glades and freshly cleared woodland and settle there in the warmer spots and on the south sides of old tree stumps (Brian, 1965).

The nuptials of termites resemble those of free-flying ants except that they end with pair formation not copulation which takes place intermittently in the nest later on. Very few are active fliers and none are known to rise out of the relatively calm air near the ground. *Allognathotermes hypogeus* are known to assemble against tree tops spaced roughly one per hectare; pairs form on the tree foliage with the male attaching to the underside of the female who then flies to the ground, breaks off her wings quickly and promenades during the site-searching period. Others that assemble include *Pseudacanthotermes militaris* (Noirot and Bodot, 1964). *Cryptotermes cynocephalus* flies to roof height only. *Allodontotermes giffardi* has females which although winged, do not fly. They climb into herbs with their abdomen raised and flutter their wings; this attracts males, either visually or acoustically and in the end no doubt chemically (Bodot, 1967).

Instead of the workers just breaking open exit holes as ants do, termites make platforms and covers for take off which are guarded by soldiers. Synchronization is achieved first by maturation at about the same season, and second by picking a certain type of weather and a certain time of day. In reviewing this, Nutting (1969) says that in general, in temperate climates termites fly in summer and in desert climates in winter, after rain. In the tropics not all species fly in the wet season, some use the dry; this depends, it seems, partly on what sort of habitat they have to dig into afterwards for rain softens wood and soil but also blocks up cracks in the latter. In fact, rather sensibly, species and even colonies have several flights each year (Nutting quotes *Pterotermes* as having 40). Some are unpredictable but others are quite predictable, e.g. five species of *Amitermes* fly on the day after rain and a species of *Heterotermes* starts 10 min after sunset and goes on for an hour! The distance flown varies a lot too: thus whereas *Macrotermes* emit thousands and some may fly several kilometres, *Calotermes flavicollis* emits 30–60 sexuals from each colony on each flight, which only fly a few tens of metres.

In a general way the type of habitat seems to be identified. Nutting (1969) suggests that desert species may identify shapes and forms that provide food and shelter. Some fly along rows of trees or streets as though reacting visually, e.g. *Amitermes* in Arizona. *Amitermes* in the Ivory Coast tends to settle near the mounds of *Trinervitermes geminatus* (Josens, 1972) and these are probably identified visually in the first place. Pairs form on the ground and the male follows the female closely keeping tactile chemical contact. Thus, in *Trinervitermes bettonianus* the male is attracted by a volatile from 10 cm away, and he palpates the tergal zones of a female, putting his head on the sixth to tenth tergal region. Both tergal and sternal glands (Leuthold, 1977) lay ephemeral trails in case contact is lost. In *Zootermopsis nevadensis* and in *Z. angusticollis* which use the sternal gland (Pasteels, 1972) females are attracted towards extracts of both sexes but males only towards female extracts and when a pair is formed the wings are broken off. To do this primitive termites rub against objects, or use the hind tarsi or get help from their mate who bites the wing base. In higher termites the wing is strained against the thorax and the blades raised until they snap off at a weak surture near the base (Nutting, 1969). In *Microtermes* the male is carried on the underside of the female as in *Allognathotermes hypogeus*. He sheds his wings and the female makes an aerial survey of the habitat features and selects a locality in which to descend and nest.

Termites that live in wood usually enter through holes made by beetles or by warping cracks; the condition of the wood, its softness and degree of decomposition are all important. Those that enter soil often use the temporary cover of leaves and debris but very quickly dig a shaft; females do most of the work but the male guards against unattached males. At the bottom of a shaft a chamber is excavated which may be lined with faeces and is just big enough to turn in. At the top of the shaft a closure is made of faeces and debris which takes various forms according to species. *Macrotermes natalensis* cuts soil with jaws and scratches out loose soil with the feet and legs. *Microtermes havilandi* makes a dome of packed soil over its shaft and the whole operation with the sexes working alternatively takes only 10 min (Nutting, 1969).

As with ants, many generalist predators make use of these flights. Migratory birds may 'top up' for their migration with termites in the tropics in spring and with ants in temperate regions in autumn. Pairs must construct a substantial shelter as quickly as possible which is not just a cover of leaves accessible to beetles and spiders and scratching birds. Although they must keep together from lack of an adequate sperm storage mechanism, they are not exposed as ants which copulate in the open and they have the advantage of co-operation in building.

12.3 Production

There are several measures of production (Brian, 1978). In social insects the ratio queens/queen/life-time must obviously give values greater than unity but requires data on longevity which is difficult to obtain in nature. Feral *Apis mellifera* colonies in eastern North America produce 0.96 swarms/colony in year 1 and 0.92 in year 2 (Seeley, 1978). In Kansas with a longer season, 3.6 daughter queens/queen/season, including afterswarms, are produced; the number is correlated with the number of worker cells, with sealed brood, but not with the number of sealed gyne cells (Winston, 1980). The size of an average prime swarm is 16 033 ± 1447, of second swarms 11 538 ± 1066 and of further swarms a mere 4000 bees. This last figure is presumably near the minimum that can succeed in surviving the next winter in the local climate. Evidently a queen living 4 years (Brian, 1965) would produce about 12 viable daughters.

The ratios queens/queen/year or just sexuals/colony/year provide useful data for comparative purposes and are more easily obtained. The latter represents the exportable surplus from a colony. Maintaining the machinery of production is essential and so the cost of replacing the workers that die during the interval chosen needs to be added. As this subject has been recently reviewed in fair detail for ants and termites by many experts (Brian, 1978), only a few examples for bees and wasps need be mentioned. Estimates of queens/queen/day gives for *Evylaeus marginatus* 0.04, for tropical *Polybia* 0.06, for *Dolichovespula* 0.07, for *Bombus pascuorum* 0.25 and for *Vespula germanica* 1.00 (Brian, 1967). These increase in value roughly according to the degree of social evolution of the species though some are temperate monogyne and others tropical polygyne.

Production by *Tetramorium caespitum* in English coastal heath has been explored for some years and measured as the weight of new ant material formed per colony/year (Brian and Elmes, 1974). This species has one queen and averages roughly 11 000 individual workers of 6.31 g (wet weight), which occupy an average territory of 43.8 m^{-2} with a perimeter of 32.4 m. Production of males and females is correlated, and when combined averages 3.9 g (this is calculated from 28 colonies over 6 years) and the total production, including workers, is 11.2 g, giving a worker replacement figure of 7.3 g: substantially more than half the total production. This is 1.77 g for 1.0 g of workers so each colony is capable of almost doubling its population each year. The biomass production is 0.26 g m^{-2}, a surprisingly low figure since the vegetation itself produces 250 g m^{-2} about a thousand times as much (Chapman *et al.*, 1975). The production of *Tetramorium caespitum* in this area is affected by annual sunshine, territory boundary, and the stage of plant regeneration after a fire has burnt off all the bushes. With gradual regrowth the influential factors change; territory area becomes less and colony weight more important. Throughout the heath, both too great or too small a biomass is unproductive; the best size is 0.11–0.14 g m^{-2} the modal value in fact, an unusual example of adaptation.

Another more primitive myrmicine in the same area but living in grassy places, is *Myrmica sabuleti* with a mean colony weight of 2.35 g over several years; it produced 3.21 g per year, that is 1.4 times their total worker weight. This is less than the multiplication factor for *Tetramorium caespitum*. Their production increased with colony weight and daily period of sunshine, but decreased with the regeneration of vegetation. When two sites, one cool and acid, the other warm and neutral were compared, colonies were found to be larger in the latter and to produce more new workers each year. Also large mature colonies gave males as a function of their size (and number), but gynes as a function of individual size (Elmes and Wardlaw, 1982a).

In the model of a *Myrmica* colony discussed earlier, when egg supply and food intake corresponded and were both sufficient, only one-third of the food units (800 units) went in maintaining the worker population whilst two-thirds (1600 units) went in the manufacture of sexuals: twice as much. This, of course, gives a total production of 2400 units, three times the value of the colony mass (800 units) and is twice as much as the wild *M. sabuleti* value. The model is, in fact, a general model of the genus *Myrmica* and is adjusted to run at maximum efficiency with optimal sexual output so the discrepancy is not surprising. The production of *Myrmica limanica* in Polish meadows has been studied in detail by Pętal (1972, 1980). Her figures were obtained by repeated sampling round the year and have been built into a numerical model (Uchmański and Pętal, 1982) which rests on two bases: 1, that production is proportional to the worker population (both as dry weight) and 2, that the fraction of production that is sexual increases with the death rate of old workers during the previous summer. This increases the proportion which are young and that care for caste-labile larvae in preference to worker-biased ones, and which do not respond actively to the queen (Brian and Jones, 1980). The foragers die more when food is short or abundant as both these states increase their activity outside the nest. This tends to stabilize the model but its return to equilibrium after a random deviation (such as winter mortality) is delayed by sexual production, and proceeds so slowly, that another episode of chance mortality is likely to intervene.

Termites probably have the highest production figures amongst social insects (Nielsen and Josens, 1978). *Macrotermes bellicosus* living at a biomass of 5001 g ha^{-1} produces 132 mg (dry)/g (wet)/day according to an estimate by Collins (1981b) who comments '. . . in general the Macro-termitinae have higher weight specific consumption rates than other groups of termites . . .'. They are aided, of course, by their fungus combs.

So far there is too little data to generalize but this is a growing branch of social insect ecology. Already we know that the greatest number of queens/ queen is not necessarily a sign of internal efficiency (*Vespula* in Archer, 1981a). Again the most advanced species do not necessarily use the habitat most efficiently as witness the advanced myrmicinae *Tetramorium* at 0.26 g m^{-2} and the simple one *Myrmica* at $3.21/4 = 0.80$ g m^{-2} (data on foraging areas from Elmes and Wardlaw, 1982a). One should take into account that organized species like the former may keep other species out of areas that they themselves do not use.

12.4 Summary

The mechanism for creating a wide range of sizes and shapes has turned out to have a similar structure in different social insect groups. Where there is limited polymorphism as in bees, wasps and primitive termites, differentiation comes late, even as late as the end of the larval period and all types follow a common stem from the unbiased egg onwards. Then the internal nutritive state is assessed and morphogenesis in the pupal stage governed by the timing and perhaps concentration of JH. The whole procedure is subject to season and to the potency of the reproductive individual (queen) or pair. These act via chemical messengers (pheromones) that change the food distribution behaviour of workers. It is interesting that in primitive termites the workers are juveniles queueing to transform into sexuals either normal or neotenic, or into soldiers, which are the first functional caste to evolve in termites but the last in ants. In both higher ants and termites a maternal physiological influence programmes larvae to mature into small workers after a quick and partial embryogenesis that cuts out inessentials, such as wings and gonads, completely. The advantage of inbuilt caste is that no restraints need be put on feeding! In ants soldiers then evolve from the class of large workers that follow the labile stem further towards the full female figure than do those from biased eggs. These polymorphisms arise from a genetic provision, subject to queen-dependent control of the society through food rationing, and from programmed development, motivated by maternal influences on egg formation.

The sexuals produced at the season that suits their need for establishment under the best environmental conditions do not at once leave their colony. They may be restrained, even imprisoned, by workers and then later become a cause of division in the society that leads to the emigration of swarms, usually after careful preparation of a new nest site. They can copulate before or later either by receiving visiting males or going out to find and attract them close by or in congregation areas. Species that rely on single queens to start new colonies usually release them after the males. They may leave and lure males for copulation and then return to their nest to form part of a bud. They may congregate at some marker or at some level above the earth before copulating. Afterwards they start colonies alone or in small co-operating groups. The lone ones may have enough reserves to build up a small group of workers in time. Choice of habitat in these cases lies perhaps in general terms with the male but in particular terms with the female. Termites belong to this class and effect a permanent pair bond before starting co-operatively to make a nest.

Early studies in production (replacement of dead workers plus sexuals) have shown how little of the essential elementary data on longevity and survival in nature is yet available. Various factors have been found to influence production: hours of sunshine, condition of habitat, size of territory, and biomass. The last proved to vary but have a mode near the optimum in one ant. In general, the base population is capable of producing a vast amount of exportable material each year either as sexuals or as swarms of workers with a few sexuals.

CHAPTER 13

Evolution of insect societies

The 'fitness' of an organism is its capacity to contribute to future populations. One widely used method is to protect a few offspring (parental care); this can be carried to the point where they learn, by apprenticeship, how to rear their siblings. They must then be released in time to apply these skills to their own progeny otherwise their fitness is reduced; but there is a disagreement between generations about the best time to part. Another way of enhancing fitness is for equals to co-operate with each other, sharing tasks and sharing reproduction; in such cases the balance is unstable and tends to shift in favour of one that dominates the egg laying and pushes the other into a supportive role. There seems to be no escape from this for though the power of the top one may decline and allow another to succeed, that may be too late for the new one to give its best reproductively. Thus the first step in social life is when an individual (an altruist) reduces its own reproductivity for the gain of another. The question is does the 'altruist' gain a subtle advantage of some kind or is he really unable to avoid helping? The first view is taken by proponents of the kin selection hypothesis, the second by those who favour the maternal manipulation hypothesis; neither excludes the other, however; both can work together to promote sociality.

13.1 Theories of individual selection

There are three main theories which are based on the natural selection of individuals: kin selection, maternal manipulation and polygyne family selection. All must offer an explanation of why the haplo-diploid system of sex determination in Hymenoptera seems to have been such a successful base from which to evolve sociality. Amongst the majority of other insects, with normal diploid sex determination (Crozier, 1979; 1980) only the termites have achieved this.

13.1.1 KIN SELECTION

It was originally thought that this theory offered an excellent explanation of social evolution, for the genetic asymmetry in the haplo-diploid system means that a female has three-quarters of her genes in common with a sister but only one-half with her mother. Hamilton (1964) argued that for a female, helping a sister would be better genetically than having offspring of her own: the inclusive fitness would be higher than the direct fitness. However, the males which are only one-quarter related to their sisters necessarily cancel out the advantage, unless as Trivers and Hare (1976) suggest this asymmetry in genetic relatedness is put to use and investment in females made three times as great as that in males (see later). An underlying assumption is that the females should all have the same father or at least that the sperm packets from different males should not mix (Charnov, 1978). For this there is at present little evidence since polyandry appears to be widespread in social Hymenoptera and although packets of sperm from different males are used unequally there is no case yet where only one is used at a time (Crozier and Brückner, 1981). Other basic assumptions on which this theory rests are: random copulation, monogyne colonies that develop as families, and queens that lay all the eggs. These difficulties have been avoided in various ways (see Dawkins, 1976; Wilson, 1971), which can now be considered. The main, rather tenuous, evidence for kin selection comes from looking at sex ratio and investment ratio.

The genetical theory of natural selection (Fisher, 1930) points out that investment in the sexes should be equal; otherwise genes that exploit a deficit in one or other will soon spread in the poulation and cancel the deviation. Investment for social insects can take the form of giving a bigger body and more stores to the female or providing the sexual female with special worker protection as in swarms. So if females are twice the size of males then roughly twice as much investment has gone into them and twice as many males will have to be produced to equalize and if one kg of workers attend a migratory queen then one kg of males should be produced to compensate. In haplo-diploid sex determination one or more loci are multiallelic and females arise wherever only one is heterozygous. Males then arise when homozygosity is complete at all pertinent loci and, of course, normally where the animal is hemizygous and develops parthenogenetically from one or other of the two possible oocytes produced in meiosis (Crozier, 1980). Diploid males have been produced either by artificial insemination or by controlling inbreeding: in *Apis mellifera* by Woyke (1980b) and in *Melipona quadrifasciata* by de Camargo (1979). In the honey-bee, diploid drones are destroyed by workers in the larval stage after only a day and if raised artificially out of the hive they are sub-fertile.

Using dry weight as measure of relative investment, and concentrating first on monogyne species in which workers do not lay eggs, Trivers and Hare (1976) found for five with enough data:

	females/male
Leptothorax curvispinosus	3.24
Solenopsis invicta	3.52

	females/male
Tetramorium caespitum	2.99
Formica pallidefulva	4.14
Prenolepis imparis	3.04

All show a ratio biased towards females three or four times. The first of these, *L. curvispinosus* has two social parasites of the same tribe (the Lepto-thoracini): *Harpagoxenus sublaevis* with 1.25 and *Leptothorax duloticus* with 0.87 females/male. Whereas normal ants appear to have ratios to suit the workers the parasites whose brood is brought up by the host's workers have ratios that suit the queens. Thus the data are compatible with the idea that kin selection is biasing investment as predicted in a male–haploid system, and that workers and queen have conflicting optima.

A further study of *Tetramorium caespitum* compares the sexual investment in two different habitats close enough together to enable the flying sexuals to interbreed (Brian, 1979b). In the moister, more richly vegetated and organic soil, colonies of *T. caespitum* maintain permanent contiguous territories with *L. alienus* living interstitially and in the drier, sparsely vegetated and sandy soil they have isolated colonies in a terrain mainly occupied by *Lasius alienus*. Both are monogyne, and in both male eggs are laid by the queen. Both species invest in more females than males in their better habitat zone and *Lasius* gives bigger gynes as well; taken over the whole area *Lasius* gives an investment ratio of 2.89:1. These data, then, support the idea that workers control investment but only in their population optima; elsewhere the queen is able to dominate investment.

A *Lasius niger* (monogyne) population in shell-sand dune slacks in the Friesian islands also does so. A gyne investment ratio of three per colony was obtained in a prisere and one per colony in a later seral stage where it was oppressed by *L. flavus* (Boomsma *et al.*, 1982). This is because the big productive colonies of *Ln* in the prisere produce more gynes. The resemblance to the Dorset ants is clear and in fact a quantitative relation was deduced that applied to both the Friesian and the Dorset ants. The weight of gynes (G) always exceeds 0.6 times the weight of sexuals (S) multiplied by the logarithm of their weight:

$$G > 0.6 \, S \log S .$$

Whether this has wide generality remains to be seen.

The best sex ratio is changed if there are several queens and if workers lay some or all of the male-producing eggs (Oster and Wilson, 1978; Trivers and Hare, 1976). In the first case each queen has a reduced fitness (if all the queens lay) since she can contribute less to the total sexual output. Since extra queens are being protected in the society, males need compensatory investment and one would expect the sex ratio to move towards the males. Trivers and Hare take *Myrmica*, a well-known polygyne species whose average queen numbers are summarized in Elmes (1980), and show that those with lower average queen number: M. *schenki* (0.97), M. *sulcinodis* (0.75) and M. *ruginodis* (*macrogyna* 0.88) have a female/male investment ratio greater than 1.5 whilst those with a higher average queen number: M. *sabuleti* (1.74), M. *rubra* (15.9) and M. *ruginodis* (*microgyna* 6.33) have a ratio less than 0.5.

These ratios are probably also biased in a male direction from the fact that workers of *Myrmica* lay reproductive eggs. According to Trivers and Hare these should give a 3:4 (female:male) and to Oster and Wilson (1978) a 1:1 investment ratio. The increased male bias in such cases arises because workers are related to sons by one-half and nephews by three-eights whereas a queen is only related to her sons by one-quarter. Since, in M. *rubra* and probably other species too, most workers lay (as nurses), they contribute substantially to the gene outflow and settle the conflict between themselves and their mother by restricting her contribution to females. Thus the two castes may have achieved a balance with the 1:1 ratio of investment. In Oster and Wilson's (1978) model of this situation they find a likelihood that the workers will take over male production completely from the queen rather than leave her a portion. This seems to be true of *Myrmica* (Brian, 1969; Brian and Rigby, 1978) and may be true for some other ants but more data are needed. These ideas have been incorporated into the *Myrmica* model, already mentioned, without violating the known facts (Brian *et al.*, 1981a). The model produces equal sexual investment at a worker population very near that giving maximum sexual output. It also predicts that males will be in excess both in poor habitats and when colonies are dividing. Thus there is some evidence that the haplo-diploid system is exploited and does produce a conflict between queens and workers. This evidence is not strong it must be admitted, and other explanations for unequal sex investment are available.

Do ant colonies that divide support this theory or not? Schneirla (1971) estimates that a colony of *Aenictus gracilis* with 85 000 workers and one queen will split into two new gyne-headed parts of 20 000 and 50 000 leaving 15 000 with the old queen. If all the workers in the new buds are considered investment in the two gynes and if 1367 males are considered investment in males with a weight equal to 25 500 workers, Masevicz (1979) calculates a female/male investment ratio of 2.74:1, in line with a worker control situation. Similar calculations for the honey-bee (*Apis mellifera*) are not so easy, as the old queen leaves with workers and with honey suggesting that the real 'bud' is the gyne or gynes left in the hive, some of which will leave shortly in after-swarms. As they have, so to speak, inherited the hive, with its stores of honey and pollen, ready-made combs, bee brood including many workers pupae as well as a sizeable population of workers, they are very well endowed compared with the old queen and her retinue. Masevicz nevertheless considered the swarm as the bud and used data from domestic bees giving a mean swarm size of 9080 workers (about a third of the population), which combined with the average male production as fresh weight gives an investment ratio between 1.5:1 and 5.1:1. Such strong female bias is surprising, since the *Apis* gyne is undoubtedly polyandrous and copulates at least a dozen times. This produces an interworker relationship similar to that between step-daughters and daughters: much less than three-quarters so there is no reason to expect a female bias in sex investment (Crozier, 1980, discusses this).

In solitary bees the sex ratio has a male bias (Trivers and Hare, 1976) but since females are slightly heavier the investment is nearer equal; the cheaper a son is to make compared with a daughter the more they can produce without breaking this 1:1 rate. In social species it is often difficult to distinguish female castes with certainty but for five species of *Bombus* the investment ratio was found to range from 1.17 to 3.13, a distinct female bias. However, since in all likelihood, males are worker generated, a ratio nearer equality would fit Oster and Wilson's model better. Moreover, Owen *et al.* (1980) have pointed out that inconspicuous small nests can be overlooked and they have shown that in *Bombus terricola* these small nests give only males and do so early in the year. They calculate an average sex investment ratio of 1:2.9 so that this species produces more males than expected on any theory! The same problem of estimation has been stressed by Herbers (1978, 1979) who collected data on the ant *Formica obscuripes* in North America and con-cluded that there could be no evolutionary equilibrium. Some colonies gave males only, some females only and some both. She found no sign of the early, seasonal production of males that others have recorded. In wasps, as in primitive bees, the castes are difficult to recognize without dissection or long, behavioural studies of marked individuals. For *Polistes fuscatus* in North America in which most viable eggs are laid by the queen foundress, Noonan (1978) found an investment ratio of 1:1. As long as one can be sure that the queen is laying the unfertilized eggs (and Noonan says no worker appeared to lay) this indicates queen control of investment. However, this is not sur-prising as the male eggs are laid and hatch before the workers become functional, only the gyne-producing female eggs remain to be laid.

Thus there is some evidence, not yet a lot, that the haplo-diploid asym-metry is exploited and causes a conflict between queens and workers. Other explanations of the unequal sex investment can do so too: one is competition for mates whilst inbreeding.

13.1.2 MATERNAL MANIPULATION

Any mother is in a strong position to influence her offspring but interference must be subtle or it will drive them away on their own rather than domesti-cate them (Alexander and Sherman, 1977; Michener and Brothers, 1974). Thus the queen has to step from simple parental care which is common in pre-social wasps and bees and which enhances her fitness as well as theirs, into control over the early development of the first adults so that they stay and help. As Craig (1979) says, '. . . parents which increase their total output of fertile offspring by first producing sterile helpers that aid them in rearing subsequent broods will be fitter than parents which do not do this . . .' and will have a selective advantage. Crozier (1979) adds '. . . workers are individuals of very low reproductive capacity programmed to sacrifice themselves to the production of siblings . . . selection acts on females to produce tractable worker offspring as assistants . . .'.

Various methods have been suggested. An obvious idea is for the mother to try to bring up too many so that most or all are substandard, but this is a risky beginning for there is no guarantee that they will help even if they are stressed in this way (in fact kin-help alleles are needed too). If, instead, the genetically substandard individuals are used by the queen does this not lead to genetic caste determination? This method suggests that the fittest eliminate the substandard as competitors by harnessing their power so that the waste usually associated with intraspecific selection is avoided. Presumably sisters once domesticated would no longer retain enough behavioural flexibility to become individualists again after the mother died; they would probably develop a new queen out of the biggest or oldest remaining individuals. Dawkins (1976) has argued that although the mother may have uncontested control over her offspring as juveniles, there must be a genetic symmetry which will enable 'cheat' or non-co-operation genes to spread in the offspring if they are not also gaining from the society.

One way in which the haplo-diploid system might facilitate the emergence of societies is through sex control which makes the queen Hymenopteran particularly powerful because she can stop her daughters copulating. Males can be excluded from early broods so that the mother then has only to 'domesticate' gynes. This appears to be easier because whole systems of behaviour (dispersal and nest foundation) need never be turned on. The saving in energy through not making males is substantial as in parthenogenetic reproduction (Maynard Smith, 1978a). Moreover a male of the solitary prototype bee or wasp is unlikely to make a good worker: he lacks building skill and a sting for defence. One way of interfering with daughters then could be to fertilize all the first eggs in the sequence. In early societies it is highly likely that the uninseminated females laid male-producing eggs. The establishment of this as a regular arrangement could be the sort of come-back which Dawkins had in mind when he argued that 'cheat' genes would evolve. These workers, invade the gene pool directly with their own progeny, though individually they might not produce many males. In the security of the nest without the risks associated with nest construction and foraging all the workers would have to do would be to evade the queen.

Another manipulation the mother might use would be to establish a gene which introduced a sensitive phase into early adult life. Each additional female could then be imprinted as a worker and left to carry on without the need for supervision. This is caste-determination in the adult stage and evidence already mentioned has shown its reality. In fact all these control systems are known to exist in primitive societies (Brian, 1980).

13.1.3 POLYGYNE FAMILIES

West-Eberhard (1978a) starts from the fact that very many social wasps are polygyne, either all the time or periodically and suggests that such a clan is the group in which sterile helpers evolved. From this state, the reproductive dominance of one female could lead to the stabilization of a monogyne matrifilial society. Drawing her examples largely from tropical wasps she points out that in *Belonogaster griseus* (Polybiini), several inseminated females (queens) co-operate with several uninseminated ones (Pardi and Marino, 1970). These gynes or workers do not, so far as is known, go on to become queens and it seems as though they are unable to copulate. Perhaps if prevented from copulating early in adult life they cannot do so later (for other examples of this in bees and wasps see Brian, 1980). Species of *Ropalidia*, and another species of *Belonogaster* appear to resemble *B. griseus* in this respect. In *Mischocyttarus flavitarsus* only 12% of foundresses live long enough to even overlap the life spans of their daughters yet true workers exist and in *M. drewseni*, studied by Jeanne (1972), there is a worker form that rears both sibs and non-sibs after their mother has stopped laying. In Colombia at 4°N inseminated females of *Metapolybia aztecoides* work in the nest when young and later go on to help in other nests though here again they may help rather than reproduce. Egg layers, indistinguishable in size and shape, 'bend' threateningly instead of dancing as non-laying workers do. Caste appears to be labile and reversible and after a swarm of queens and workers settles down, the number of queens declines: some are driven off, some go for no apparent reason in new swarms and some are converted into workers. Then, the group may have just one queen for several months but when she vanishes several young ones take her place and elicit worker dances. Eventually males are produced but only in big permanent polygyne societies. The queens that are eliminated after laying some eggs (that give workers) are usually inferior and represent the exclusion of less fit individuals. Female accretion followed by reproductive monopoly could thus be a social selection process.

West-Eberhard suggests four possible stages in social evolution:

(1) solitary but with degrees of reproductive success that are genetic in origin and associated with differences in female size and aggressiveness.

(2) primitively social, nests shared, the weak joining the strong, thus foreshadowing rudimentary caste differences along these lines, e.g. *Polistes*.

(3) castes formed by young which first help and then go on to lay elsewhere when older, e.g. *Belonogaster*, *Ropalidia*, *Metapolybia*, and probably many Stenogastrinae.

(4) highly social with sterile worker castes and a system of alternating

mono- and poly-gyny, e.g. *Mischocyttarus, Polistes* and swarming Polybiini.

From the last situation which probably adapts them to disturbance by ants and birds in the tropics, a permanent monogyny evolved in temperate regions. This hypothesis differs from others in postulating a gradual evolution of reproductive dominance after a period of undifferentiated group living and in not requiring the overlap of generations for the formation of family groups, though some degree of relatedness must be present. The polygyne system enables the fittest queens to be selected socially.

13.2 Models of these theories

Oster and Wilson (1978) have modelled a society of haplo-diploid insects with both individual and colony-level selection. Relative sexual investment at colony-level is that which a hermaphrodite colony in a community of colonies interbreeding fully at a steady state would require, assuming that all the genotypes are derived from a single founding pair. They took into account the difference in cost of manufacture of each sex and they maximized efficiency of the total production process. Their model indicates that either the workers or the queen lay all the male eggs but in neither case do the three components: queen, workers and colony, 'agree' on the best allocation of resources between the sexes. If workers lay all the male eggs the best ratio for queens would be 1:1, for workers 1:1 also but for the colony 1:2. In this case, although the female castes agree on equality, the colony will do best if there is a male bias. If, on the other hand, the queen lays all the eggs the best ratios are 1:1 for the queen, 3:1 for the workers and 1:1 for the colony which thus supports the queen in reducing the female bias required by workers. In these circumstances the system may settle for a steady state, that is a compromise, or it may fluctuate with the environment.

Several population geneticists have constructed models in which alleles in the mother for maternal manipulation or alleles in the offspring for helping the mother (kin selection) have spread through a population and converted it from a pre-social to a eusocial one. Craig (1979) found that the actual relatedness is not that predicted by pedigrees but varies with gene frequency and degree of genetic dominance. Moreover, for dominant and co-dominant alleles, the 3:1 sex investment ratio is actually a disadvantage to the allele for helping, whereas a male-biased ratio is an advantage. Craig includes an allele for manipulation by mothers that forces daughters to help irrespective of their genotype; workers are then completely sterile. In his model, infinite populations mate randomly and copulate once only and there are two broods a season or in the tropics two cycles a year. Here then, instead of a gene in the offspring causing them to help (kin selection), there is a gene in the mother causing her to force them to help. These, of course, reinforce, the difference being that no sexual bias in investment is needed by the 'manipulation' model; from a genetic point of view the offspring in this situation gain nothing by helping since both the parental (P) and the filial (Fl) generations have half their genes in common and are reciprocal and symmetrical. Craig found that parental manipulation genes are selected whether dominant or recessive as long as the gain to the mother exceeds the loss to the offspring. In Hamilton's terms this means $k > 1/r \geqslant 1$ where r is the pedigree degree of relationship and k the gain to a beneficiary relative to the cost to the donor (nett gain).

Craig maintains that a condition in which $k > 1$ is quite realistic. He suggests that (1) females which do not lay eggs have energy to spare for rearing sisters; (2) sub-fertile daughters may be more likely to work thus favouring manipulation alleles and leading eventually to obligatory service; (3) a colony with several adults can defend itself better; (4) since females survive better in established colonies so will the gene for manipulation. The gene for maternal manipulation succeeds even if the gain to the mother is less than the loss to the offspring donor ($k < 1$); in fact in much less profitable conditions than are necessary for the kin-help gene to spread. Craig found zero selection at $k = 1/2$, which is in conformity with theory since if daughters only raise half as many of the queen's daughters ($k = \frac{1}{2}, r = \frac{1}{2}$) as they would produce grandchildren ($k = 1, r = \frac{1}{4}$) the advantage just balances. Judging by the low values of k needed to establish manipulation genes, maternal control is a more likely first step in social evolution than is kin selection. Moreover, multiple insemination is irrelevant since relatedness does not matter as it does in kin selection.

Charlesworth (1978) shows that both kin and manipulation models favour the evolution of complete helper sterility in suitable conditions. In his model of kin selection the spread of an allele for helping behaviour (A) is traced in a population that is initially all non-A, infinite, mates randomly, and has discrete generations. Not all A carriers manage to help; this depends on environment and luck, for they only occur in the first brood and they help both A and non-A types in the second brood. In his model of maternal manipulation, a gene A causes offspring to work, operates from either the male or female parents and enables them to dominate offspring, either physical or chemical. He shows that the probability that an A genotype will actually serve the group increases with the gain from doing so and '. . . in every case there is a critical value of k which must be exceeded if an evolutionary stable strategy is to exist'. With parental manipulation there is '. . . a powerful tendency towards the evolution of altruism . . .'. Charlesworth concludes that social evolution has been slow and proceeds by small stages and he points to the interesting idea that a worker/queen ratio in a colony can be selected '. . . in a rather mechanical way'. His models agree with those of Craig in showing that the manipulation allele spreads at lower k values than the kin selection allele and that evolutionary stable strategies (ESS) exist with high levels of service independently of the number of copulations a gyne makes.

The usual assumption of random mating is, in fact, doubtful and female-biased sex ratios can be caused by competition for males (Alexander and Sherman, 1977). With inbreeding, a sex ratio of 2:1 (female:male) is best for both sexes (Trivers and Hare, 1976). In this state with the actual proportion of males one-third this means that i, the number of brother/sister mating is $(1-i)/2$ also one-third (Maynard Smith, 1978b). Since most colonies are hermaphrodite, extensive inbreeding is possible though obvious mechanisms for avoiding too much exist, including lack of sexual attraction in the nest. This breaks down in many ants towards the end of the season and in those species in which females lure males just outside the nest considerable inbreeding could occur in isolated colonies. This includes social parasites. Inbreeding increases the relatedness between a queen and her daughters, normally one-half, by a factor as big as 3. For if f is the number of female offspring that a queen has, then the relatedness is given by $(1 + 3f)/(1 + f)2$ (Oster and Wilson, 1978).

With the termite diploid sex determination, kin selection is an even more remote possibility but parental manipulation is quite feasible. As already discussed they rely, in the primitive families, on the activity of juveniles that always have the chance later on of becoming either substitute sexuals or winged sexuals. Their classic fitness is not in the least reduced, and may be increased by their early experience in their parents' nest. Parental manipulation probably plays a part in a later stage of evolution in the Termitidae when they are matured and their caste lability destroyed. From then on their evolutionary fitness is always inclusive and proportional to the degree of assistance they give to their parents in rearing their sibs.

13.3 Group selection

The last theory of social origins to be discussed is that which postulates 'group selection'. In this the population is split up into sections (demes) that only intercommunicate in a genetic sense briefly and occasionally. Internally they mix but the effective population size is so small that random drifts in allele frequency can occur and become established. Later they get the opportunity to spread into areas where unsuccessful demes have died out. In this way a good population, one that is well organized and does not overexploit its habitat, survives. This system enables the establishment of characteristics which though favouring a group may actually be unfavourable to the individual (such as sterility). Williams and Williams (1957) start with families in which some offspring tend to serve and though at a disadvantage as individuals do their groups good in that their ant resources are not wasted. Levin and Kilmer (1974) model a population split into demes connected by migratory movement but subject to periodic extinction and re-colonization of the empty resource. The chance of surviving is a direct function of the proportion of helper genes in the deme. Interdemic selection for a helper allele overrides the effects of individual selection against it. Restraints are severe, however: gene flow has to be below 5% per generation and effective population size less than 25. This aspect has been further developed by D.S. Wilson (1980).

Group selection is particularly applicable to polygyne ant and wasp colonies with unrelated queens. Successful colonies in equilibrium with their resources will need to regulate their queen number using the mechanism based on the worker/queen ratio. Recruits may be taken in selectively and later even further refined. This enables a trial under social conditions during which queens have the chance to compete for worker attention and allegiance. Success will then depend on how closely correlated attractiveness and fitness, in the evolutionary sense, are. Colonies will go through a cycle of queen intake, worker population growth, sexual production, copulation and again queen intake. Thus the polygyne colony could be a deme in the group selection process.

Sturtevant (1938) points out that a queen which produces more gyne-biased eggs than another will at first leave more descendants but later run down the society. She may end up lacking the power to lay worker-biased eggs at all, and for these she will be dependent on others: a social parasite in fact. If these fail the local population or colony will die out and leave the area open to colonization by demes or colonies with a better mix of queens. Workerless parasites have only established themselves in a few cases such as *Anergates atratulus* that uses and destroys colonies of *Tetramorium caespitum*.

13.4 Conclusions

With such a wealth of theorization all of which points to ways by which societies could have started and successfully evolved, the only question left is: why have they appeared so rarely in insects as a class? Can it be that the male–haploid system of Hymenoptera confers an advantage? This seems unlikely since there are a number of other animal classes with haplo-diploid sex systems (such as rotifers and mites) that are not social and there are the diploid termites. These may well have started through manipulative parents or group selection. None of the theories are mutually exclusive and it seems that wasps and bees may well have used kin selection and maternal manipulation as a starting point.

The haplo-diploid system does have a few advantages nevertheless. One is that sex can be determined by controlling sperm access to the ovum. This enables males to be excluded from the society and used for gene dispersal.

Males are also useful as sinks for unsatisfactory recessive alleles including those that are produced all the time by mutation: Smith and Shaw (1980) show that in parasitoid Hymenoptera a differential haploid mortality of 10% can exist. Even in pre-social Hymenoptera males do not have the qualities of constructiveness, parental care and intelligence, and they cannot produce, as diploid females can, the variety of genotype that adapts to different situations. That males can arise parthenogenetically makes a semi-sterile worker caste possible. They can help the female-producing queens but still contribute directly to the gene pool. This step is half way to the goal of complete sterility.

CHAPTER 14

Colonies

Populations of the same species are divided into colonies which may have one or many queens but which can be recognized by their mutual hostility. A colony is an organized food-converting and reproducing system that sets up and defends barriers against entry by individuals of the same species as well as many other organisms. A colony depends on the ability of its members to recognize each other instantly. This is not just defence against predation, social parasites and other intruders, it disperses the population over its resources. Fighting is often avoided by learning where attacks may occur and, in some species, it has been ritualized. Other dispersive behavioural mechanisms, sensitive to population density, have also been postulated (Wynne-Edwards, 1962).

14.1 The Colony Barrier

The basis of distinction is chemical since even dead members of a strange colony can be recognized. In very small groups it has been shown that individuals are able to learn each other's smell. Thus, the bee *Lasioglossum* has colonies consisting of a queen, a guard and a few foragers; they do not acquire distinctive odours from the environment, and guards will accept younger bees of any nest but a special colony odour develops 2 days after emergence (Barrows *et al.*, 1975). A guard can remember a nest mate for at least 7 days but forgets after 12 days. Although the queen can do all the work, she is not a good guard and her first daughter (as there are no males to begin with) takes on this function, blocks the entrance hole and fights if pressed, resisting enemies and strange members of the same species (Breed and Gamboa, 1977; Breed *et al.*, 1978). Guards are slotted into their role when about 2-days old. This process is a kind of habituation or imprinting with an age-dependent sensitivity during which the guards learn to inhibit attack when they perceive their queen's odour and later on that of their sisters. In tubes in the laboratory they will reject any bee from another colony unless it is under a day old, the age presumably at which their exocrine glands begin to produce a characteristic aroma. Guards isolated for 5 days and then returned even reject their own sisters, again presumably because they have not learnt their individual smells and cannot recognize them as relatives. However, in another instance, degree of acceptance by guards is correlated with relatedness and they can recognize sisters even if they have never met them before (Greenberg, 1979). Presumably the odour of a bee is genetically determined rather than acquired and guards, once they have learned the odours of nest-mates, can recognize their relatives too (Holldöbler and Michener, 1980). Later on bees not only recognize each other as individuals but also according to their occupational categories: foragers, the guard, the queen. Even males can learn and recall the odours of individual gynes (Barrows, 1975).

This faculty for individual identification cannot be expected in colonies of hundreds or thousands except as far as the queen or a few queens are concerned. Only three possibilities exist: either individuals acquire a chemical mix from their surroundings or from their queen(s) or they have a glandular odour of their own inherited from their parents. In the first case the odour could come from food via faeces or the tracheal system or it might come from odiferous soil fungi and other micro-organisms nearby. In the second case it could be generated by the queen(s) and passed round on the cuticle or even internally; each colony would then have a mixture of chemicals derived from their queen(s) directly or via other workers. Thirdly, exocrine glands in the workers secrete an aroma governed by their genotype. Since this derives from both parents it will be more variable than a smell transferred from the queen (Crozier and Dix, 1979). For acceptance to take place at least one allele at all odour loci (of which there can be many) is shared by colony members. Unless aromas blend each worker must learn the spectrum of phenotypes that a single parental pair can generate: feasible in *Lasioglossum* but what of bigger societies that may be polygyne?

The most studied social insect is the monogyne honey-bee (*Apis mellifera*) yet even in this case the situation is unresolved. At the hive entrance the guard bees are extremely sensitive to bees of different colonies trying to enter by mistake, or to rob, and they grapple immediately. At feeding places they leave a scent that is more attractive to hive mates than to other bees (Baroni Urbani, 1979; Ferguson and Free, 1979). At sugar dishes when syrup dries up, bees of different colonies, though normally peaceful, will start to fight (Kalmus and Ribbands, 1952). Clearly they can recognize colony differences outside the hive and therefore carried on foragers. At the hive entrance they leave a foot odour, generated, not by tarsal glands (as in the queen) but by glands on the body that secrete a colony-specific mixture. Yet genetic variability of the sisters in one society must be considerable as they have collectively many fathers. The acquisition of odours takes place too. Workers drift into other hives of an apiary; this tendency can be increased by feeding all hives on similarly scented food or stopped within 3 years by changing to different scents for each hive (Köhler, 1955). Bees from different hives can also be joined together as long as it is done gradually, especially if a covering odour such as aniseed is added and a diversion provided such as a powder that makes them groom themselves. Colonies from different areas transferred to ling (*Calluna vulgaris*) lose their enmity (Kalmus and Ribbands, 1952) and colonies divided into two parts and fed differently show hostility if reunited after some weeks.

Perhaps bees can acquire the smell of their queen? They can recognize their own queens and even diagnose the smell of a paint mark (Seeley, 1979) and the queen's absence is known to the bees within minutes of removal. Moreover, they will only accept another queen if their own is removed and if the new one is added gradually in a way which gives time for the bees to learn the new odour or to acquire and spread it through the colony. Experimental evidence is thus in favour of acquired odours being important in the honey-bees even though the colony-specific secretion that passes down their legs from their bodies is basic.

Myrmica rubra, although polygyne, has definite colonies between which hostile behaviour occurs (Winterbottom, 1981). Strange workers can be identified by others chemically from any part of the body and need not move or behave at all. The colony odour, absent in larvae and pupae, appears in young workers; when this happens in a foster colony they are nevertheless accepted without hostility. At first sight this implies that their odour has mixed with that of the foster workers. However, the chances are that the fosterers, though they refrain from attack, can distinguish the unusual workers since members of the same foster colony that have never met them before (having been kept separately), when presented with a mixed set attack only the strange young ones, not their con-colonial fosterers. Probably the gradual emergence of the odour from the transplanted workers allows their fosterers to habituate to them. Yet the transplanted workers must also have changed for their original colony no longer accepted them! The only interpretation possible is that they have a genetically determined chemical base and an acquired chemical veneer; the inexperienced workers of the foster colony attack them because of their base and the inexperienced workers of their own colony because of the foster veneer. What is certain is that acceptance does not imply an inability to discriminate only habituation to the aroma.

When Winterbottom looked for the colony-specific chemicals she could not find them. She naturally searched in Dufour's gland since this has been shown to give different chemicals in different species of *Myrmica*, and Cammaerts *et al.* (1981b) have shown that it is used to mark blank territory to familiarize foragers and, presumably, to deter strangers. This secretion does not at first involve colony specific features, these only appear after several days and it is not certain that they come from Dufour's. However, Winterbottom could not find any satisfactory chemical differences in these glands from different colonies and the mystery awaits either more sensitive chemical techniques or a wider search. Perhaps a non-volatile material spreads over the body and down the legs to give foot smears as in *Apis* and *Vespula* (Butler *et al.*, 1969).

An endogenous odour spread on blank territory would be expected in the nest. When *Solenopsis invicta* workers are given soil to dig in, either from their own nest or another nest or just plain soil, they chose their own soil (Hubbard, 1974); this is also similar with *Pogonomyrmex badius*, according to Hangartner *et al.* (1970). In *Acromyrmex octospinosus* colonies can be divided, fed differently and re-combined with no more than mild aggression and again a basic genetically determined odour can not be completely disguised by one acquired from the food (Jutsum *et al.*, 1979). Using *Formica fusca*, Wallis (1962), after an experimental study, came to the conclusion that endogenous odours were more important than exogenous ones but using *F. polyctena*, Mabelis (1979b) concluded that a divided colony developed hostility when the two parts were fed differently, one on *Lasius fuliginosus* which has a strong terpenoid smell, and one on other members of its own species. *Oecophylla longinoda* workers mark territory with colony-specific fluids from the rectal sac; this is not just defecation outside the nest. So far then, ants and bees show both a basic inherited odour and an acquired superficial odour.

14.2 Queen number and species ecology

The strength of a colony barrier is influenced by the number of queens that it contains. Monogyne colonies always have strong barriers and defend their nest ruthlessly against all animals except those parasites that have the skill to enter. Polygyne colonies, on the other hand, vary a great deal; sometimes they defend generally and sometimes only against animals of a different species. In wasps the number of queens is related to climate: the genus *Polistes* is largely polygyne in the tropics and monogyne in temperate regions (Yoshikawa, 1963). The usual explanation offered is that in the tropics predators like ants and birds are ubiquitous and that having several queens and many workers enables the wasp colony to break up under pressure, fly off and start again elsewhere. In temperate regions the necessity to survive a long winter gives single queens an advantage as they can go into diapause ready for hibernation in dry, cool places and start up independently afterwards. Moreover, in spring, food is thinly spread and dispersal over it important. A species of *Mischocyttarus* living in Florida, which is, of course, warm and only slightly seasonal, is monogyne in spring but polygyne in autumn. Yet, even in the tropics, swarming Polybiini can be periodically monogyne, e.g. *Metapolybia aztecoides* and it looks as though there may be some more subtle reason for this such as the need for natural selection at the colony level. The need for sexual reproduction, dispersal, and selection of the fittest families may be periodic. Polygyne *Polistes* can achieve 20 000 individuals compared to 200 for tropical monogyne *Polistes* and the polygyne species has evolved communication to effect swarm movement (Jeanne, 1981). Hostility between queens of *Vespula germanica* only arises if they have come out of diapause after a cool winter; Spradbery (1973b) noticed that queens in Tasmania where winters are warm remain amicable and form a giant colony.

Bombus is like *Polistes* in that in the tropics species are known (though relatively few) which develop polygyne colonies, e.g. *B. atratus* though single queens also found colonies alone (Sakagami *et al.*, 1967a). But the genus *Apis* is definitely monogyne and the species *Apis mellifera* which lives throughout the world in both tropical and temperate regions is, in all places, strictly monogyne. Similarly with the Meliponini in which gynes are sealed in cells until a contest or until a new nest site is ready. Evidently one queen with considerable powers of expansion can supply all the eggs a colony needs and vast colonies are disadvantageous; it is better to divide and disperse.

Whereas in wasps monogyny is part of the technique for invading temperate climates, in ants the problem is not climatic. Mono- and polygyne ants occur in all regions of the earth, in fact often as sibspecies or as a polygyne patch in a generally monogyne population. The flying monogynes are able to find new places and start new colonies alone with the help of their reserves; they can, therefore, pioneer into new habitats especially priseres. Polygyne forms can build up the fecundity of an existing colony quicker so that it grows and spreads out but since the queens are dependent on flightless workers all the time they are limited to pedestrian travel. So, whilst the monogyne queen is independent, self-supporting and able to pioneer habitats by air, the polygyne queens are dependent on worker support, and limited to such new habitats as workers can reach on foot. Although pedestrian they have the substantial advantage that in densely populated areas workers can test and assess the intensity and nature of competition, and then if necessary oust it. One takes a risk, the other plays safe.

Ideally, a species retains both options and uses them as occasion demands just as plants both seed (monogyne) and spread by rhizomes (polygyne). However, evidence increases to the effect that a species frequently divides along these lines, i.e. by ecological separation into sibling species, one with aerial dispersal and the other with ground dispersal, one a pioneer into vacant habitats and the other thrusting slowly into occupied habitats. Now some examples.

14.2.1 TWO SPECIES OF PSEUDOMYRMEX

In tropical America both live in *Acacia* stems: one, *P. ferruginea* (Pf) has one queen and the other, *P. venefica* (Pv) has many (Janzen, 1973). Males of Pf congregate by a conspicuous tree to which females fly around midnight whereas males of *Pv* do not congregate at all but search the foliage of trees a few hours before sunrise for luring females. Both copulate once only: *Pf* on the ground by the assembly tree, *Pv* on the lure tree. Then the queens break off their wings and start to walk; neither of these species flies in search of a new place. *Pf* queens search for a seedling *Acacia* and often wander hundreds of metres away from the assembly ground; their success may depend broadly on the males assembling in the right sort of habitat. *Pv* queens, by contrast, try to get back into their own tree but may be opposed by workers, who refuse them entry through existing holes and stop them cutting new ones. After some attempts they drop to the ground and start a search for new shoots, not seedlings. Clearly the important situation for them is whether the workers will take them back or not and the conditions governing this are not known but they probably depend on the worker/queen ratio in the colony and on the genotype of the young queen. Queens of *Pf* are very hostile towards each other and will fight over *Acacia* seedlings which they refuse to share even temporarily. Workers, too, are aggressive towards any queen other than their own mother and towards workers of a different colony; they fight by biting and smearing poison. Inter-colony combats are frequent and relentless where trees of *Acacia* grow to overlap. *Pv* queens, on the other hand, do not attempt to nest in seedlings and have clearly evolved a distinct selection behaviour: they seek out root suckers on vegetatively propagating plants. Their workers are much less hostile, not only towards other insects but towards their own species which they do not fight even when growth of their bushes brings them into contact. They simply mix and the groups fuse.

Colonies of *Pf* grow slowly in their seedlings but give them complete protection from other insects thus increasing the likelihood of renewing the ants resources quickly. Mature colonies with some 30 000 workers produce males when the population reaches some one or two thousand adults and then both sexes. The queen and the workers which live in specially large hollow thorns move to new ones about three times a year; they are not totally static. The queen stops laying before this migration so that she can walk easily and get into the new thorn. Workers accompany her and they may even cross over the ground between trees. This, of course, is risky for if the queen is eaten by a bird the colony dies out since it will neither accept or make a replacement; the fragments produce males for a time. *Pv* populations grow more quickly, increase their reproductive potential by taking back queens and moving eggs about, and are able to keep pace with the faster growth of the established tree though they protect it less. They have no boundaries between populations, and it is impossible to estimate numbers; Janzen suggests that four million can live amicably in several thousand old *Acacia* trees. Though they have no colony barriers they are somehow able to segregate queenless zones of a few hundred workers for sexual production.

Further ecological differences that Janzen noticed were that the queens

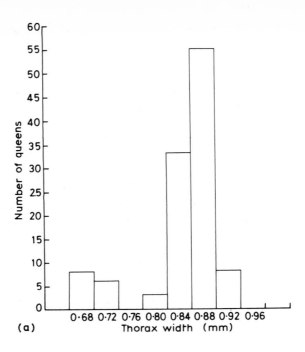

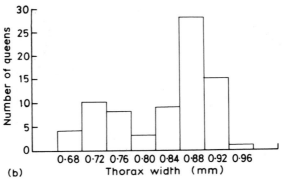

Fig. 14.1 The thorax widths (mm) of (a) gynes of *Pseudomyrmex venefica* from inside thorns of *Acacia*, and (b) queens. (After Janzen, 1973.)

differ in size as well as behaviour. Those of *Pf* are strong fliers with big wing muscles, useful as a food reserve during the period of colony foundation. Their jaws, too, are very strong and easily cut into a thorn. Obviously size and strength, particularly of wings and jaws, favour success. Queens of *Pv* are smaller and weaker and their success depends on more subtle properties such as attracting mates, pacifying worker guards, and obtaining food from workers. In this way they will be able to contribute more eggs to the pool, create more worker progeny and influence the growth and reproduction of the population. It is unlikely that the workers actually assess their egg laying capacity but quite likely that some linked feature, perhaps activity and energy, perhaps docility and attractiveness which is correlated with their fitness as queens, is reviewed periodically. *Pv* queens have a bimodal size distribution and Janzen noticed that the smaller ones could re-enter more easily than the larger ones (Fig. 14.1). Such frequency distributions are not uncommon in ants and the smaller queens are always about the same size as the largest workers; they are called 'microgynes'.

Thus, these two distinct species both live in *Acacia* thorns in tropical America but do not interbreed, indeed could not interbreed, their habits and distribution are so different. One inhabits a seed-reproducing *Acacia* and the other a vegetatively-reproducing *Acacia*. To achieve this each has adopted a way of reproduction analogous to that of the plant: individual dispersal of big, store-laden queens for the seeding plants and worker-dependent small queens, spreading in swarms, for suckering plants. Janzen has pointed out that the plants themselves inhabit different habitats, though, with consider-able overlap, and inter-digitation: the seed dispersing ones tend to live in moister areas than the suckering ones.

14.2.2 SPECIES OF MYRMICA

Pairs of sibspecies specialized in similar ways are now commonly known in ants though they are usually adapted to more generalized habitats than those we have just discussed. A well-studied double species is *Myrmica ruginodis*. One part has normal queens that can found colonies alone and will settle permanently in groups of up to six (*macrogyna*, M). The other part has microgynes, which cannot found colonies alone but must move in swarms with workers. Their colonies spread out by budding and they restock with queens periodically (*microgyna*, m). M produces intolerant workers that will not even take in a new queen when queenless; m is tolerant and will mix with its own kind freely but rejects all M. M stages outbreeding nuptial flights, often on the rocky ridges between valleys, with males congregating by conspicuous rocks and trees to which females fly for copulation. m may use these places too but they mostly copulate near the nest and only produce sexuals every few years. This periodicity combined with the fact that there is a definite tendency to mate preferentially with their own kind, prevents hybridization (Brian and Brian, 1949, 1955).

These reproductive differences cause the two types to live in different stages of their wet acid heath habitat (Elmes, 1978b). M has a wider range and can colonize many quite isolated spots like the centres of ling clumps (*Calluna vulgaris*) that have spread out and grown moss or small sunny glades in forest, whereas m lives in sites in which the vegetation is frequently grazed. Elmes concluded from a survey of 18 sites in England, 14 of which had both forms, that habitats which provided support for a few big colonies were occupied by M whilst those which provided support for many small, well-spaced sub-colonies were occupied by m; these latter were richer in grasses and regularly grazed. This form no doubt has a socially organized dispersal.

This double species, has been regarded as a single one because the two types cannot be distinguished with certainty even by queen size; however, workers can recognize two distinct queen types presumably by their smell even though they have the same head width. Both forms occur together throughout the palaearctic from the British Isles to Japan and show no difference in geographical distribution; they are not subspecies. The small queens breed true and their workers are only slightly smaller than those of the big queens. A likely explanation is then that m originates in M nests which have been established long enough to have produced by mutation and inbreeding the microgyne form and that these start as temporary social parasites; later, if the habitat endures and the M die out, m takes over (Pearson, 1981; Pearson and Child, 1980). Elmes' survey shows that other species of *Myrmica* are attracted to the same sort of habitat as the m, so that they are therefore living in competitive conditions to which their swarm style of reproduction is suited. *M. ruginodis* then has the appearance of speciating sympatrically and even continuously: M is adapted to turbulent, ephemeral places to which it flies as lone queens; whilst m is adapted to steady habitats into which it spreads in swarms. Presumably the rarity of intermediates indicate that they are at a reproductive disadvantage compared with either extreme form and are almost completely selected out.

The larval development of a microgyne has been charted once (Fig. 14.2). This individual showed several growth arrests but managed to exceed the size normal for worker larvae though not to reach the size normal for gyne larvae at a crucial stage in caste development. It is thus possible that quite a simple genetic change affecting larval growth could produce a microgyne and this might be expected in an established population which had become isolated and was inbreeding. Provided the microgynes manage to stay in or re-enter the original colony there seems to be no reason why temporary social parasitism should not persist indefinitely.

Myrmica rubra, a species very similar to *M. ruginodis*, is consistently polygyne with an average of 16 queens in a colony and a negative binomial distribution of frequencies, Fig. 14.3 (Elmes, 1973). Single queens can found colonies alone, at least in the laboratory (Brian, 1951) but it is more likely that, in nature, small groups accompanied by workers, do this. They certainly restock with their own queens and possibly those of other colonies as well. Elmes (1973) has shown an inverse relation between queen number and queen size that might be due to overcrowding during production or to a genetic deterioration in size in large, long established colonies or to a tendency to accept small queens back rather than large ones (as in *Pseudo-*

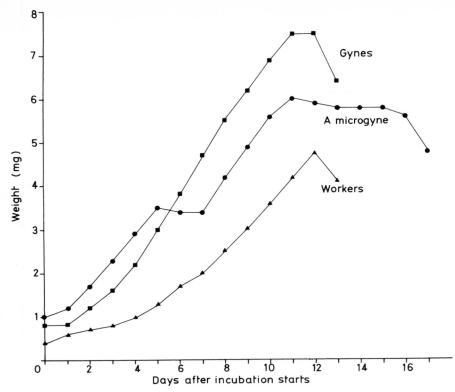

Fig. 14.2 The growth of larvae after hibernation, some produce gynes, some workers; one produced a microgyne. (Data from Brian, 1955b.)

myrmex vinecola). Thus, unlike the double *M. ruginodis* it combines in a single species with normal queens both the ability to fly and start colonies unaided and the ability to restock with young queens and form buds which probably set up barriers and become colonies as they age. *M. rubra* produces microgynes widely and sometimes many can be found in a single colony along with normal queens. They have never been found alone (Elmes, 1976) and appear to be unable to establish viable colonies. One reason for this is that microgynes have difficulty in producing workers: their eggs are not worker biased and they cannot control worker brood rearing. Microgynes are thus condemned to live as parasites on the normal type.

This is not the end of the story, for *Myrmica sabuleti* has a workerless social parasite now known as *M. hirsuta*, which was at first thought to be a sporadic microgyne. Moreover, the whole range of workerless social parasites now grouped in the genus *Sifolinia* may have originated in this way for the genus is morphologically very close to *Myrmica* (Elmes, 1978a). This seems to be the logical consequence of letting in smaller queens preferentially, as in

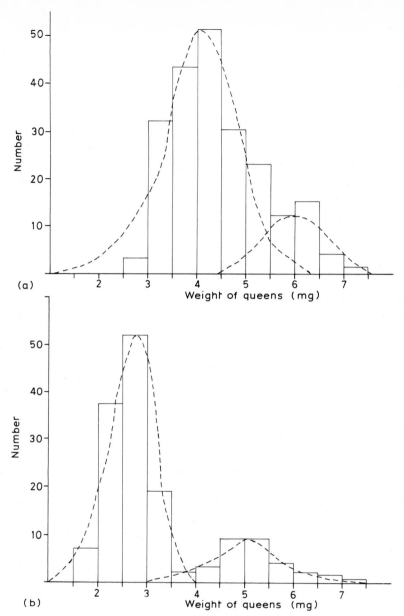

Fig. 14.3 Numbers of queens with a given fresh weight (mg) for (a) *Myrmica ruginodis*; (b) *M. rubra*. Each is selected to show the contrast in weight of microgynes (on left of distribution) and normal queens. The broken lines indicate hypothetical distributions within the bimodal overall distributions. Notice that the mode of *M. rubra* normal queens is smaller than those of *M. ruginodis* by about 1 mg and that *M. rubra* microgynes are slightly more than 1 mg smaller which makes them very small; indeed so small they cannot produce workers. (From Elmes, 1975.)

Pseudomyrmex vinecola. They may have difficulty in producing workers and end up as social parasites. The microgyne form of *M. ruginodis* is no doubt exceptional in being able to lead a relatively free existence. Polygyny may degenerate into social parasitism and so extinction through over specialization.

In the same subfamily, *Leptothorax* and *Monomorium*, both common genera, contain polygyne species but very successful genera such as *Tetramorium*, *Solenopsis* and *Pheidole* as well as all the Attini are usually monogyne.

The imported fire ant in North America, *Solenopsis invicta*, is now known to have polygyne patches. One especially large colony with 20 000 mated queens (Glancey *et al.*, 1975, 1976) shows signs of inbreeding from the existence of a high proportion of diploid males: 96% of males examined have little or no testes and no sperm whereas in normal colonies with haploid males less than 1% are sterile. No doubt this inbreeding led to polygyny as well as diploid male formation. The increase in diploid males is compatible with the hypothesis that sex is determined by genes in one sex locus though in this case the *Solenopsis* species introduced from South America could have undergone a drastic reduction in genetic variability so that in some colonies homozygosity at several major sex-determining loci could exist (Crozier, 1971). Fifteen polygyne colonies (with 3 to 63 queens) were found in Mississippi but more in Georgia; 90% of queens were inseminated and laying eggs, though not as many as the large physogastric normal (monogyne) queens (Fletcher *et al.*, 1980). It is thought that polygyny may be increasing in this species which when originally imported is likely to have been single queened.

14.2.3 DORYLINAE AND FORMICINAE

In a special study of the Dorylinae in America in which 144 species were examined, Rettenmeyer and Watkins (1978) found only one that had more than one queen; *Neivamyrmex carolinensis* had between 3 and 13. They suggest that this helps in their migrations underground in a very periodic climate but there are many species of hypogaeic Doryline that are not polygyne.

In the Formicinae, many genera are strictly monogyne (e.g. *Oecophylla*) but others (e.g. *Lasius*) contain species with polygyne patches and others which are wholly polygyne. *Lasius sakagami* (*Ls*) which is very close morphologically and behaviourally to *L. niger* (*Ln*) has from 0 to 309 queens (Yamauchi *et al.*, 1981). *Ls* copulate in, or just outside, their nest but are not produced every year unlike *Ln* who fly and assemble or stratify. Queens of *Ls* either re-enter the nest or go off, in the latter case apparently at low density, and as with *Ln* start a new nest. The queens of *Ls* are less fecund but are tolerated by their workers and accepted back into the nest. As a result, colonies become vast and run into millions instead of tens of thousands. The habitat of *Ls* is a narrow one: they live in sandy soil with sparse herbs and make underground nests that unlike those of *Ln* are resistant to flooding (cf. *L. flavus*); *Ln* is a much more eurytopic species. Thus *Lasius sakagamii* retains both reproductive options like *Myrmica rubra*, whereas *L. niger* is strictly monogyne.

14.2.4 THE GENUS FORMICA

This genus is very interesting from the queen number point of view. The ecologically basic species, *F. fusca*, is entirely independent of the others but along with its allies or sibspecies (e.g. *F. rufibarbis*, *F. neogagates*) is the host to many of the *F. rufa* group as well as the dulotic *F. sanguinea* and *Polyergus rufescens*. *Formica fusca* itself is oligogyne in Europe according to Pamilo *et al.* (1978, 1979), and in Japan it varies between monogyne near Tokyo and permanently polygyne in another part of Japan (Higashi, 1979). The monogyne form is seasonally polygyne due to the return in late summer of freshly mated queens (Kondoh, 1968). After winter these queens are forced out, often with the workers, to start new nests which yet again, can be retracted in winter. This seasonal polygyny is reminiscent of that of some of the tropical polybiine wasps. In North America, *Formica pallidefulva* has long been known to be both mono- and polygyne (Talbot, 1946), and is discussed by Hölldobler and Wilson (1977b, p. 13). Most amazing is the *F. microgyna* group of species which have regular microgyne queens that start new colonies by temporary social parasitism using *Formica fusca*, *F. neogagates* or *F. pallidefulva* as host. *F. microgyna* ranges widely and varies in hairiness but shows no other very satisfactory taxonomic diagnosis (Letendre and Huot, 1972). Did it evolve from a host like *F. fusca*? A series of increasing dependence is completed by a workerless social parasite even smaller and hairier than the typical microgyne: *F. talbotae* (Wilson, 1976c). It uses *F. obscuripes* (of the *F. rufa* group) as a host; the parallel with *Myrmica hirsuta* needs no stressing.

In *F. rufa* there is a full range of sibspecies from monogyne to polygyne which includes microgyny and workerless social parasitism. *F. rufa* itself has a monogyne form on the mainland of Europe (Collingwood, 1979) but is oligogyne in England. The morphologically indistinguishable *F. polyctena* is restricted to the mainland and may have a thousand or more small queens in 50–100 nests in a colony covering 30 ha (Raignier, 1948). This species is more receptive to new queens in late autumn and early spring and also just after a bud has been started (Elton, 1958). There is some disagreement about whether it has indubitable colonies which fight or not; this will depend on population density and the history of the local habitat. Mabelis (1979a) describes this fighting in detail and points out that it is especially common in spring when the colonies are re-establishing their territories after winter dormancy. Colony division to give one or two new nests at a time may take over a month and is initiated by disturbing factors such as shading or the discovery that the nest is not as near the centre of its territory as it could be or by the intrusion of other species (Mabelis, 1979b). Both workers and brood are carried to and fro indecisively but on average more go to the new nest; the queen either walks or is carried, but predation of queens in polygyne ants is not a disaster as it is in monogyne ones. Nest survival is not very high and a turnover of between 8% and 32% is recorded in Klimetzek's summary (1981). Welch (1978) records substantial turnover for *F. rufa* in England.

Formica opaciventris, a member of the *exsecta* group in North America, buds new colonies (Scherba, 1964), and *F. exsecta* in Poland is polygyne and has been studied in detail for many years by Pisarski (1973). *F. pratensis* is largely monogyne in Eurasia but in Siberia there is one centre with a territory around, into which tunnels pass where workers go to await food recruitment signals; at certain times of year the separate tunnel-groups acquire queens and move off (Reznikova, 1979). In part of Germany and the Netherlands it is polygyne and lives in shaded woodland which delays sexual ripening compared to that of the monogyne which lives in open places; this reduces interbreeding (Collingwood, 1979). *F. lugubris*, in Ireland, is monogyne or only slightly polygyne, with slightly overdispersed nests, which have a longevity of 23–38 years (Breen, 1977). A specially large colony exists in the Juran forests (France/Switzerland) at over 1300 m where snow lies from October to the end of May and there is 2 m precipitation a year; the average annual temperature is only 4–5°C. The colony covers 70 ha, contains 1200 nests, and the inhabitants are hostile to other polygyne groups on their borders (Cherix, 1980; Cherix and Gris, 1977). A similarly extensive colony of *Formica truncorum* is known in Japan; in this too, workers walk or are carried between nests but individuals rarely travel more than 4 m (Higashi, 1976, 1978).

The population genetics of European species has been studied by isozyme analysis (Pamilo *et al.*, 1978). Geographical variation as in *F. sanguinea* is very small but the disjunct species, *F. acquilonia*, polygyne and boreal, has different alleles fixed in one locus in Finland and in Scotland; so does *F. lugubris*. Differences between local populations are usually small: 5.1% in *F. polyctena* rising to 24.7% in *F. fusca*, but in *F. acquilonia* a figure of over 50% exists. Gene diversity is very low, at 0.043 compared with 0.150 in diploid insects (Pamilo *et al.* 1978, cited Selander, 1976), no doubt due to the extinction of recessive subviable genes in haploids; it has the useful effect of eliminating the harm due to inbreeding often extensive in *Formica*.

Ideally, a species will have a mixed strategy with some queens big enough to found colonies alone, after flying in search of suitable prisere spots, and others, smaller, able to return to base and bud with workers where competition is intense. This may be the situation in *Myrmica rubra*. More usually a species appears to give way either to monogyne flier or to polygyne swarmer species (as in *M. ruginodis*). Is a mixed strategy unstable in an evolutionary sense? This dilemma faces many flowering plants which have the choice between dispersing seeds and vegetating; although some opt for one or other some retain both potentialities and use then at appropriate times in suitable environments (Harper, 1977). Many *Formica* species may have a stable solution in that their queens can either re-enter their own nests or take over those of another species and spread their range; solitary foundation is never required.

14.3 Queen interaction and queen relatedness

There is no lack of evidence that in monogyne species the queens are hostile to each other and the workers inherit their hostility, towards all but their sisters and stepsisters. It spaces the colonies out.

14.3.1 INTERQUEEN AFFAIRS

Hostility between queens can be temporarily reduced as in *Formica fusca* and *Apis mellifera* where supersedure is accompanied by the co-existence for some time of the old and the young queen. During the preliminary swarming of the honey-bee the colony is again temporarily polygynic but the queens are kept away from each other by the workers as they are during the division of colonies of *Eciton*.

In all these cases the workers are acting with and for their queens but an interesting case of tolerant workers and hostile queens has recently come to light in the genus *Leptothorax* (Provost 1979, 1981). If two colonies of a monogyne species are kept in the laboratory with access to a common arena, in 40% of cases one starts to move over to join the other and carries first, the brood and workers and finally, the queen. She at once fights with the resident queen using stings as weapons not jaws. Hostility spreads briefly to the workers, which grip each other with their jaws but make no attempt to sting and the combat is not lethal. Afterwards the workers continue to hold down the queens for a long time (reminiscent of the 'sealing off' treatment *Eciton* workers give to their young gynes). Both are pinned down in this way until one, always the immigrant, is slowly released, and the resident queen cut to pieces and thrown out. This looks as though the active set are invading the passive set. Workers blend and lick the new queen but whether they are habituated to two sets of odours which they distinguish covertly or they have an umbrella pheromone from the immigrant queen is unknown. This sort of fusion was not possible with *Camponotus lateralis*; if one queen was removed fusion might take place in half the trials but if both queens were taken out mergence was regular. These experiments underline the importance of the queen in establishing colony identity. With a queen, a monogyne species becomes defensive in relation to its nest and brood though it may share an arena of food in the laboratory.

Workers of *Apis mellifera* will stop gynes fighting; it is called 'balling' and often leads to the destruction of one gyne or queen by the workers. So 'sealing off', holding down or 'balling' may be a slow way of neutralizing and finally destroying the queens that workers do not want. The contest between queen and gyne *Trigona subnuda* has already been mentioned; again it involves the participation of the workers and the process seems to be more active than that in *Leptothorax*.

Supernumerary termites are known to be eliminated in a similar way. They are produced in great numbers during a phase of replacement after experimental removal of the founder pair. They fight in a way which has been described in detail by Lüscher (1974) and those which become wounded in the process are finished off by workers so that in the end a single pair is left to take over the reproductive function. Some cases of natural colony fusion have also been described amongst termites; after the two colonies have joined, one of the reproductive pairs is destroyed as with queens of *Leptothorax*. At least one termite, *Nasutitermes corniger*, though normally monogyne, is facultatively polygyne. A colony may have 2 to 22 primary queens (and males), all friendly, all laying, living in a single multichambered reproductive complex. As in ants these polygyne colonies form by queens joining buds but cooperative colony foundation is also possible. Thorne (1982) found them in areas of Panama that were regenerating secondary forest.

In polygyne species, the inter-relations between the queens is very interesting. In *Camponotus herculeanus* each individual queen has a territory in the nest (Hölldobler, 1962). The same is true for *Bombus atratus* (Sakagami *et al.*, 1967a). Where queens are not territorial one would expect a dominance hierarchy. This is more likely to take a subtle form in which status is indicated by nest zonation, food transfer and brood relationship, e.g. the polygyne wasp *Metapolybia*. In *Leptothorax curvispinosus* up to three queens can be present and they appear to live amicably and lay eggs together, but in each of two groups studied by Wilson (1974a, b) one queen was found to treat her own eggs with much more care than those of the others! This top queen would 'accidentally on purpose' break and eat the eggs of other queens and of workers too and they could be seen struggling over them though workers apparently never eat eggs themselves. Queens got most of the food by regurgitation of worker crop contents and by helping themselves from the mouths of larvae. Workers without food to offer avoided queens or even ran off from them apparently recognizing them after antennating their heads; are these queens low on the worker preference order? Wilson could find no indications that the queens recognized each other, though they clearly distinguished the eggs of others from their own. If the top queen succeeded in destroying eggs of others this would be a case of facultative monogyny such as has been recorded in *Myrmecina graminicola* by Baroni Urbani (1968). In the genus *Formicoxenus* and the species *Leptothorax gredleri*, Büschinger (1979) has noted similar situations. In *Lasius sakagamii*, the polygyne sib of *L. niger*, queens are aggregated within the nest into dense brood-rearing regions but within each aggregation they tend to occupy separate chambers; only when new daughter queens are recruited is there a tendency for more than one to live in each chamber (Yamauchi *et al.*, 1981).

Studies of interqueen relations in *Myrmica rubra* show that, though they average 16 a colony and range from 0 to over 70 (Elmes, 1973) they are by no means clustered except perhaps in the spring when they come up to warm themselves. They tend to have 'home' ranges, not really territories within which they live and there is a good deal of variation between those which are fairly sedentary and live in a chamber deep in the nest and those which are more restless, frequently coming to the surface and even breaking out on to the soil surface and running away (Fig. 14.4). Such queens, when they meet, fight briefly with their fore legs and antennae, rarely with their jaws, and never their sting. It seems that the sedentary ones attract worker attention more, lay more eggs, are preferred and probably better fed and can be said to dominate the society with the minimum of aggression (Evesham, 1982). Groups of this species only recently become queenless will accept queens

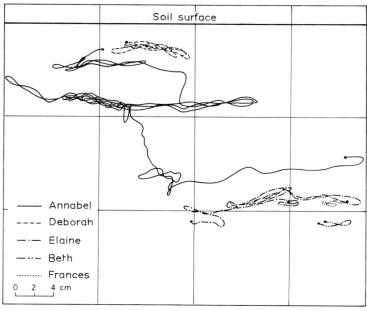

Fig. 14.4 *Myrmica rubra* in a self-made soil-slice nest in the laboratory; individual queen movements were recorded for 0.5 h. Notice the general dispersion and the localized movement; queens avoid each other and may fight. (After Evesham, 1982.)

from their own colony readily and even from strange colonies as well. Acceptance has several degrees, for if presented dead a queen from a different colony is rejected and put on the rubbish area whereas one from the same colony similarly presented is taken in and put on the brood, i.e. accepted. Living strange queens will move broodwards and may either be held down by workers or allowed to remain there and lay eggs. These eggs are probably fewer than a queen of their own group would lay and their oviposition does not necessarily cause workers to lay trophic eggs, in fact they lay a substantial proportion of reproductive ones under these conditions (Brian *et al.*, 1981b; Winterbottom, 1981). Thus these queens are tolerated but not allowed to influence worker behaviour and indirectly ovarial physiology. Groups of queenless workers that have been many weeks without a queen will not even accept one taken from their own colony and are, for a time at least, mildly hostile holding her down for hours or even days. This hostility within the nest can extend in some ants to queens in the territory outside and be mediated by workers, e.g. *Lasius niger* and *L. flavus*, both of which show greater aggression towards queens of their own species (Pontin, 1961b). In *Pogonomyrmex* queens discovered in the territory are carried to the edge of the nest without damage: exquisitely civilized (Hölldobler, 1976a)!

14.3.2 QUEEN RELATEDNESS

It is likely that a high degree of relationship between queens (e.g. sisters) helps in their acceptance and amicability inside the nest. *Tetramorium caespitum* queens will only co-operate if they are actually sisters (Poldi, 1963). In a number of cases winged gynes have been marked and found subsequently as queens in their parental nest and even in nearby nests, e.g. *Myrmica sabuleti* (Brian, 1972; Jones, personal communication) and *Myrmica rubra* (Elmes, 1973). With wasps, too, West-Eberhard (1969) and Klahn (1979) have both shown that associate queens are sisters, and Ross and Gamboa (1981) that queens prefer to associate with former nest-mates and can do this even after 100 days isolation.

Metcalf and Whitt (1977a, b), using isozyme analysis on *Polistes metricus*, have also found that a similar degree of relationship exists and revealed many other interesting features of their social lives. Colonies are started by one or two sisters (or half-sisters) each of which may copulate twice and so receive sperm biased 9:1 towards one of the males. The a-female produces 78% of the colonies' females and 87% of the males, the rest in each sex being due to the associate female. Workers rarely lay eggs unless the queens die and then one dominates and lays 95% of them. Inbreeding is negligible and the relatedness (r) between gyne progeny is 0.66 if there is only one queen or 0.49 if there are two. Gynes are related to their mother by 0.50 if she is the a-female and by 0.32–0.36 if she is not. To males relatedness is similar.

Nests started by two founders were only 2.25 times as successful as single founder nests but the inclusive fitness of the a-female exceeded that of a single founder. Even the b-female, though a joiner and a subordinate does not fare worse than if she starts alone! Nevertheless two-queen nests have a greater chance of success and usually only fail if the joiner dies early before workers can take over effectively; queens die from predation or accident away from the nest. Metcalf and Whitt conclude that these results accord with the kin selection theory of social origin but do not encourage females to join since they do almost as well alone.

Isozyme analysis of workers and queens in wild colonies can produce additional information. Pearson (1980) studied two populations of *Myrmica rubra* in adjacent valleys, one of which showed no relationship between queens and the other showed relationship in one year but not in the next: a surprising variability. In the polygyne *Myrmecia pilosula*, Craig and Crozier (1979) found that queens are only about as related as cousins, and workers only distantly related. Within each caste the frequency of genotypes fits the Hardy–Weinberg distribution and thus suggests outbreeding without selection. With *Aphaenogaster rudis*, Crozier (1977) found that a once-inseminated queen and her worker progeny were full sisters and that males arose from the queen or from the workers if there was no queen. Workers are full sisters in a form of *Iridomyrmex purpureus* (Dolichoderinae), in *Rhytidoponera chalybaea* and in another species, both of which are monogyne (Crozier, 1980). In *Rhytidoponera impressa*, a polygyne species, workers are less closely related than sisters.

In the dulotic *Formica sanguinea* in Finland, study of an enzyme locus encoding for malate dehydrogenase gave four segregating alleles (Pamilo and Varvio–Aho, 1979). Monogyny can be ruled out since 75% of queens are heterozygous (expected is 35%) and 10/92 nests have workers with three or four different genotypes. The authors suggest that several related queens, perhaps mother–daughter pairs are laying in each nest with the mother producing most of the eggs. Each queen copulates once. No significant geographical variation is present and only 2.4% of the total variation lies between populations. Protein segregation methods should open up this genetic field. Relationships can be explored and the mechanisms used in social systems, elucidated.

The infraspecific organization of social insects into 'colonies' thus covers a wide range of exclusive practices, from the strictly family group of the monogyne to the much more receptive polygyne group with low relatedness between individuals. The former is primarily a dispersive device and regulates population density by territorial means though it certainly keeps social parasites and other undesirable creatures out too. At high densities founding queens (or pairs) may cluster and regulate egg production by dominance orders. Where these polygyne groups endure and requeen cyclically a dispersive system exists between queens of a similar kind, and this, with the ability of workers to select and regulate queen density in the population establishes a population that is economically adjusted to its resources. Not much is yet known of the details of the mechanism used and how effective it actually is at promoting the best reproductives but there is no doubt that the social organization of insects is sensitive to the environment. There is increasing evidence that the pioneer monogyne state can pass into a competitive polygyne state as establishment proceeds.

CHAPTER 15

Comparative ecology of congeneric species

Species which are distinguishable by taxonomists usually live differently. Many of their 'species' if examined cytologically for chromosomes, and biochemically for allozymes or pheromones, turn out to be divisible into sibling species (sibspecies for short). Since these methods are not easy to apply in the field there are as yet few comparative studies. The account which follows where possible starts with the relations between subspecies just capable of interbreeding and passes on via sibspecies to congeneric species.

15.1 Ant and termite races

The study of chromosomes has, until recent technical advances, been rather limited. Moreover, even where differences are found the possibility exists that one is dealing with a single but cytologically variable species (Crozier, 1968). In an extensive study of 105 Australian ants the haploid number of chromosomes ranged from 3 in *Ponera scabra* to 42 in *Myrmecia brevinoda* with a mode of 11 and a median of 15. It is impossible to say what the primitive state is but Imai *et al.* (1977) infer from their analysis that the principle rearrangements are not polyploid but in the main change in number by fission and fusion of existing chromosomes. There appears to be little or no correlation between whether the species is morphologically primitive or advanced and its karyotype evolution.

Chromosome number can vary between races within a taxonomic species. *Aphaenogaster rudis* of the Myrmicini has monogyne colonies in mixed woodland in eastern North America which nest under stones, in soil or in logs. It has colour variations: dark brown in the Appalachian Mountains, reddish-brown on the coastal plain. These come within 1 km in one place in the foothills where Crozier (1977) finds that the 'plains' form has 2 haploid chromosomes and the 'mountain' form 22. No intermediates occur. Isozyme analysis also gives predominantly different esterase and malate dehydrogenase frequencies, and very little evidence of gene interchange. Evidently sibspecies are evolving sympatrically at different altitudes which govern the climate and a whole galaxy of dependent vegetational factors. On one mountain, in addition to the typical population with a haploid number of 22 chromosomes there exists, especially near the peak, another population with 18. Again no intermediates occur. Where the forms overlap they are said to use different nest-sites, the former under stones and the latter in logs though such specificity is unusual in ants. Slight colour differences and isozyme differences also occur. Males assemble nearby, and have distinctive allele frequencies as determined from allozyme data. So the two varieties are really distinct species using different microhabitats in the same area, but how they arrange their ecological separation is still obscure, as is the way in which they avoid cross-copulation at the common congregation point. Crozier concludes that the 18–20- and 22-chromosome races are reproductively isolated wherever they occur together and are distinct species despite the close similarity in shape, coloration and appearance of the phenotypes; evidently their behaviour differs in some vital respect. A recognizably different species, *Aphaenogaster fulva*, has 18 haploid chromosomes and perhaps an identical karyotype to the 18-chromosome *A. rudis* but it differs allelically. Karyotype evidence may not always reveal clusters of sibspecies but multi-locus studies probably do.

Amongst termites the dry wood species *Cryptotermes brevis* (Kalotermitidae), which occurs widely, has physiological races that differ in the way they replace reproductives that are taken away experimentally. In Australia at 27°S many substitutes were made and then all but one killed; in America at 27°N the selective culling appears to take place earlier before the 'neotenics' reach sexual maturity (Lenz *et al.*, 1982).

15.2 Desert ants and termites

Desert provides a fluctuating food supply distributed patchily in two dimensions. One would expect congeneric species to find difficulty in avoiding each other unless the physical instability and meagreness of the habitat prevented populations growing into contact and competition.

In the Chihuahuan desert southwest of N. America, two species of *Novomessor* differ nonetheless in feeding habits and food. *N. cockerelli* food is half insects (ants and termites) compared with 7% for *N. albisetosus*. The seeds of perennial shrubs and grass are also eaten: *N. cockerelli* eats less *Acacia* and less grass but more *Prosopis* (Whitford *et al.*, 1980).

The American genus *Pogonomyrmex* includes a number of species that live in acid, scrub/desert areas. Some such as *P. rugosus* (Pr) and *P. barbatus* (Pb) do not appear to have much ecological difference, both feed on seeds and termite sexuals as available, covering a similar range: in June, 60–70% of food is grass but in July only 30–40%, then more herbs, especially buckwheat (*Eriogonum*), are included. Only further analysis will show any further differences not yet recorded. Colonies are dispersed randomly with no obvious habitat difference and even the nests are similar shaft and chamber constructions in the centre of a clear disc (Whitford *et al.*, 1976). Their tracks, though they interdigitate, do not cross and forager contacts between species are infrequent even though they wander from the trails. They must avoid each other and the same two species, in mesquite-acacia desert in Arizona and New Mexico defend their nest area and trunk trails vigorously. Both stop foraging at dusk though *P. barbatus* forages at night in the hotter Texan desert (Hölldobler, 1976a). Are there feeding differences that have not come to light? A third species, *P. maricopa* is less efficient, eats a smaller size range of seeds, is less territorial and moves over the territory of the other species.

At least five species of *Pheidole* may live together in some areas of the south-western deserts of North America. Three of these use tracks and have species-specific trails (*P. militicida, P. tucsonica* and *P. rugolosa*) whilst two others do not use tracks and their trails are not species-specific (*P. desertorum, P. vallicola*). Tracks are not always distinct, sometimes they cross each other as in *P. tucsonica* and *P. militicida* but these two use them at different times. The former starts in the morning but stops at a temperature of 33–36°C, then comes out again in the afternoon as the temperature drops, and ceases altogether at about 20.00 h, whereas *P. militicida*, is nocturnal. Soldiers of *P. tucsonica* accompanying a set of workers were once seen to attack *P. militicida* foragers, otherwise encounters, as between *P. militicida* and *P. desertorum* where their tracks overlap, are usually avoided. In general it looks as though desert ants that cannot stratify because of lack of vegetation and lack of steady food resources, disperse themselves in space by maintaining territories and in time by having different hours of work and temperature tolerances.

15.3 Ants and termites in grassland

Even a herb/grass layer provides strata into which social insects can specialize: roots, soil surface and foliage. Grassland can be held back from development into scrub and forest by grazing animals of many kinds or it can exist through lack of water or continual mismanagement (especially burning) by man (savannah). Both sorts are included here.

Solenopsis exists in N. America as a complex of six species, two of which were introduced into the southern US and have since spread. The first, *S. richteri*, (*Sr*, the 'black imported fire ant') from Argentina and Uruguay, arrived in 1918 or thereabouts and the second *S. invicta*, (*Si*, the 'red imported fire ant') arrived from Matto Grosso, Brazil in 1933–1945. Both can be traced back unchanged to South American populations and rarely hybridize (Lofgren *et al.*, 1975). *Sr* is now limited to north Mississippi and Alabama whereas *Si* is spread widely from Texas to North Carolina. Usually there is one queen, but sometimes two, and at least one highly polygynous colony is known. It is suggested that *Si*, an aggressive species, replaced *Sr* in most parts, by means of a superior ability to hunt out and destroy *Sr* queens about to start colonies. *Si* also suppressed the indigenous *S. xyloni* (*Sx*) for both were grassland species, but a fourth species, *S. geminata* (*Sg*), persists in woodland though originally it too lived in grassland. However, *Sx* is not entirely expelled, for Baroni Urbani and Kannowski (1974) found it with *Si* in grassland; but it is restricted to the shade of trees in which it forages and in this way avoids *Si*. Even when baits are arranged to increase contact they still avoid each other and never come to grips. *Si* colonies are aggregated which is surprising since it is monogyne normally and has well-marked territories; presumably some habitat condition itself varies randomly. *Si* and *Sg* have been compared by Wilson (1978). He measured the time each caste of worker spent performing its repertory of actions. Both species are polymorphic but *Sg* has much larger major workers than *Si*. In both the smallest workers stay in the nest, whilst larger ones carry soil and food pieces that match their size. Whereas the major workers of *Si* chew seeds and store oil, those of *Sg* are exclusively seed millers; they will not even attack prey. In nature *Si* are more omnivorous than *Sg* which eats many types of seed.

Myrmica is a holarctic genus with many species (Fig. 15.1). A low temperature optimum (20°C) adapts it to cool summer and a requirement of at least 6 weeks around 10°C in winter for the development of gyne-potentiality means that in the south they are restricted to mountains. The common species pair in uncultivated grass–heath scrubland in Europe is *M. ruginodis* and *M. scabrinodis*, quite distinct in size and shape (Elmes has studied them biometrically, 1978b, 1980a), they show microhabitat differences that can be related to behaviour. *Myrmica scabrinodis* (*Ms*) nests in short grass–herb vegetation or under lichens between ling (*Calluna vulgaris*) clumps and makes nests that include a lot of mud in the walls and are divided into small chambers. These nests are ideal for picking up the heat of the sun, for resisting desiccation and can be used very effectively as fortification when pushing into the nests of *M. ruginodis* (*Mr*) so that (*Ms*) can advance chamber by chamber. *Ms* hunts near the soil surface amongst stem bases and its compact build enables it to creep through thick grass. Thus its habits enable it to live in the hottest areas of the short turf patches in the grass–heath complex. In contrast, *Mr* nests in tall herbs and grasses often near shrubs and even in loose moss. Its nests are less substantial, more of an envelope or tent covering a large vestibule with a few thin partitions. This design enables them to reach and absorb sunshine flickering through taller vegetation and to move to a new site if shading becomes severe, but it is not resistant to desiccation or invasion. *Mr* feeds in the leaf stratum above *Ms* and runs up bushes and for this their long legs and slender out-reaching antennae are very useful. Like *Ms* it feeds on a variety of small animals alive and dead as well as honey-dew but higher up the

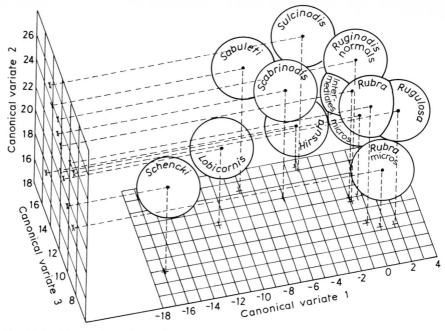

Fig. 15.1 Biometric analysis of 10 European species of *Myrmica* with microgynes of *M. rubra* and *M. ruginodis*. (From Elmes, 1980b.)

plants. Thus the ecology of the two species differs enough for both to survive in the same type of habitat: they have been able to evolve structural and behavioural differences that prevent them clashing. Conflict only occurs over the short turf that *Ms* settles in; here it can both hold its own defensively and eject *Mr* offensively. *Ms* lay seige and eventually the whole *Mr* colony leaves for a new place and the *Ms* go in and alter the nest style or, if the *Mr* do not go, the *Ms* push into it chamber by chamber. Dry weather helps since its own nests protect it from evaporation. Thus contact between the two species is reduced by foraging and nesting in different levels of the habitat. Whereas one (*Ms*) can dominate the other completely in warm dry soil and make better use of it, the other (*Mr*) is able to survive outside these zones where there is no pressure from *Ms*, but the right microclimate is patchy. Elmes (1978a) has studied *Myrmica* in *Calluna* scrub in southern England and confirmed these differences in distribution between *Mr* and *Ms* noting that *Mr* is pushed out into the '. . . poorer, more heterogeneous *Calluna* areas . . .' by *Ms* which dominates '. . . the richer, more uniform turf type areas . . .'. He also suggests that the microgyne form of *Mr* '. . . enables the species to make full use of the heterogeneous areas . . .'.

Further east, in both a warmer climate and on a more neutral soil this pair of complementary species encounters two more: *M. rubra* (*Mra*) and *M. sabuleti* (*Msi*). A glance at the figure of biometric distance shows that each member of the northern pair has a counter part in the south; *Mra* overlaps *Mr* and *Msi* overlaps *Ms*. This could make identification difficult were it not that to the taxonomic criteria are added behavioural peculiarities that naturalists learn to rely on. *Mra*, though built to forage in herbs like *Mr*, lives in neutral, and so normally agricultural soils, which are grazed but not managed intensively, i.e. thistles grow and support aphids above the soil, stones fall off walls and give trampleproof ready-made nests. In limestone grassland it lives patchily but extensively in shady places away from the competition of *Lasius flavus* which needs more sunshine. Even so, *Mra* probably lives in warmer places than *Mr* which is forced into open scrub and woodland (Elmes and Wardlaw, 1982b). *Msi*, with a build like *Ms*, forages near the soil surface and lives in grassland that is either grazed (if neutral) or grows patchily (if acid) and in either case enables them to nest in the soil with full sunshine or build small mounds. It needs more warmth than *Ms* to judge from its more southerly geographical distribution.

The two co-exist in an area of acid grassland (pH 5.2) in south-west England. *Msi* does best in more open vegetation whilst *Ms* occurs where tall grasses abound; only a few centimetres difference is needed. In calcareous grassland (pH 7.8) at about the same latitude only *Msi* occurs but here is less abundant in grazed and trampled areas. It is possible to distinguish the vegetation around their nests from other local vegetation; the best nests are in open places and have *Hieracium pilosella* (compositae) common nearby (Elmes and Wardlaw, 1982b). DeVroey (1979) shows that in Belgium *M. sabuleti* and *M. scabrinodis* are spatially dissociated and, in a study of the same two species on the Welsh marine island of Ramsey, Doncaster (1981) shows that *Msi* is more eurytopic and has a much wider range in two principal components representing exposure and drainage than *Ms*. In a third, relatively unimportant component representing altitude, bracken and maritime plants, *Msi* is drawn away from *Ms* and characterized as an inhabitant of low, rocky ground (cliff edges) and shallow soil with lichens that is well insolated and warmer. *Ms* lives higher up the hills where it is cooler and is quite closely associated with *Lasius flavus*.

This quartet of *Myrmica* species, *Mr, Mra, Ms* and *Msi*, thus separates into two pairs each with a north and a south and each with a short and a tall herb representative. When cultured at a fixed temperature the time taken for winter larvae to pupate follows a similar sequence; *Mr* are quickest, *Mra* take 1.2 times as long, *Ms* take 1.4 and *Msi* 1.5 times as long (Elmes and Wardlaw, 1982c, d). All show a Q_{10} of 3. Integrated temperatures at 5 cm below the soil surface give *Mr* 13.1°C, *Mrs* 13.9°C, *Ms* 14.3°C and *Msi* 16.0°C for their habitat, exactly the same ranking.

The genus *Messor* may have a similar design. In South West France *M.*

capitatus, with big workers and vast colonies, is monogyne and probably starts new colonies from single queens, whilst *M. structor* is polygyne with up to 17 queens (Delage, 1968). *M. capitatus* is restricted to dry, calcareous well-vegetated grassland slopes in which it lives very successfully (with *Pheidole pallidula*) whereas *M. structor*, much less specialized, can live in a wider range of habitats including valley and limestone plateaux, where it makes a deep, ramifying nest and escapes desiccation during drought; it often lives around *M. capitatus* areas. In fact one can ask: is the *M. structor/M. capitatus* pair ecologically equivalent to *Myrmica ruginodis/M. scabrinodis*? There are many species of Messor, but two, studied in savannah at Lamto, Ivory Coast, are especially interesting (Lévieux, 1979; Lévieux and Diomande, 1978a) *M. galla* lives in open, short grass herb and *M. regalis* in taller vegetation again, a difference recalling *Myrmica*. This leads them to collect the seeds of different species of grass. As if to further ensure that they do not meet, they forage at different times of day. *M. galla* has foraging columns in which 60–70% of returning workers are laden and collects all the year round in savannah. *M. regalis* forages singly and sporadically except when food is abundant in the dry seasons and only 40–50% of the workers return laden. The nests are also different: *M. galla* makes diffuse ones (and may be polygyne) whilst *M. regalis* makes a dome with about three openings, each 10 cm diameter. These congeneric species thus differ in habitat, food, time of activity, method of collection and nest structure: ample for ecological differentiation.

Crematogaster, a very diverse genus of myrmecine ants has two species in the savannah at Lamto, one *C. heliophila* (*Ch*) in the tree-tops and another *C. impressa* (*Ci*) in bushes near the ground (Delage-Darchen, 1971). Both are very similar in structure. *Ch* makes carton nests and *Ci* lives in twigs though it does make flimsy, carton nests as well. Both feed on plant sap and scale insect products but the insects they eat differ since they are stratified. *Ch* has one queen and is very aggressive and may occupy several adjacent trees. *Ci* colonies break up easily, especially after fire and the fragments take in several new queens though they do not seem to be always polygyne and may reduce numbers periodically as the wasps do. The stratification is at least partly enforced by the hostility of worker *Ch* though some degree of habitat selection by the queens is also likely. There are obvious resemblances to *M. ruginodis* and *M. scabrinodis* here in that *Ch* occupies the warm canopy of the forest and builds durable nests whilst forcing the *Ci* into cooler, less stable strata in which it survives through greater mobility achieved by a less costly nesting behaviour.

The genus *Lasius* (Formicinae) has been studied in temperate grasslands in Europe. Pontin (1961a, b) worked on the interrelations of the black epigaeic *Lasius niger* (*Ln*) and the yellow hypogaeic *L. flavus* (*Lf*) in England and found that the former eats the latter and has a detrimental effect on its sexual

production. The use of *Lf* by *Ln* is restricted, however, by its poor performance as a nest builder, for it can only live where flat stones provide a natural
ready-made nest. They are also stratified feeders, for *Lf* exists almost entirely
on soil arthropods and cultivates many types of aphid on plant roots whilst
Ln ascends herbs and shrubs, collects 'wild' honey-dew, catches prey and
scavenges. In any rough pasture, with flat stones and some scrub, the two
stratify and co-exist with *Ln* eating *Lf* at high densities.

This pair are important members of the community on Wadden island
(Netherlands). *Ln* pioneers into the freshly vegetating shell-sand dune slacks
and is followed by *Lf* and other underground species only when sufficient soil
has built up above water level to avoid salt stress and encourage grasses like
Festuca rubra (Boomsma and van Loon, 1982). Then *Lf* can outnumber the
Ln considerably and exert strong negative effects on its worker size, and gyne
emission but not male production. This complements Pontin's research in
England (Boomsma *et al.*, 1982).

L. alienus (*La*) is much closer to *L. niger* structurally and behaviourally. It
has a more southerly distribution where it can live freely without molestation
by the larger, more aggressive, workers of *Ln* but in southern England where
they both occur together it can only avoid *Ln* and find the warmth it needs by
living in dry, exposed, insolated slopes (Brian, 1964; Brian *et al.*, 1976). Here,
it nests underground in a sunny spot and makes foraging tunnels which
protect it from wind, rain and periodic heathland fires. It subsists mainly on
soil arthropods, particularly after fire but is closely associated with *Aphis
ulicis* on the recumbent branches of dwarf gorse (*Ulex minor*). In the absence
of *Ln* foragers it comes up from the soil and ascends ferns and bushes for
nectar and honey-dew but *Ln* workers collect round the *La* exit holes and
attack those that try to break through. Nuptials are in July–September for *Ln*
and September–October for *La* so there is some risk of hybridization and
Pearson (1982a) in fact has isoenzyme evidence that this occurs. He has also
shown that *La* may have a chromosome polymorphism with either 28 or 30 in
the diploid.

Ln queens appear to avoid the dry heath that is favoured by *La* perhaps
because they find it too hot and dry in July though they choose bare soil in
gardens (Pontin, 1960) like the American *L. neoniger* which avoids woodland if placed there after breaking its wings off and refuses to dig in (Wilson
and Hunt, 1966). In southern England, Elmes (1971) translocated *Ln* into *La*
habitats and found that *Ln* was slowly eradicated. Thus the species are
designed for different habitats, but sometimes overlap; where this occurs,
stratification is given a finishing touch by immediate interactive behaviour
which reduces the sexual production of both and so their fitness. This sets a
premium on habitat discrimination by young queens.

In grassland grazed by rabbits on Ramsey, *Ln* nests inland closer to old
walls whose large stones provide it with ideal nest sites (Doncaster, 1981).

La, in contrast, is almost entirely coastal and lives on cliff-tops in underground tunnels. Its diffuse nests are built under very short herbs or lichens, exposed to strong winds but also to sunshine and are liable to desiccate periodically in early summer. Using principal component analysis, Doncaster was able to show no overlap at all in the two dimensions which account for about 89% of the numerical dispersion: *Ln* was confined to sheltered, moderately moist and *La* to exposed, rather dry areas. The *La* habitat is thus much the same as that in England; the *Ln* one is probably more like the garden/scrub habitat which complements the marsh habitat in England. *Lf* is the most widely distributed of the species on Ramsey; its ability to build mounds of its own and to forage underground enables it to survive well where rabbit grazing is light enough to reduce the growth of the herbs though rabbits, if too dense, can destroy mounds by scratching. *Lf* can be brought out in the principal component analysis and both *Ln* and *La* by a third component representing altitude, that features much bracken (*Pteridium acquilinum*) and maritime vegetation. The three species are then clearly ecodifferentiated with 'distances' in the three-dimensional component space as follows: *La* to *Ln* 3.14, *La* to *Lf* 2.61 and *Lf* to *Ln* 2.66.

Camponotus is probably the most common genus of ants (Wilson, 1976d), and ranges from tropical to temperate climates. There are 50 described species in the Ivory Coast of W. Africa alone, some of which have been studied ecologically by Lévieux (1976). *C. acvapimensis*, which lives in savannah soil, is the commonest ant with a biomass of 0.52 g m^{-2} in black soils, 0.47 g m^{-2} in ferralitic soils and 0.2 g m^{-2} in hydromorphic soils (dry weights). Queens, after flight, are destroyed by workers if they land in occupied territory. Another species, *C. solon*, with workers over 1 cm long, nests in tall tree trunks that are well spaced (1–2 ha^{-1}), and eats plant exudates in the tree canopy up to 60 m away.

Each foray is concentrated into a few square metres of tree foliage but over a year they spread around to cover several trees. Scouts alert colonies at nightfall and a mass emerges, climbs a tree and comes back with distended gasters an hour later. *C. vividus* lives in nest groups constructed from enlarged beetle borings in branches and collects its food from an area of hundreds of square metres of canopy changing its emphasis gradually, sometimes even foraging on the ground. It visits aleurodid plant lice for honey-dew and collects exudates, especially a white, dry gum: in fact 90% of its forage is of plant origin. They have activity peaks at 06.00–08.00 h and again at 16.00–18.00 h; activity drops to 20% in between. Only rain, never temperature, stops them foraging. *Camponotus* thus shows a variety of lifestyle in the tropical rain forest savannah ecosystem (Lévieux and Louis, 1975).

Termites of a single genus are equally well differentiated ecologically. Five species of *Trinervitermes* (Nasutitermitinae) have been studied by Sands (1965) in West African savannah where 112–122 mm of rain peaks in August and grass is burnt in December. He surveyed three biomes: woodland with grass 1–2 m high, scrubland with similar grass and an area in which dense tall grass alone existed. The abundance of grasses on which they all fed made food competition unlikely but he found zonation and differences in nest building. *T. geminatus* lives in open, sunny areas in aggregated mound-nests and is well adapted to habitat and climate. *T. trinervius* does not forage actively and is more at home in moister savannah with a two-peak rainfall for which its life cycle is better suited. *T. togoensis* is clearly unable to make strong mounds of its own and uses old mounds of *T. geminatus* and *Cubitermes*. *T. oeconomus* and *T. occidentalis*, both pale in colour and sensitive to sunlight, forage continuously for grasses, whereas the others have a store which lags the nest against heat from the sun and even fire. More than half the nests of *T. oeconomus* occur in *Macrotermes* mounds. Sands concluded that the differences in nest-making skill, foraging and feeding method (though not necessarily in food), and climatic adaptation enable the species to co-exist in grassland with only a few trees. Differences in habitat selection behaviour space the species populations and only when these termites accidentally encounter another species are they aggressive.

Microtermes (Macrotermitinae) is a genus of small hypogaeic fungus cultivators that make chambers (3–4 cm diameter) linked by galleries in soil. At Mokwa (Nigeria) 9°N, seven species studied by Wood (1981) are reproductively isolated. Most hold nuptials after rain but either 1, 2 or 3 to 4 days afterwards, at specifically characteristic times of day or night. Some assemble round trees and in others the female lures males from vegetation.

15.4 Forest ants and termites

Trees provide additional opportunities for species diversification in spite of weakening or destroying the ground herb stratum where their canopy closes. The genus *Formica* is divided into several sub-genera which include the slave-making social parasites (*Raptiformica*), their free-living slave groups that make small mounds or none at all and collect food in bushes (*Serviformica*) and the wood ants with well-developed mounds, heat conservation and tree foraging (*Formica*). A study of 13 European species shows a remarkably close genetic relationship. The average genetic distance is 0.023 between con-specific populations compared with 0.283 between species or 0.332 between sub-genera. All these distances are smaller than normal in animals (Pamilo *et al.*, 1978, 1979). The *F. rufa* (Fr) group of six species (Kutter, 1977) can be distinguished from others (see dendrogram (Fig. 15.2)) but are either a group of species with high genetic similarity or a group that is now interbreeding all the time after divergence in different geographical areas. Virtually nothing is known about the genetic differences required to stop gene flow and initiate speciation. Hybridization is possible for sexuals often come

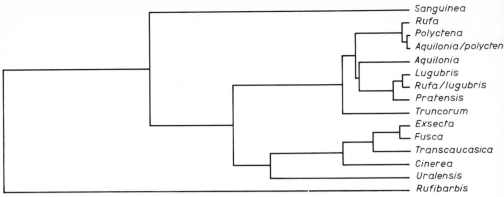

Fig. 15.2 A dendrogram showing the relationships between the *Formica* species of Eurasia. (From Pamilo *et al.*, 1979.) The construction is based on genetic distances and described in the above paper (p. 68).

out at the same time in Europe and cross-species copulation has been observed. *F. truncorum* is most different from *F. pratensis*, and can, in fact, be distinguished on morphological grounds (Dlussky, 1967). For the others no diagnostic loci were found, though the similarity between conspecific populations is just a little greater than between species. Curiously, *F. acquilonia* (*Fa*) in Scotland, now an isolated population fragment, resembles *F. lugubris* (*Fl*) from slightly further south, more than it does north Finnish *Fa* (which is distinct from the Scottish form at two loci). In its turn the Finnish *Fa* resembles the south Finnish *Fl* more than it does Scottish *Fa*. From the genetic evidence these two species appear to have diverged independently at opposite ends of Europe.

In the British Isles *Fl* is northerly and *Fr* southerly in general distribution. They meet and interdigitate in Central Wales, the Lake District, Yorkshire Moors and perhaps the Pennines but only the former, *Fl*, exists in Ireland (Breen, 1977, 1979). In Wales *Fr* shows its need of warmth by ranging from sea level to 200 m in sunny, south-facing slopes (mostly abandoned smelt furnace areas) with stable gaps surrounded by oak scrub (Fig. 15.3). Where forest is closing and encroaching this species vanishes. *Fl* tolerates more shade, is found on north-facing slopes up to 305 m (the tree line) and ranges more widely even entering conifer plantations. In one zone they occur within 80 m of each other: the *Fr* nests on the site of a derelict copper mine now a warm sunny, sheltered rock-strewn glade, the *Fl* nests completely shaded under closed oak canopy, cool and damp. No interaction is recorded (Hughes, 1975). Hence, where they meet there is an ecodifferentiation that expresses their differences in climatic adaptation, though how this affects foraging activity or nest-site search is quite unknown.

A similar temperature dependent zonation exists in the species pair,

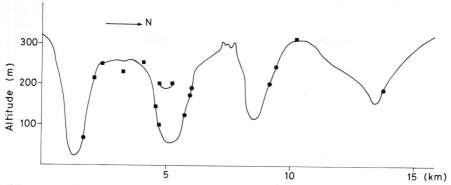

Fig. 15.3 Transect (north/south) through Gwydir Valley systematized to show wood ant position. ●, *Formica rufa*; ■, *F. lugubris*. (After Hughes, 1975.)

Formica lemani and *F. fusca* in the British Isles: the former lives much further north than *F. fusca* and where they interdigitate it tends to live on cool northern slopes whilst *F. fusca* lives in warm open places with a southerly aspect. Perhaps *F. lemani* uses scarce sunlight better through constructing its nest just under the soil surface.

In tropical rain forest (in Papua New Guinea) two species of *Leptomyrmex* (Dolichoderinae) live together. *L. lugubris* (*Ll*) has large workers that feed from definite, though temporary, trails on a variety of insects (mainly other ants) and plant material. *L. fragilis* (*Lf*) is a smaller individual, with smaller colonies that have many ill-defined trails from which they collect smaller food items especially winged insects. In addition to *Ll* being a bigger insect with a bias towards plant and ant food it has clear territories that are not shared with other colonies or with *Lf*. Their nests are distributed on the edges of these exudate-providing trees, of which the right kind are scarce. *Ll* can take over the nests of *Lf* (which tend to be central to an ill-defined territory) without fighting and literally 'drive' them away. (Plowman, 1981). The key resource is probably the correct species of exudate-providing tree, *Ll* can acquire and dominate any available but this still leaves plenty of space and food for *Lf*.

The way in which two species of *Macrotermes* use tropical rain forest in west Malaysia and together comprise more than half the termite biomass forms part of a study by Matsumoto and Abe (1979). Both species have fungus combs which they feed with pieces of litter mostly taken from leaves fallen from trees but whereas *M. carbonarius* (*Mc*, 0.33–0.43 g m^{-2}) approaches from above and has mound nests, *M. malaccensis* (*Mm*, 1.06 g m^{-2}) approaches from below and nests below ground. *Mc* prefers freshly fallen leaves, which it approaches at night in columns guarded by soldiers which run over the surface often as far as 10 m from the nest. *Mm* prefers rotting leaves though these may be scarce but primarily feeds on the wood of logs and only collects litter near these when it makes a soil canopy and attacks from underneath. The two thus have one common food item, namely leaves on the ground but choose different stages of decay and collect in different ways, either from above at night or from below under shelters. An active population of *Mc* can deprive *Mm* of its litter food but *Mm* will still have logs.

15.5 Wasps and bumblebees

Wasps and bees, both travelling by air, might be expected to differ from the pedestrian ants and termites in their interspecific relations. Both *Polistes fuscatus* and *P. metricus* nest close together in North America and both forage into the same tree, searching foliage for soft-bodied larvae which they chew to a pulp and carry into the nest. In a search for differences that would minimize conflict, Dew and Michener (1978) found that *P. metricus*, the larger individual, went further afield (average 102.4 m) and foraged in tree tops whilst *P. fuscatus* only travelled half as far (average 48.1 m) and foraged lower down. *P. annularis* is said to forage entirely in trees, and it is clear that stratified feeding exists in *Polistes*.

The competition between species of *Vespa* for exuding plant sap has been described by Sakagami and Fukushima (1957). An interspecies dominance order exists: *V. mandarinia* ≥ *V. dybowskii* > *V. analis* > *V. crabro* > *V. xanthoptera* > *V. tropica*. This is slightly oversimplified for *V. analis* is vague in its relationships and does not fight. *V. mandarinia* is rarely offered resistance and then only by the next one down, *V. dybowskii*. This last species is ferocious and, though it can start colonies unaided, often displaces *V. crabro*; its thick cuticle and strong foul smell help it to penetrate their nest. The two common European species of *Vespula germanica* (*Vg*) and *V. vulgaris* (*Vv*) are especially interesting. Small food differences have already been listed. Both nest underground in cavities but *Vv* makes nests with soft decomposing wood that is worked into a friable carton whereas *Vg* uses dead but sound wood which is worked into a grey and supple carton. As nests are extremely important structurally in defence and for thermal regulation, this difference may well reduce conflict for raw material though the general impression is that dead wood, both decaying and sound, is not in short supply, even today. There is also a stratification. *Vg* always nests underground, *Vv* usually, *V. rufa* near the soil surface (which it cannot excavate), *V. sylvestris* on or above the surface and *V. norwegica* in bushes and small trees. The last two species also make substantial envelopes.

Bombus females undoubtedly select different stages of a vegetation sere: grass, herb, scrub or forest for nesting and foraging in; the males too identify the general stage of plant regeneration when making scent marks. In Europe *B. lapidarius* marks tree tops, *B. pascuorum* and *B. pratorum* mark bushes or small trees, *B. terrestris* and *B. lucorum* mark herbs and *B. hypnorum* marks the ground (Haas, 1949). Correspondingly *B. pascuorum* and *B. pratorum* feed from flowers in open woodland whilst *B. lucorum* and *B. hortorum* use the flowers of grassland and low shrub (A.D. Brian, 1957). The former pair also nest on the soil surface or even in bushes and the latter pair below ground safe from heavy grazing animals. Within each stratum differences in tongue length enable the bees to use different types of flower: thus *B. pratorum* has a shorter tongue than *B. pascuorum* and *B. lucorum* than *B. hortorum*. The latter pair are very different. *B. lucorum* feeds on honey-dew and on a variety of open flowers with short nectar tubes, it is also strong enough to cut into the nectaries of flowers, down which it cannot probe directly. *B. hortorum* specializes in flowers with very long corollas and is prepared to go right inside. *B. lapidarius* favours clusters of shallow nectary flowers and all these species have characteristic soil temperatures for spring emergence (Prŷs-Jones, 1982). These habitat and flower biases almost eliminate interference between species, but *B. pratorum* is aggressive and attacks other bees in flowers displacing *B. pascuorum* but not *B. lucorum*. The former may be attacked in flight and knocked to the ground, (A.D. Brian, 1957).

B. lucorum (*Bl*) is a small white-tailed form of *B. terrestris* (*Bt*) with a northerly shift in its general area of distribution; where they overlap they may hybridize. *Bl* emerges and starts nest making before *Bt* each year, as it responds to a lower soil temperature. This enables it to live further north and presumably in cooler parts of the common area but also exposes it to temporary social parasitism by *Bt* whose greater size helps physically in the takeover of the nest. *Bl* has a morphologically indistinguishable variety whose male uses a different scent (Bergström, 1975) and *Bt* has dihydro-farnesol as a main component, unusual in this genus (for a summary of these pheromones read Bergström *et al.*, 1981).

The interrelations of three species of *Bombus* in the Rocky mountains of Colorado have been investigated by Inouye (1978). *Bombus appositus* (*Ba*) with a longer tongue feeds principally on *Delphinium barbeyi*. *Bombus flavifrons* (*Bf*) with a shorter tongue uses monkshood (*Aconitum colum-bianum*) for the most part, whereas *B. occidentalis* (*Bo*) cuts corolla and steals nectar. Experimental removal of *Bf* from *Aconitum* causes smaller individuals of the same species, which have not been seen before, to come out and forage, but it also leads to an increase in *Ba* which suggests that the presence of *Bf* is normally enough to make it unprofitable for the *Ba* to use this flower. Removing *Ba* from *Delphinium* allows the nectar tubes to fill up and brings the shorter tongued *Bf* in. This sort of inconspicuous oppor-tunistic competition (exploitation) may be more common than is realized. The competition between two bees for a rose flower (*Rosa carolina*) that provides pollen and yet has room for only one forager at a time proves that workers of *B. terricola* can be displaced by workers of *B. vagans* of the same or greater size. The bees appear to forage randomly and encounter each other every 15th flower, yet there is also evidence that they avoid any occupied flower and so reduce the frequency of collision (Morse, 1978). This complements the observation that *B. pratorum* may attack bees in occupied flowers for variations in aggressivity between species are well known.

Although these wasps and bees do not dominate blocks of territory they have many ways of reducing their overlap in space and time. They emerge and start work in different temperatures and peak at different seasons with different length cycles (Prŷs-Jones, 1982). They forage after flying different characteristic distances often in different strata and they may nest like this too. They may select different stages in seral regeneration and feed on different materials or have a dominance order for the same materials. Flower use by *Bombus* shows some attracted to shallow flowers or clustered shallow flowers and others attracted to deep corolla flowers. Males may reinforce the attraction of females to certain types of seral habitat stage. Interspecific hostility is rare but occurs exceptionally in small limited food sources (sap exudates, flowers).

15.6 Advanced bees

The Meliponini and Apini are conveniently taken together. *Trigona fuscipennis* (*Ts*) and *T. fulviventris* (*Tl*) are bees of similar size both of which feed on the pollen of *Cassia* bushes by cutting open the anthers (Johnson and Hubbell, 1974, 1975). These plants flower in the dry season of the Costa Rican coastal plain. *Ts* tends to forage in groups on clumped bushes, average nearest neighbour 1.5 m, whereas *Tl* either works single plants or sparse flowers on clumped plants, average nearest neighbour 8.1 m. This difference is partly due to interaction; *Tl* can be seen to avoid *Ts* which is blatantly aggressive and hovers within 1 cm of *Tl* workers until they fly away. *Tl* even leaves vacant flower clusters alone, although whether it learns to do this after molestation or is avoiding a scent mark left by *Ts* is unknown. *Ts* has a species-specific pheromone and arrives in groups led, perhaps, by a scout; other insects are driven off and it reaches densities of 22/m². In contrast, *Tl* makes no attempt to interfere with other insects and only attains 3/m². In short, *Ts* is an aggressive social forager and spends energy monopolising and retaining good flower clusters whereas *Tl* is a lone forager and spends energy searching for new flowers; either tactic succeeds.

Hubbell and Johnson (1978) also tried sugar bait grids and attracted about nine species of bee. In this experiment *Tl* from several different colonies fought when they met in the grid centre: not lethally, merely by threat chase and butt. *Ts* arrived later in a cloud led by a scout and monopolized a few baits. *T. sylvestriana*, a big black bee, easily buzzed *Tl* off but was not very effective at using the baits it gained; in fact a sort of blustering bully. *T. testaceicornis*, in total contrast, insinuated its way on to occupied baits and fed alongside only leaving if detected; yet it can build up a monopolizing density. So again we see different techniques for dealing with competitive situations: outright attack, bluster, persistence and recruitment, or insinuation and retreat under pressure. Johnson and Hubbell suggest that these strategies have evolved through co-existence to minimize interactive waste of energy.

The genus *Apis* provides some fine examples of interspecies ecological differences. Two, *A. mellifera* (*Am*) and *A. cerana* (*Ac*) are very close taxonomically and, prior to human interference, spread geographically west and east, respectively, of the south Asian area where they probably originated. They are not just geographical subspecies for there is unequivocal evidence that they do not interbreed (Ruttner *et al.*, 1972, 1973). The reason lies with the males which congregate in the same places and are attracted to the same pheromone (9–ODA) which all *Apis* gynes produce in their mandibular glands. *Am* males within a few kilometres will prevent *Ac* males from inseminating *Ac* gynes. This is so in spite of the fact that *Ac* drones fly faster and at a lower temperature. It may be that the *Am* sperm packet, which is larger and more fluid than that of *Ac*, destroys the *Ac* packet after copulation, though this is unlikely since in Pakistan normal *Am* copulations are blocked if *Ac* are present in the area (Koeniger and Wijayagunasekera, 1976)!

Ac is the native species in Japan but *Am* has been introduced by human beings as it stores more honey and absconds less often. Sakagami (1959, 1960; see also Brian, 1965) has described their interaction. *Ac* have strong jaws that they use to enlarge their cavities thus increasing the size range of cavity suitable for nesting; this may contribute towards the better hibernation success that *Ac* shows. *Am* are quicker to spot enemies and sting than *Ac* and can hold *Ac* away and prevent it stinging back, and *Am* recruits more quickly to sugar dishes and robs *Ac* colonies in groups; evidently it is a quicker communicator and acts socially more effectively. However, *Ac* has an evasive tactic: they abscond whilst a raid is on and return when it is over. This is a big advantage when attacked by a predatory hornet (*Vespa mandarinia*) and saves the colony from destruction; in fact, *Am* resists and is often wiped out. Thus, the ecological balance between the two *Apis* species could be held by the predatory hornet which eats *Am* and the robber human which cultures it. At least *Ac* appears better adapted to feral life in mountains.

In Pakistan *Ac* meets *Apis florea* (*Af*) and *A. dorsata* (*Ad*) in the mountains to which it is confined (Ruttner *et al.*, 1972). In Sri Lanka it also co-exists with these species as well as with the meliponine, *Trigona iridipennis* (*Ti*) (Koeniger and Vorwohl, 1979). Food appears to be abundant in the regenerating fallow which follows brief cropping especially as in the dry summer months both *Ad* and *Af* emigrate. In competition for artificial food, interactions between individuals do not always go the same way, i.e. there is no clear dominance hierarchy, but on balance the *Ti* dominates *Ac* and *Af* but never meets *Ad*, whilst *Af* dominates *Ac* and *Ac* dominates *Ad*. The average order is: *Ti* > *Af* > *Ac* > *Ad* and is in inverse order of individual size. The authors point out that this arrangement enables the smallest, which cannot fly far, to forage around the nest without difficulty (100 m or so), whilst those which can fly further, e.g. *Ad* (5 km) do so and thus improve interspecies dispersion. Only *Ac* shows intercolonial conflict. Nest sites differ completely as *Ti*, a small bee, uses smaller cavities than *Ac* and the other two nest in the open: *Af* around small branches and *Ad* from big branches. Although all gynes of the *Apis* species use 9–ODA as a luring scent and all drones go for it, hybridization is avoided by flying out to copulate at different times of day (Fig. 15.4). *Af* fly first around 12.00–15.00 h, *Ac* next around 16.00–18.00 h and *Ad* last around 18.00–19.00 h.

Apis mellifera has, with or without human aid, spread over most of the world to form many subspecies that today interbreed. Several African subspecies exist, Fletcher (1978b). Thus *A.m. adansonii* to the south of the Sahara Desert, lives in thorn tree and tall grass savannah and another, *A.m. scutellata*, in central Africa lives in a similar habitat though with more tropical evergreen and deciduous forest. The former subspecies was introduced in Brazil in 1956 where it competes with the European *Apis mellifera*

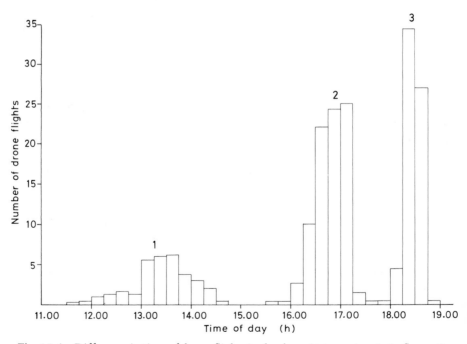

Fig. 15.4 Difference in time of drone flights in the three *Apis* species; 1, *A. florea*; 2, *A. cerana*; 3, *A. dorsata*. (From Koeniger and Wijayagunasekera, 1976.)

introduced earlier and by 1971 the 'Africanized honey-bee' covered most of the country (Michener, 1975).

The peculiarities of these subspecies are that they have smaller workers and smaller cells and nest in any weatherproof shelter, e.g. termite nests, crevices in rocks or buildings, boxes, under trees and even in the soil, features which provide a greater habitat flexibility. The hibernation behaviour is absent: there is no slowing down of tempo and only brief cluster formation; they readily abscond when food is scarce. Both these features indicate a selection for tropical rather than temperate climates. They are active, industrious foragers that fly quickly in and out and between flowers: and their recruitment system, which gives direction down to 50 m and sometimes as close as 10 m, calibrates at a lower waggle rate per unit distance than that of *A.m. ligustica*, the main European subspecies. They also collect and store a lot of honey. In defence they are highly co-ordinated and issue in a mass that is strongly deterrent, though individual stings are no worse than in normal honey-bees. In short, they are an effective tropical variant of the typical *Apis mellifera*.

In their replacement of the European type in Brazil they have used a number of devices. One is the building of a large feral population through their ability to endure sub-standard nesting conditions. This enables them to replace colonies of *Apis mellifera* directly by robbery and attack and it gives their drones more chance to hybridize viably. These hybrids have a mixture of features from their parent races: they will use standard hives and foundations, are active foragers with good sensitive direction indication but still abscond easily and are sensitive to disturbance and aggression.

From this survey of congeneric species ecology there seems to be no end to the number of ways in which species avoid interacting at a short resource whether it is food or nest-site. They can have different periods of seasonal and diurnal activity usually keyed by different temperature triggers. They can have differences in size as individuals that enable them either to collect bigger pieces of food or travel further from their nest in search. They can nest or feed at different levels in highly structured biomes with the result that they often eat different food items. Hybridization between sibspecies can be reduced by having nuptials at different times of day or year. Some species appear to be dividing into mono- and polygyne forms which are, respectively, pioneer and intrusive in their ecology. Yet others migrate into other geographical areas to avoid competition during the bad season. In the next chapter the aim is to put these groups of congeneric species into their community and look for wider patterns of interaction.

CHAPTER 16

Communities

Now that the differences between congeneric species have been explored it is possible to see how these species groups fit into natural communities. In this chapter communities are followed geographically from grass and woodland in high latitudes including dry steppe and savannah to deserts and then on to tropical rain forest.

16.1 Temperate zone communities in grass and woodland

16.1.1 GRASSLANDS IN EUROPE

Many excellent faunistic studies have been made in temperate regions but very few analyse the interrelationships between species. The community in which the two *Myrmica* live at Strathclyde 56° N in a maritime climate 100–200 m above the sea has two additional ants: *Leptothorax acervorum* (*La*, Myrmicinae) a small species, and *Formica lemani* (*Fl*, Formicinae) a large one. As neither of these can nest in grass they are restricted to scrubby places where there is some bare soil or decomposing stumps and branches. *Fl* nests under the bark on the south side of stumps with *Myrmica scabrinodis* (*Ms*) in slightly cooler places east and west and *M. ruginodis* (*Mr*) right in the shade. This temperature zonation is established by the queens themselves as they settle to found nests and is later reinforced by the workers: *Fl* will evict or kill and eat *Myrmica* workers and brood. *La* escapes destruction by tunnelling into the hard undecomposed wood where the bigger ants cannot follow. As the tree stump degrades *Fl* move into short grass nearby and though they may interfere with its growth they cannot build mounds, so that eventually shade forces them to move. If they cannot find a place they die out slowly. *Myrmica* survive in a regenerating forest if there are patches of sunshine on the floor. Prior to the deforestation of Scotland last century one could have expected *Formica lugubris* or *F. acquilonia* to take over as young trees established but they are not very mobile and may take decades to travel a few kilometres.

On top of this zonation is a foraging stratification which spreads the species out over the resource space. *Fl* climbs up bushes and young trees; *La* forages several metres away from the nest and goes into shrubs but its main source of food is unknown. Any wood ants, would, of course, go still higher up into the trees. *Fl* also forages more during the middle part of the day in sunny weather just as Talbot (1956) found in a similar community in Michigan. This reduces its likelihood of contact and competition with the two species of *Myrmica* which forage morning and late afternoon and in dull, though not wet windy, weather. None maintain a field territory; their nest and a few clusters of aphids sheltered by soil canopies are vigorously defended but between there is only a vague system of trackways. This simple collection of ant species is a community in that they share a habitat, and divide it into spheres of influence which only overlap a little in space/time. Their mutual avoidance when resources are plentiful can turn to animosity when they are scarce. Then, the dominance order can become a food chain. Mutual benefit may arise from co-operative defence against vertebrates. Five degrees latitude further south (Dorset) and still only 20 m above sea level and in base-poor soil the plants, still predominantly ericaceous shrubs, are interspersed with gorse (*Ulex minor*). *U. minor* which probably fixes scarce nitrogen with its root nodules contributes seeds, of which *Lasius alienus* (*La*) eats the caruncle, and carries aphids (*Aphis ulicis*) on recumbent shoots which *La* attends. *Lasius niger* (*Ln*) lives in the sheltered, cool, wet valley bottoms and in scrubby areas which are comparatively shady and has colonies and territories that range over many square metres. The interrelationships of this pair of species have already been discussed. A third species dominates the core of the community namely *Tetramorium caespitum* (*Tc*, Myrmicinae). This comes from the south and take the warmest zones where it is highly organized on a territorial basis using underground tunnels that hold ants always available for contingencies like invasion or discovery of food (Fig. 16.1). Moreover, it exploits the seed resources of the habitat using not only those of the ericaceous plants but also those of various grasses, which it collects both off the ground and from the capsules in late summer and eats in spring after winter storage. It also hunts small arthropods and cultures aphids and, in fact, has many 'tentacles' ramifying deeply into the heath biome. So the core of the habitat consists of the advanced myrmicine and the two species of *Lasius*. The genera present in the north are still found but on the fringe of the community: *Formica lemani* is replaced by a rather similar southern variant, *F. fusca* which is not dominant, and in fact lives in association with *Tc*, nesting in its territories but avoiding contact by having a small, single entrance to its nests and foraging further afield and higher up shrubs; they may benefit each other. *Myrmica scabrinodis* is replaced in dry heath by *M. sabuleti*, which needs more warmth and can live in neutral as well as acid grassland. *M. ruginodis* is replaced by *M. rubra* in most grass scrub that is neutral but survives in scrub and woodland and in bog. In the neutral grassland *M. rubra* meets *Lasius flavus* but occupies a slightly different sub-habitat, shadier, with flat wall stones for nests, and feeds at least in part,

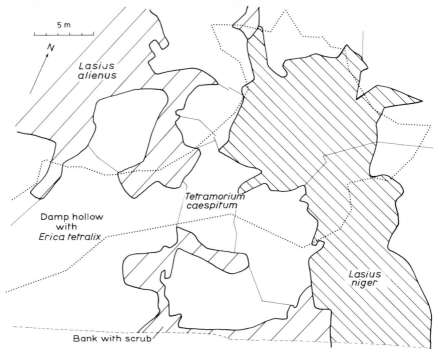

Fig. 16.1 Dorset heath community; thick lines divide different species, fine lines different colonies. *Lasius niger* lives in the shade of the bank and in a damp hollow (dotted line), *L. alienus* on higher ground to the west and *Tetramorium caespitum* in between.

on aphids on herbs. Where sunshine creates enough warmth the *Lf* can drive *M. rubra* away and cover its stone with soil (Elmes, 1974). The wood ant is again sporadic but represented by the southerly *F. rufa* rather than the northerly *F. lugubris* or *F. acquilonia*.

The establishment of woodland is stopped by human fire-raising. Only a small proportion of the ants are killed directly by the heatwave that passes down from the soil surface but the food shortage that follows can be devastating. The soil-living ants then eat arthropods and collect the small animals that fall to earth during dispersal flights but the expansive search for food soon leads to contact with neighbouring colonies and species. *Lasius alienus* expands temporarily at the expense of *Tetramorium* but as the plants return and regenerate so does the ant and in terms of the fire regeneration sequence these two species see-saw and never establish a steady state. Principal component analysis shows that in each sub-habitat one species is a formicine and the other a myrmicine (Brian *et al.*, 1976). Evidently both subfamilies have radiated into all the habitat types.

The ants on Ramsey island, have already been discussed as species of

Lasius and *Myrmica*. An important difference between the genera is that the nest densities, highest near the coast, differ in degree of aggregation: *Myrmica* are less clumped than *Lasius* and of the three *Lasius* species, *L. flavus* is least clumped. In spite of this aggregation the species density distribution is not far off random with most quadrats containing either no or one species and with four or five being rare. This could be due to two opposing tendencies: both to collect in mutually favoured spots and yet to restrict each other's density by competition (Doncaster, 1981). The Ramsey community represents an intermediate state between those of west Scotland and southern England for though *F. lemani* is replaced by *F. fusca*, *M. ruginodis* (*Mr*) and *M. scabrinodis* (*Ms*) are present as well as *M. sabuleti* (*Msi*). *Tetramorium caespitum* (*Tc*) occurs in only two of the most sheltered coastal situations but is nowhere extensive enough to dominate the ant fauna generally as it does in southern England. The same can be said of *Lasius niger* (*Ln*) which with *L. flavus* is predominant in an area of grass (*Festuca rubra*) sheltered from gales; it also lives by stone walls. The *Myrmica* are, however, on the wet side of the principal component for 'drainage', living in moist, flat old fields and seemingly indifferent to the first component representing 'exposure' which separates *Ln* and *Lf* (sheltered) from *La* (exposed). So, if one looks for a dominance order one finds only three local ones: *Tc* heading one system, *Ln* another, and *La* a third. *Lf* especially high up in areas with bracken, establishes high density populations and excludes other species through superior use of resources for it is not an aggressive ant. Where it has a low biomass the other species co-exist, in higher soil strata. The adaptable species, *Lf*, *Msi* and *Ms* are widely spread whilst the specialized dominants *Tc*, *Ln* and *La* are restricted and aggregated. The genera *Myrmica* and *Lasius* both contribute species to all habitats. *Myrmica* more often occurs in quadrats with *Lasius* than not, and one-third of quadrats of *La* also contain a *Myrmica*. *Myrmica* are able to co-exist with *Lasius* and *Formica* by occupying moister intersticial areas as in Dorset. *Formica* may be restricted by lack of deep soil which it needs for nesting and which encourages shrubs for them to forage in. Thus the Strathclyde and Dorset communities blend in Ramsey.

In the high altitude grasslands of the Appenines on calcareous rocks at 1700 m and 42° N, Baroni Urbani (1969) has also studied *T. caespitum* (*Tc*) and *L. alienus* (*La*). Here, above the tree line the main grasses are *Brachypodium pinnatum* and *Festuca ovina* each with bare patches between them and the main ants are *La* 55%, *Tc* 36% with *Formica fusca* (*Ff*), *F. rufibarbis* and *Myrmica sulcinodis*. *La* is more abundant in the barer, *Festuca* soil but no other zonation is apparent and nests of *La* and *Tc* are distributed randomly. Baroni Urbani could find no sign of the territorial habit so clear in Dorset but he noticed a tendency for *La* to forage alone by night in August. Normally both start at sunrise increase to midday then rest until evening when peaks again coincide. However, *La* is more periodic than *Tc*. Whether the non-territoriality of *Tc* in the Appenines is due to climate or to the grassland habitat is not established; making the necessary horizontal tunnels in grass turf may be too difficult.

Ants have colonized mires. In southern English heaths, waterlogged areas are inhabited mainly by *Lasius niger*, *Myrmica ruginodis* and *M. scabrinodis* but locally *Formica transkaucasica* (*F. picea*) makes its nests in grass tussocks and other islands.

Lasius niger is undoubtedly adapted to waterlogging. In Hungary, (47°N) it lives in grooves in sandy soil dissected by the wind whilst *L. alienus* lives on the ridges in between (Gallé, 1980). In the sandy coastal plain between the North Sea and the Wadden Zee, *L. niger* pioneers places that become water-logged in winter especially where only a few grasses (mainly *Festuca rubra*) are established and the other vegetation is sparse. *Myrmica rubra*, but not its northern sib *M. ruginodis*, is zoned lower on dunes and limited to *Festuca* turf and *Linum catharticum* that is sandy but not too salty. *M. scabrinodis* is rare in the dunes but occurs in small hillocks between them that are vegetated with *Festuca* turf where there is more silt and salt in the soil and halophytes occur. *Lasius flavus* is absent, probably due to lack of vegetation and a fine sparse soil yet very common nearby (Boomsma and De Vries 1980). *L. flavus* can survive salt water at 5°C for days (Nielsen, 1977), and in southern Michigan, Talbot (1965) found that it occurred along the borders of water, nesting near the surface, never deep, and showing no mound of excavated soil. Other evidence that *L. flavus* can live on the borders of salt marsh and develop a *Festuca rubra* vegetation grazed by rabbits, on top of its nest comes from the work of Woodell (1974) in eastern England. He found that *Frankenia laevis* normally a Mediterranean shrub, was zoned on the south side of these soil mounds. The ant brings up more middle size grains from subsoil levels and makes a very porous nest structure.

16.1.2 STEPPE AND SCRUB

The grassland communities of the British Isles are poor in *Formica* but in the steppeland of Western Siberia, 55° N, communities dominated by *F. pratensis* (*Fp*) and *F. uralensis* (*Fu*) occur (Stebaev and Reznikova, 1972, 1974). The large territories of these two main species allow smaller ones in between, e.g. *F. subpilosa* (*Fs*) and *F. transkaucasica* (*Ft*). Each day the territory is ex-panded and then contracted in the late evening and *Fs*, a less industrious ant than *Ft*, tends to forage only when *Fp* have retired for the night (at 22.00 h) whereas alone they peak at 18.00 h. The interstitial *Ft* is a temporary host to *Fu* queens (Collingwood, 1979); and aids *Fu* in foraging for hidden meat baits but not for exposed baits. Clearly the *Fu* must have formed the habit of checking what *Ft* are doing or of responding to their trail and recruitment pheromones. In fact, *Fu* and *Ft* are not hostile to each other; presumably the latter gains an amount of protection from *Fu* in exchange for launching its queens and finding its food: symbiosis (mutualism) rather than parabiosis (sharing resources).

In western Siberia where birch groves and open steppe mix, another community comprises *Formica pratensis* (*Fp*) and *F. cunicularia* (*Fc*), a species close to *F. fusca*, as well as the widespread *T. caespitum* (*Tc*), *Lasius alienus* (*La*) and *M. scabrinodis* (*Ms*) (Reznikova and Kulikov, 1978). They feed very largely as predators and scavengers on meadow and herb-eating invertebrates, e.g. grasshoppers, crickets, flies, leaf-hoppers and other bugs but above all on other ants in a seasonal succession in which the young forms predominate. *Ms*, in particular, collects 70% dry remains of which 40% are other ants, especially *Formica*. However, 80% of *Formica* food is fresh invertebrates in the juvenile stage, the variety taken by *Fp* being greater than by *Fc* which eats small inhabitants of the grass layer. Its food territories may include 4–7 colonies of *Fc* as well as 10–12 of *Ms*. *Fp* operates in groups to attack living crickets and grasshoppers which *Fc* can find but cannot deal with, and Reznikova (1971, 1975) has shown experimentally that *Fc* increases the effectiveness of *Fp* foraging by finding these big insects for it. Possibly *Fp* robs *Fc* of many of its larger prey for if *Fp* is fenced out it collects more bigger items of food for itself. Yasuno (1964a, b, 1965a–c) points out that in Japan species with big workers and big colonies are independently distributed from those with small workers and small colonies. The latter groups comprise *Tetramorium caespitum*, *Pheidole fervida* and *Paratrechina flaviceps*. As all the big species belong to the subfamily Formicinae and two out of three of the small belong to the subfamily Myrmicinae there is a distinct resemblance to the pattern of co-existence between subfamilies in subhabitats in the grass-heath of Dorset.

In an ant community in a Louisiana pasture (31° N) Baroni Urbani and Kannowski (1974) found that mounds and territories of *Solenopsis invicta* (*Si*) predominate but *Monomorium minimum* (*Mm*) and *Paratrechina arenivaga* are also common, whilst *Cyphomyrmex rimosus* is sparsely but independently distributed: *Solenopsis xyloni* (*Sx*) is restricted to the shade of trees. *Mm*, many of which inhabit *Si* mounds, interfere with *Si* using a repellent spray that forces the *Si* workers to clean themselves instead of recruiting help; this tactic works even when the *Mm* do not inhabit the *Si* mound. With *Sx*, *Mm* is less successful; they build up a population but are then displaced by workers of *Sx* who crush them in their jaws; *Mm* is thus at a particular disadvantage in shade. In open daylight even though it largely avoids *Si*, which is more active at night, it has its aerosol repellant to compensate for its small size.

A little further north (39° N) in the humid climate of Maryland, deciduous woodland (*Quercus, Fagus, Carya*) with a rich understorey contains 24 ant species; the four main ones are *Aphaenogaster rudis* (*Ar*, Myrmicinae) with 27%, *Prenolepis imparis* (*Pi*, Formicinae) with 23%, *Paratrechina melanderi*

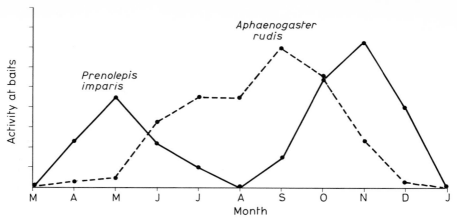

Fig. 16.2 The complementarity of two ant species in seasonal foraging activity on the hardwood forest floor in Maryland. (After Lynch *et al.*, 1980.)

(*Pm*, Formicinae) with 20% and *Camponotus ferrugineus* (*Cf*, Formicinae) with 6%. Lynch *et al.* (1980) found that *Pi* lives in big, fixed colonies deep in permanent nests and that activity above ground peaks in May and again in November; it avoids 26°C and over but is effective as low as 10°C (Fig. 16.2). During July–October, *Ar* and *Pm*, in contrast, are active on the surface by day whereas *Cf* is active at night. Thus time zonation of foraging keeps three of the species apart and leaves only *Ar* and *Pm* to overlap. Of this pair, *Pm* is a small individual living in small colonies, and though quick to find food, needs to recruit help to collect it. *Pi* is much the most skilled recruiter. Food sources are, of course, best defended by *Cf* which can kill *Pi* but *Pi* is very aggressive and comes next in the dominance order. Only contests between *Cf* and *Pi* cause severe damage; others merely involve chasing, biting and hanging on. *Pi* may occupy half a grid of baits within 0.25 h and then slowly extend to the others; if given crickets 66% go to *Pi*, 26% to *Ar* and only 4% to *Pm*. *Pi*, are unable to stop *Ar* getting some food in diffuse sets of small baits but can monopolize large patchy ones. In brief, the dominant ant (*Cf*) is reclusive and forages at night; the next (*Pi*) avoids the heat of summer and lives deep in the soil leaving the surface in summer to *Ar*, a eurytopic, eclectic, versatile myrmicine with a formicine satellite (*Pm*).

This pattern is world-wide. In S. Australian communities *Iridomyrmex* species are diurnal and dominant and space themselves out but leave gaps between territories in which other species of genera like *Camponotus* and *Melophorus* can live (Greenslade, 1979); the former genus are large, evasive, versatile and may feed at night whilst the latter forage in great heat (Fig. 16.3). *Iridomyrmex purpureus* (the meat ant) lives only in open places in S. Australia; it needs sunshine and cannot walk well on vegetation. It makes a

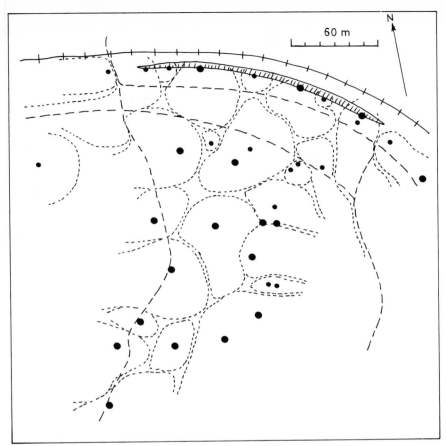

Fig. 16.3 Colonies of *Iridomyrmex purpureus* bounded by broken lines, nests as black circles. (After Greenslade, 1975a.)

stiff legged walk on its territory boundaries and avoids lethal conflicts (Greenslade, 1975); in this, it resembles *Myrmecocystus mimicus* (Fig. 16.4, Hölldobler, 1976a).

Dulosis (slave-making) in ants is only known in three tribes: the Tetramoriini and Leptothoracini (Myrmicinae) and the Formicini (Formicinae) (Büschinger *et al.*, 1980). It is only in the north temperate grass–scrub communities with dense slave populations that this intense interdependence has evolved: 2% of species are social parasites and 17.5% are dulotic. All are monogyne and use monogyne species as hosts though one, in fact, *Polyergus lucidus*, can be polygyne for a while but its surplus queens go on raids and stay with a few workers to start a new nest. Strong pointed jaws are used by *Polyergus* and *Strongylognathus* to pierce the heads of resistant slaves. There

Fig. 16.4 *Myrmecocystus mimicus* workers in stilt-walking and head-on confrontation display at territory boundary. (After Hölldobler, 1976a.)

are many hypothetical stages by which dulosis can be achieved: first the queens may go after nuptials into nests of a closely related species; frequently this social parasitism is only temporary and then the invader builds a colony of her own. But if her own workers are specialized fighters she will have to 'send' them out to raid local 'host' colonies for pupae.

F. *sanguinea* (*Fsa*) appears to dominate the grass–scrub forest habitats of Eurasia. It is aggressive towards F. *rufa* (*Fr*) and F. *pratensis* (*Fp*) taking pupae first and later destroying the workers, but some pupae taken back to the *Fsa* nest emerge and are integrated and even refuse their own queens. F. *fusca* (*Ff*) and F. *rufibarbis* (*Frs*) whose workers show less hostility to *Fsa*, may escape destruction. Aggression towards L. *niger*, L. *flavus* and C. *herculeanus* also occurs and may lead to their elimination but the *Fsa* can still survive independently, feeding on honey-dew and insects like wood-ants (Marikowsky, 1963, cited Brian, 1965). The more specialized social parasite and slave maker, *Polyergus rufescens* (*Pr*) raids F. *fusca* (*Ff*) and F. *cinerea* (*Fca*). The conduct of a raid depends on differences in reaction between the hosts (Dobrzańska, 1978): *Ff*, whose nest entrances are concealed, carries its juveniles out and ascends plants to escape; *Fca*, which is polycalic with worker interchange between nests, builds barricades. These are effective with *Fsa* but *Pr* attacks so suddenly and overwhelmingly that time does not allow defence. The slaves at the receiving end react differently during a raid too: *Ff* enlarges its entrance holes and reduces the traffic jam that returning *Pr* create, but *Fca* simply picks up the pupae as the *Pr* drop them outside; it may even take them off *Pr* forcibly. Part of this behaviour is learnt and comes gradually to perfection. *Pr* with F. *rufibarbis* slaves begins activity earlier in the season. The evidence is against subdivision of the *Pr* into races with different 'prey';

in fact Dobrzańska records a few cases of one colony of *Pr* having a species-mixed slave population.

Spreading polygyne forms of *F. rufa* displace other species as they extend and may even eliminate *F. fusca*, the host of monogyne queens. Some ants make use of wood ants: *Formicoxenus nitidulus* (Myrmicinae) nests in the mounds and *Diplorhoptrum fugax* predates the brood, whilst *Leptothorax* inhabit the mound cortex. The spreading *Formica truncorum yessensis* on Hokkaido, probably excludes many smaller species of *Lasius, Aphaenogaster* and *Myrmica*, but *F. japonica* is agile and lives in open spots out of the reach of *F. yessensis* whilst *Diplorhoptrum* and *Leptothorax* both co-exist with *Fy* (Higashi and Yamauchi, 1979). The spreading population of *F. lugubris* in the Jura mountains does not destroy the myrmicines, *Tetramorium caespitum, Leptothorax acervorum, Manica rubida*, and three species of *Myrmica* (Cherix, 1980). *F. opaciventris* (Fo of the *F. exsecta* group) lives only in western North America and Scherba (1964), in Wyoming at 2065 m in a sagebrush meadow surrounded by poplars, willows and pines, notes that its nests are spaced some 6 m apart: it appears to be keeping *F. fusca*, which is restricted to open spots in the wood, out of the meadow. Where they meet *Fo* lays siege to *Ff* nests, kills their workers gradually and removes brood but it can only do this in meadows; in woodland *Ff* can resist effectively often by leaving their nest temporarily.

A final example comes from Japan at 41° N up to 600 m altitude where Yasuno (1964a, b, 1965a–c) analysed the community in grassland surrounded by beech forest. *Formica exsecta* dominates some areas and *F. truncorum* (*Ftm*) others. The latter (*Ftm*) is settled in warm banks on the forest margin and if removed is replaced by *F. fusca* (*Ff*) from open grassland or *Camponotus herculeanus* (*Ch*) from grass scrub. *Ftm* is polycalic and breeds new colonies but may use *Ff* as a temporary host as well.

Polyergus rufescens (*Pr*) avoids *Ftm* and even leaves *Ff* nests in *Ftm* territory alone, which means that in effect *Ftm* protects *Ff* where it can. Immediately round its nest *Pr* eradicates *Ff* but this is a very adaptable species, and quickly returns if *Pr* is removed. Thus *Ff* is basic in the ant community; it is protected by the dominant (*Ftm*) for use as a temporary host and used by a subordinate (*Pr*) as a host and a slave. *Ftm* also keeps *M. ruginodis* inside the cool interior of the wood.

Slave-makers subjugate the host queen by preventing sexualization of her brood (e.g. *Strongylognathus karawegi*): presumably all the male eggs she lays are destroyed and the female eggs are suppressed if larvae show signs of growing gynewards. Such a colony can live as long as the host queen. If the host queen is destroyed, it becomes necessary if the colony is to last longer than the host workers live, to collect replacement workers (as pupae) from different nests from time to time. Dulosis thus evolves from competition, or inefficient predation in which the pupae collected are not all consumed.

Emerging workers are then able to learn to tend what they find in their immediate environment. Only species exhibiting early learning can become slaves; those with wholly innate behaviour (e.g. *Lasius niger*) are immune (Le Moli, 1980).

Dulosis leads to the rapid decay of normal independent behaviour by the slave-maker. In *Leptothorax duloticus* of North America the ability to gather and prepare solid food has gone altogether though it can still build nests and care for brood. The removal of its slaves (*L. curvispinosus*) actually increases its brood care but not enough to survive (Wilson, 1975). *L. duloticus*, which collects larvae and pupae in a raid, has certain minor morphological traits that assist in carrying big objects (also seen in dulotic *Formica*). Wilson points to behaviour that may be advantageous for the slaves; thus their workers lay and add eggs to the cluster of the *L. duloticus* queen. Although some are eaten, the slaves do, perhaps, produce a few males which will, presumably, copulate with gynes from free nests and disperse genes for slavery. They would thus gain protection from the slave-makers in exchange for nourishing their young: a mutualism. This mutualism is reminiscent of 'parabiosis' in which the search capability of foragers is exploited in exchange for protection.

This cursory review of the ant communities in temperate areas of the world shows that a general design exists based on an interspecies dominance hierarchy. The dominant holds the hot nutritious zones through individual strength and ferocity, organization and economic strength but leaves plenty of suboptimal space around in which species with adaptation to lower temperature or with special skills in feeding or nest-building or just with evasiveness, can make a living. In stable areas polygyny evolves leading to social parasitism and dulosis but this need not be one-sided if the slave can reproduce parthenogenetically, as it gains protection from its master species.

16.2   Desert communities

Termites are prominent members of arid ecosystems (Table 16.1).

16.2.1   TERMITES

The distribution of termites in arid zones of Africa and Arabia has been reviewed recently by Johnson and Wood (1979). The genus *Psammotermes* (four species) of the family Rhinotermitidae is restricted to arid areas. It lives in Sahara oases and desert fringes nesting very deep, down to 5 m, to avoid fluctuating temperatures, drifting sand and desiccation. It will collect water from 40-m deep water tables to moisten the walls of its nest. It forages, beneath protective sheeting, on a wide variety of organic food: dung, fossil soils, litter, and wood during moister periods. In the Santa Rita mountains of

Arizona at 950 m, five hypogaeic species feed on different foods in different ways (Haverty and Nutting, 1975, Haverty et al., 1975). One, *Heterotermes aureus* (Rhinotermitidae) eats dead wood on the surface (56% *Opuntia*, 17% *Acacia*, 13% *Prosopsis*) whilst other termites prefer *Acacia* to *Opuntia* and eat *Cercidium* or grasses. In South Australia, Lee and Wood (1971) list nine underground species in 'mallee' with 250 mm of rain a year. In African savannah with the tussock grass *Loudetia*, Bodot (1966) found 20 species, mostly underground, but some were mound builders. Again, their food is different; many cut dry grass and store it (*Trinervitermes*) whilst others eat grass roots (*Amitermes evuncifer*). There are also humus eaters, like *Cubitermes severus* and two species of *Trinervitermes* which eat a similar range of grass but exploit it in different ways (Bodot, 1966). Bodot also describes how soil type, moisture and plant cover affect termite distribution in the Ivory Coast savannah. In this area *Macrotermes natalensis* is disappearing gradually perhaps due to encroaching human activity perhaps due to a climatic trend or both. *Amitermes evuncifer* settles in the old, decomposing mounds but also invades the mounds of *Cubitermes severus* and species of *Trinervitermes* which it converts into its own design and structure (Bodot, 1966). Sexuals apparently drop and tear their wings off near these mounds, especially those of *Trinervitermes geminatus* and then enter gradually (Josens, 1972).

Josens found that at Lamto (Ivory Coast) the Macrotermitinae probably compete for food but not for nest space, since they are mostly diffusive hypogaeic nesters; those that make mounds may be limited by interaction, for the grasses on which they feed are abundant and in this area 90% of epigaeic termites feed on grass. Josens noted 26 species of which 11 were very rare. The remaining 15 could be classified ecologically into: epigaeic harvesters genus *Trinervitermes* (four) and a wood eater, *Amitermes evuncifer*; hypogaeic fungus cultivators, *Ancistrotermes cavithorax* (*Ac*), *Microtermes toumediensis* (*Mt*), *Odontotermes* (*O*, of unknown species) and *Pseudacanthotermes militaris* (*Pm*), one wood-eater without a fungus, *Microcerotermes parvulus*, and five humivorous species (Josens, 1972). The four Macrotermitinae were not identifiable as colonies, indeed there is only indirect evidence that colonies exist and they may have fused, but Josens sampled the numbers of comb chambers in an average hectare, and got 21 500 of *Ac* and 24 400 of *Mt*, 3300 of *O* and 8000 of *Pm*. Details of the composition of these populations, their biomass and energy conversion were also obtained (Figs. 16.5 and 16.6). After the annual fire, vegetable refuse is sparse even though the bushes and trees drop their leaves; then they must compete even though they have diffuse nests, are stratified and feed differently. Moreover, on a grid of 9920 stick-bundle baits only 40 were used by two species (*Ac* and *Mt*) simultaneously. Since *Ac* uses 2485 alone and *Mt* uses 2059 alone, if they are indifferent to each other there should be 511 joint

Table 16.1 Abundance and live weight biomass of termites in different ecosystems. (From Wood and Sands, 1978.)

Ecosystem	Species or group of species	Type of nest	No./m²	g/m²	Reference
Warm temperate woodland					
Sclerophyll forest, South Australia	Nasutitermes exitiosus	Mound	600	3.0	Lea & Wood, 1971b
Semi-arid Grassland, North America	Gnathamitermes tubiformans	Subterranean (1) Mean over three years	0–9127 2139	0–22.21 5.2	Bodine & Ueckert, 1975
Shrub-grassland, North America	Heterotermes aureus	Subterranean (1)	431		Haverty, Nutting & Fage, 1975
Sahel savannah, West Africa	All species	Subterranean+mounds (2)	229	1.0	Lapage, 1974b
Tropical savannah					
Savannah woodland, North Australia	All species	Subterranean (3)	2000		Lee & Wood, 1971b
Grass savannah, Central Africa	Cubitermes exiguus Apicotermes 'gurguliflex'	Mound Subterranean	612–701 70	1.3–1.9	Hébrant, in Bouillon, 1970 Bouillon, 1962, 1964
N. Guinea savannah, West Africa	Trinervitermes geminatus	Mound	110–2860		Sands, 1965a, b

Location	Species	Sampling type			Reference
S. Guinea savannah, West Africa	All species	Subterranean+Mound (4)	4402	11.1	Wood et al., 1977 and unpublished
Secondary S.G. savannah, West Africa	All species	Subterranean (5)	2966	3.6	Wood et al., 1977
'Derived' savannah, West Africa	All species	Subterranean+Mound (6)	861	1.7	Josens, 1972[†]
Tropical forest					
Semi-deciduous forest, West Africa	All species	Subterranean (5)	3163	8.0	Wood & Johnson, unpub.
Riverine forest, Central Africa	All species	Subterranean+mound	1000	11.0	Maldague, 1964
Rainforest, West Indies	All species	Subterranean (7)	4450		Strickland, 1944
Rainforest, South America	Nasutitermes costalis	Mound	87–104	0.1	Wiegert, 1970
Rainforest, Malaysia	Four species	Mound	1330	3.4	Masumoto, 1976
Agro-ecosystems					
Grazed pasture, West Africa	All species	Subterranean (5)	2010	2.8	Wood et al., 1977
Maize (first year), West Africa	All species	Subterranean (5)	1553	1.7	Wood et al., 1977
Maize (8–24 years), West Africa	All species	Subterranean (5)	6825	18.9	Wood et al., 1977

Depth of soil samples: (1) 30 cm, (2) 75 cm, (3) 8 cm, (4) 200 cm, (5) 100 cm, (6) 50–60 cm, (7) 7.5 cm.
Sampling methods for subterranean species: hand-excavated pits, Josens, 1972; Lepage, 1974b; core samples used in all other studies.
*For full references see Wood and Sands (1978).
[†]Mean of facies D1 (open savannah woodland) and D2 (moderately open savannah woodland).

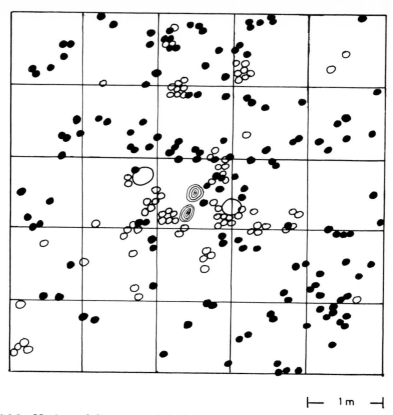

├─── 1 m ───┤

Fig. 16.5 Horizontal dispersion of the fungus combs of two termites: (a) *Ancistro-termes cavithorax* white circles (b) *Microtermes toumediensis* black circles. Two tree bases are present in the centre (concentric rings). Two reproductive chambers of *A. cavithorax* are present and probably represent two separate colonies. (After Josens, 1972.)

ones. Where both species do use the same bait they make an earthen screen to separate each other.

Whilst examples and descriptions of termite communities are numerous (see Krishna and Weesner, 1970) actual tests of termite interaction are few. Different genera, at least, normally avoid each other according to Noirot (1959) and a good example of interaction has been given by Bouillon (1970) in which a mound nest of *Macrotermes natalensis* is clearly preventing *Cubitermes sankurensis* from nesting within 5–10 m. Competition of this sort does not necessitate combat, but results from a superior exploitation of food that is more effective near the nest base. *Hodotermes mossambicus* and *Trinervitermes trinervoides* both eat grass in the same locality and normally avoid each other, but will fight if forcibly mixed (Nel, 1968). Species of

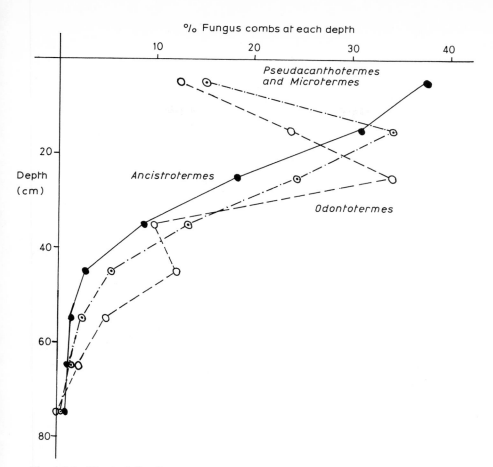

Fig. 16.6 Vertical distribution (as % of total) for four species of termite; *Ancistrotermes cavithorax*, *Odontotermes* species with *Pseudacanthotermes militaris* and *Microtermes toumediensis* together. (After Josens, 1972.)

Coptotermes live in different logs in South Australia (Calaby and Gay, 1956, cited Lee and Wood, 1971). *Nasutitermes exitiosus* attains densities of 8.6–9.2 mounds/ha alone but only 2.4/ha with *Coptotermes lacteus* present. They are dispersed over the area and three genera of grass eaters also show dispersive tendencies. The mounds of *Nasutitermes triodei* are more often abandoned in the presence of *Hemitermes laurensis* both of which feed on grass; the former eats dung and plant litter as well but this evidently does not save it (Lee and Wood, 1971).

Similar communities of partly epigaeic termites were studied by Josens (1972) at Lamto. He found 40.5/ha of *Trinervitermes geminatus* (60% of nests and 80% of the total termite biomass) 14.7/ha of *Trinervitermes togoensis*, 3.2/ha of *Amitermes evuncifer* and 2.6/ha of *T. oeconomus* with smaller amounts of two other species of *Trinervitermes*. In the grassy savannah (*Loudetia* mainly) there are hillocks that arise from old, crumbling *Macrotermes* mounds and on these *T. togoensis* establishes itself and is able to live in wet grassy places by finding these better drained islands. *T. geminatus* lives on more open, but bushy places, but *T. oeconomus* peaks in very dense bushy savannah. The nests of the hypogaeic Macrotermitinae are so diffuse that Josens does not think they compete except for food, but in the case of these epigaeic species, of which more than 9 out of 10 are harvesters, his calculations indicate that of 4–5.5 tonnes total production (dry)/ha only a few tens of kilos (dry) are eaten.

Sand's (1965) study of *Trinervitermes* in Nigerian savannah was orientated more towards a comparison of congeneric species than community study yet it is interesting to recall that he found *Trinervitermes* nests were commonly shared with other termites or with ants or both. Normally there seems to be no contact between the different species but if a mound is opened up, there is 'immediate and indiscriminate conflict'. One species of *Microtermes* lives in nests of *Trinervitermes trinervius*, another in nests of *Macrotermes natalensis*, but both are ignored by their hosts. Since there is a great discrepancy in size this seems to be a typical case of parabiosis.

In a study of termite interrelations in the desert near Tucson (Arizona) paper baits were put down in a grid on or under the surface. *Heterotermes aureus* made narrow spotty galleries into 53% of them and *Gnathotermes perplexus*, which used 96% of the baits, made bigger, cleaner galleries; in shared baits they avoided each other's galleries (LaFage and Nutting, 1978).

Inquilinism is common in termites. The guest is often much smaller than the host and quieter and may either be of a related or a very different species (Bouillon, 1970). Sometimes this parabiosis is obligatory and both species can use the same galleries and have sexual pairs quite close together or they can be sealed off by the host. Josens (1972) shows a chamber of the termite *Ancistrotermes cavithorax*, with galleries of a much smaller humivorous termite around it and in its walls, he also shows such galleries in the walls of the reproductive chamber of *Pseudacanthotermes militaris*. Apart from these obvious cases of one species living under the umbrella of another, the larger mound builders loosen the soil and help other species to enter. In South America *Comitermes cumulans* makes nests in grassland that can be 4 m high. When abandoned these are used by social wasps, ants, reptiles even, as well as by other termites. Sands has, in fact, shown experimentally that some species select soil that has been in another termite mound in preference to normal soil. Not only is soil broken up by termites, but wood can be broken up to provide humus for humivorous termites, creating a succession from wood to humus feeders.

Ants frequently prey on termites; *Camponotus*, *Crematogaster* and *Pheidole* eat *Trinervitermes*. *Megaponera foetens* not only predates many

Macrotermitinae but nests in their old mounds. At Mokwa (9° N) Longhurst and Howse (1979a, b) found that of 83 nests examined 37 were in deserted *Macrotermes bellicosus* mounds, 3 in *Trinervitermes* mounds, 36 in the ground under bushes and 7 in open ground. Nest sharing with *Trinervitermes geminatus* varies with the habitat and increases where old mounds are common. In parabiosis, ants normally avoid termites and certainly do not attack them as long as their nest is intact. Many doryline ants predate termites whose soldiers either resist them step by step as they penetrate the nest or move off in columns overground in a body until the attack ceases. *Anomma kohli* may take several days to work its way into a large *Cubitermes* mound. *Macrotermes bellicosus* mounds at Mokwa which are abundant and randomly distributed in savannah show a very high death rate due to predation, and Collins (1981a) takes the view that their density is more likely to be controlled by predation in the early stages than by intraspecific competition.

In a hectare of soil at Lamto one may find 1.6 million foraging or harvesting termites along with 4.5 million humivorous termites and 5 million fungus-growing termites (Josens, 1972). The foragers or harvesters consume 30–50 kg dry grass a year, the humivores consume about 30 kg cellulose a year and remove at least 15 tonnes of soil a year, whilst the fungus growers incorporate about 1.3 tonnes of litter (dry) a year.

16.2.2 ANTS

Ants, to survive in deserts, must be catholic feeders and include direct plant products in their diet, e.g. seeds and leaves. In the Sahara 55 out of 97 species exploit underground water sources (Délye, 1968), and compared with similar European species have more waterproof cuticles, and actively avoid extremes of heat and aridity. They nest wherever plants occur, as around oases, and collect and store seeds during the short moist season, insects are also eaten when available. A transect from coast to mountains across Israeli desert measured soil and climate as well as species of ants (totalling 49, Ofer *et al.*, 1978). These habitat types could be separated statistically in order of warmth, coastal desert, hill mountain, northern spring, and it emerged that there is a characteristic species in each zone and a distinct tendency for congeneric species to live in different habitats. Even *Lasius alienus* was present in the coolest areas! A survey in the Chiricahua Mountains of Arizona showed that a cool, moist forest in the north, changed to a dry, scrub with succulents in the south (Eastlake *et al.* 1980). Myrmicines were common in the north-west and rare in the south-east whereas dolichoderines were rare in the north, common in the south-west and abundant in the south-east. Of seven species of seed-eaters, five were *Pheidole* (all small ants), but on the north slope three common seed-eaters differed substantially in size and probably ate different seeds.

The correlation between ant size and seed size has been established in a transect of the Mojave and Sonoran Deserts (California and New Mexico)

chosen to represent a rainfall trend (Davidson, 1977). In California with <
100 mm rain/annum and 29–34° C in July (mean) three species were found
but in New Mexico with 225–275 mm rain and 26–27°C, eight species were
found. No species ranged throughout but some were present in six out of ten
stations. In California, *Veromessor pergandei* is common along with two
species of *Pogonomyrmex* whereas in New Mexico *Novomessor cockerelli*,
Solenopsis xyloni, *Pogonomyrmex negosus* and *Pheidole desertorum* are
predominant. Davidson found that there are not only more species in moist,
productive habitats but that colonies are more closely packed than in drier
habitats. The additional species are mainly ones with small workers (< 3 mm
long) which forage only 3–4 m from their nests but there are more species
with large (> 9 mm) workers too, e.g. *N. cockerelli*, *P. rugosus* and *P.
barbatus* with foraging ranges up to 40 m (as in Hölldobler, 1976a).
Veromessor pergandei ranges from 3.5–8.4 mm in a total range of 1.8–9.8
mm but intense competition from other species restricts it size range.

All this suggests that food limitation is more important than nest-site
limitation but Davidson could not find much evidence in these seed-eaters of
either zonation or stratification or other habit partitioning. Although seed-
size specialization helps species assortment, Davidson (1977) has evidence
that species of different body size only co-exist if they differ in foraging
method too. Species whose workers forage singly exploit dispersed, low-
density seed sources and species whose workers forage in groups exploit
patchy high density resources. So their style of foraging suits the nature of the
food distribution. The group foragers collect strenuously when the weather is
right and store their seeds whilst the others work through under poor conditions. Davidson found that large species can forage both in groups and as
individuals (e.g. *P. rugosus*) but that small species are all group foragers (e.g.
Pheidole).

Quadrats in a series of latitudes from 44° down to 36° N in the Great Basin
Desert and at a series of altitudes in the Mojave Desert, have been surveyed at
35° N up to 1500 m (Bernstein, 1975, 1979a, b; Bernstein and Gobbel, 1979).
Novomessor pergandei occurs only at the four lowest stations 500, 610, 700
and 830 m and *P. rugosus*, only at the three highest, 870, 1150, and 1500
whilst *P. californicus* ranges from 610 to 1500 m, though patchily. In the
lowest altitudes 75 and 500 m there are some six species including *V.
pergandei*, *P. californicus* and *Conomyrma insana*. At 1500 m *V. pergandei*
and *C. insana* are absent but *Pogonomyrmex californicus* is still present with
P. rugosus, *Pheidole xerophila*, and *Iridomyrmex pruinosum*. *P. rugosus* and
P. californicus forage singly, not in groups though unless there is a genotypic
difference between *P. rugosus* populations it must be assumed from Hölldobler's (1976a) observations that they can do so when food distribution
encourages this. *V. pergandei* can vary between group collection where the
seeds are scarce and patchy, to individual collection during seed production

times when they are well distributed.

The temperature range at which ants forage increases up mountains and further north so that foraging tends to overlap more in time. Conversely, in the hotter, drier desert conditions which are unproductive, each species collects at a short season when it is profitable and when the food is abundant. This, of course, differs between species. There was no sign of interference between foragers of different colonies. Over-dispersion of nests in these ranges of altitude and latitude is common amongst those of a single species arising, probably, from interference between foragers, i.e. defence of foraging areas. As one changes from communities in high altitudes and latitudes, to those in low ones, spacing of all colonies of all species changes from the clustered arrangement typical of broken perennial vegetation which produces a zoned and stratified habitat towards a dispersed, thin, more uniform pattern. Nest sites between species are also overdispersed (Byron *et al.*, 1980). In general, the more species there are the less the overlap in foraging areas at whatever latitude or altitude. Bernstein (1979b) has also studied diet. The ants either collect seeds, prey or nectar. With a reduction in altitude in the Great Basin Desert food variety decreases and overlaps more between species. The competition between *N. cockerelli* and *V. pergandei* intensifies in the Sonoran Desert when food is short and the former then subsists more on prey and corpses and less on seeds.

Two Chihauhuan Desert habitats: bajada, an alluvial fan at the entrance to an old lake containing creosote bush (*Larrea tridentata*) and playa, an area of sandy soil around a dry lake with mesquite-yucca vegetation (*Prosopsis-Yucca*) have been investigated by Whitford (1978). In bajada, seed eaters predominate and include five species of *Pogonomyrmex* and four of *Pheidole*; they are largely diurnal and rain stimulates foraging. In playa, general feeders like *Novomessor cockerelli*, *Formica perpilosa*, *Iridomyrmex pruinosum* and *Myrmecocystus* prevail if a variety of insects are available at night. The attine *Trachymyrmex smithi* that collects detritus for its garden and the doryline predator, *Neivamyrmex nigrescens*, range through both sub-habitats, though obviously differences in food, feeding method and diurnal conditions are important. Whitford also found that species coexistence was aided by differences in body size and seasonal activity.

An interesting form of interspecies competition has been brought to light (Möglich and Alpert, 1979). In the Chiricahua mountains of Arizona the dolichoderine ant *Conomyrma bicolor* (*Cb*) picks up small stones and other objects and drops them down the holes of other ants including three species of *Myrmecocystus* (*M*), *Novomessor cockerelli* and *Pogonomyrmex desertorum* with which they compete for aphids on yucca and for termites. This prevents *M* but not *Pogonomyrmex desertorum* from foraging. When the *Cb* find a single nest hole with guards in it they collect stones and drop them in at a rate of one or two a minute; fewer than five are needed to stop foraging. The stones have no mechanical effect if dropped in by the experimenter nor do they seem to carry a pheromone but they constitute a threat of hostility. *Cb* work all day long except for midday when it is too hot so they are able to interfere with *M. mexicanus* by night and two other species by day. When it rains (albeit rarely) *Cb* stop foraging and all the *M* come out. The aggressive behaviour of *Cb* is much greater near its many nest holes. The authors also show that this is a special case of a general reaction to smells for *Cb* will put stones on cotton wool soaked in chemicals. This is a very general habit with ants in fact. *Myrmica* will cover many solutions (but not water) to a degree that declines exponentially with distance from their nest hole (Frith, 1979). The use of small stones and soil for blocking the holes of other ants is also reminiscent of the barricades that *Lasius flavus* build when fighting underground.

16.3 Tropical rain forest

16.3.1 WASPS, BEES AND TERMITES

Continual warmth, a daily rainfall and a steady 12 hours of light build up enormously productive ecosystems in the tropics. Social insects are at their species peak in this rain forest and though social parasitism and slavery have not evolved many transpecific collaborations have. Thus queens of *Mischocyttarus immarginatus* (*Mi*) in relatively dry forest in Costa Rica compete for sites near a nest being newly constructed by *Polybia occidentalis* (*Po*) (Gorton, 1978). The queens actually locked, biting and kicking and once nests had been made spent their time robbing and interfering with each other. Out of 24 *Mi* nests, 22 were near *Po* and perhaps getting some protection from lizards and birds. Ant trees frequently serve as umbrellas too: they collect more than a chance number of wasps and even bird nests (Evans and West-Eberhard, 1973). The Richards (1951) have the record perhaps: eight nests of *Polybia rejecta*, two of different species of *Mischocyttarus*, one of *Protopolybia*, one of *Metapolybia* and an oriole nest. The wasps are docile and retreat to their nests whilst the ants defend the tree against allcomers! Though co-operation is a characteristic of the tropics there is plenty of competition too. This may be quite subtle: 'Africanized' honey-bees (in French Guiana), reduce the time that stingless bees spend on flowers collecting pollen (Roubik, 1978); the total number of foragers is not affected, only the time spent collecting. *Apis mellifera*, introduced some centuries ago into America, has a tongue about the same length as the local *Bombus*, whereas in Eurasia, where *Apis* originated and evolved, many *Bombus* have longer tongues and competition is reduced (Pleasants, 1981; Prŷs-Jones, 1982). Can it be presumed that *Bombus* have had to lengthen their tongues in Eurasia in order to survive? Are there not enough differences between *Apis* and *Bombus*?

Termites in West Malaysia exist in 57 species of 23 genera of six sub-families (Matsumoto and Abe, 1979). One hectare can yield 52 species! The termite biomass runs at 3–4 kg m^{-2} (wet) or 2.2–2.6 g m^{-2} (dry), The majority make either diffuse nests in the soil or towers up to 2 m high, sometimes supported by a tree base (Fig. 16.7). In trees and bushes either dead branches are excavated or carton nests made in and around branches with covered ways to the soil surface. Nest spacing varies from species to species but taken together the epigaeic ones are random. Thirty-seven species feed on dead wood on the forest floor, 24 species on leaf litter, but there are 17 species of humus eaters and two species that feed by scraping lichens from the trunks of trees often 40 m above the soil surface. The two main species of *Macrotermes* have already been discussed. *Longipeditermes longipes* come from the trees above, and guarded by soldiers cut pieces from leaves lying on the ground underneath. *Bulbitermes* nesting 1–9 m above the soil surface also come down covered ways and take bits of decomposing twigs and leaves from the surface. Some hypogaeic humivores come up to the surface, which is clearly a very reactive zone. This indicates that the stratification of nests bears little relation to the feeding stratification since the termites are prepared to walk with soldier escort or under trackways between the nest and the food source.

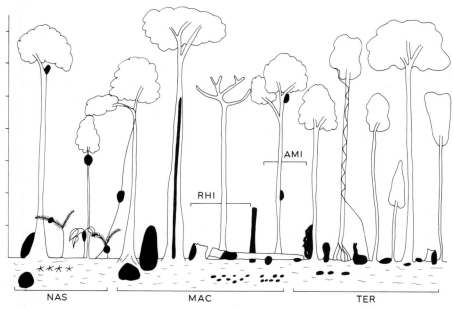

Fig. 16.7 The stratification of termite nests in Pasoh Forest, a West Malaysian tropical rain forest. (After Abe and Matsumoto, 1979.) RHI = Rhinotermitidae, three genera AMI = Amitermitidae, two genera TER = Termitinae, five genera MAC = Macrotermitinae, three genera NAS = Nasutitermitinae, five genera.

In general, one can say that the tropical rain forest is very rich in species, that termites are the main component of the soil microfauna, that they nest in the soil, up the trees and in bushes, that they all eat fibre, often from a different stratum and that more than half have symbiotic fungi. There are 23 species in undisturbed Savannah woodland: 33% are *Microcerotermes*, 26% *Ancistrotermes* and 25% *Microtermes*. The last genus, with dry diffuse nests survives agricultural impact and enters the roots of maize (Wood *et al.*, 1977, 1980).

Termite nests attract many other invertebrates. In an area in the Donala-Eden Forest Reserve 3° N by the shore of a lake, Collins (1981b) found that 52% of nests were built by *Cubitermes fungifaber* and 37% of these were secondarily used, but cohabitation was unusual. Usually, there is no change in the outward appearance and shape but *Noditermes* covers the original with fresh material. Collins (1981b) lists 31 species using the nests of five mound builders.

16.3.2 ANTS IN FOREST

The number of species that inhabit rain forests in the tropics is stupendous. There are 29 genera common to all tropical land areas. Moreover the neotropics has 65 endemic genera and Africa which is less isolated, only 31. In the Australia–orient region some genera (there are 42 endemics) are slightly more primitive than the African ones suggesting that the African genera are younger. The species of pantropical genera differ between Africa and S. America too. Brown (1973) suggests that at least four genera have spread over the earth since oligo-miocene times: *Tetramorium, Pheidole, Crematogaster* (all Myrmicinae) and *Camponotus* (Formicinae). He points out that *Crematogaster* (*Cr*) unknown before the miocene is now pantropical and has entered the 'adaptive zone' (niche) previously occupied by *Iridomyrmex* (*I*) a dominant ant in the older oligocene era. Today they are complementary: in Africa, *Cr* 175 species, *I* none; in the orient, *Cr* 125 species, *I* 5; in Australia, *Cr* 30, *I* 80. Brown (1973) also points out that in the northern hemisphere since the miocene the myrmicine genera *Pheidole, Crematogaster, Tetramonium* and *Myrmica* have largely replaced dolichoderine genera. *Pheidole* appears to have evolved since the oligocene (Baltic amber) into hundreds of species in the warmer earth zones. *Tetramorium* is also a complex genus; very common in the south palaeartic, Africa, orient and north Australian zones but with only one species in America.

Lévieux (1975, 1977) has recorded that 30 ant species occur in the soil alone in 'gallery' forest floor as against 120 in savannah soil nearby; only 20 of these are common to both habitats. Of course, the forest floor lacks herbs and many of the ants which forage there descend like termites from tree nests. Lévieux points out the contrast between species with large individuals and big foraging ranges, e.g. *Megaponera* and smaller ones such as *Crematogaster* which have local foraging areas. For a graphic account of tropical ants read

Bequaert (1922) who describes ants in the Congo region of Africa. The latitudinal impoverishment of the predatory ant fauna has been indicated vividly by Jeanne (1979a) who baited for such with wasp larvae, in low wet habitats at five latitudes: 2° S, 10° N, 19° N, 31° N, 43° N in America. He captured only 22 species in the most northerly site, but 74 in the most southerly. Species were usually more varied in forest than field and at all latitudes forest had more species on the ground than in the trees. *Pheidole* was the commonest genus at all latitudes, and *Solenopsis geminata* the most widespread species in the field south of 43° N where it replaces *Formica obscuriventris*. *Eciton*, though an abundant predator, never came to the baits.

The ant subfamily Dorylinae is pantropical and enormously important as predators. *Neivamyrmex* can ascend mountains and endure short winters but most are confined to hot, though not necessarily wet, climates. Many are hypogaeic, others are epigaeic and arboreal. *Aenictus*, a genus of small eyeless monomorphic individuals with many species in tropical Asia and Africa lives mostly below ground but Schneirla (1971) made a special study of the two epigaeic species *A. gracilis* and *A. laeviceps*. Although raids start at dusk they go on without reference to light and the ants may rest on the surface; each raid lasts an hour or two and extends for 25 m. During the stationary period when brood is pupal and eggs are being laid they go underground. They eat much ant brood, termites, wasps and cockroaches and are quick, agile, and effective with their sting pulling in immobilized prey together; even big ants, like *Polyrachis* and *Camponotus* can be captured and piled temporarily at trail junctions. Schneirla also describes the stratification and raiding differences between *Eciton burchelli* and *E. hamatum*. *Eciton vagans* that hunts on the soil surface probably eats more ants than the others. He says that *Eciton* and *Neivamyrmex* colonies are hostile and will not use each other's trails. To summarize, the dorylines have workers of different sizes; colonies of different sizes, and raid in different ways at different levels in the ecosystem, all these things enlarge their ecological grip but we have very little information yet on how they space themselves out except that, within a species, division of colonies is followed by movement in opposite directions.

Raiding parties of army ants are frequently followed by birds of several families, thrushes, owls and ant-birds of the neotropical Formicariidae. They eat the insects that are flushed out. Three or so *Eciton burchelli* colonies per square kilometre with a raiding front 20 m wide which moves at 15 m h^{-1} on Barrow Colorado Island in Panama all have a regular following of birds which obtain at least half their food in this way (Willis and Oniki, 1978). The authors list 50 'professionals' which include 4 cuckoos, 13 wood creepers, 5 tanagers and 28 ant-birds. They form a dominance zonation with the top bird nearest the centre of the swarm and subordinate smaller ones to the outside. Some of the last are very agile and are not at all deterred by the aggression of the top birds. Zones between professional birds can often be used by less skilful species. Whilst following a swarm of ants these birds call constantly and attract others.

The Attini are described as tolerant (Weber, 1972). Weber reports seeing *Acromyrmex* in Guyana cutting grass from the same clump as the termite *Amitermes*. He also points out that their nests interdigitate and says 'the material for their gardens is generally sufficient for all and they forage by one another without hostility'. Yet they usually avoid contact with other species. Weber saw a file of *Atta cephalotes* meet *Eciton burchelli*, intermingle without hostility, separate and move off; this is particularly surprising as doryline ants eat many *Attini*. As with the termites, many other species of ant often nest on mounds of the large *Atta*. Weber cites *Trachymyrmex*, *Sericomyrmex* and *Myrmicocrypta*, the workers of which forage over each other. Two colonies in the lab used the same rose flowers but cut pieces separately; yet Weber says that *Trachymyrmex* 'clearly dominated the cutting of substrate' and was the aggressor when occasion demanded. Thus interspecies hostility definitely exists in Attini but is normally avoided. Jutsum (1979) has confirmed that biting is quite normal between different colonies and species, though avoidance, if possible, is preferred. In Trinidad, *Atta cephalotes* is usually found in forest and *Acromyrmex octospinosus* in clear ground and it is likely that the queens select these habitats after nuptials. Jutsum's work leaves no doubt that small adjustments are made after the queens have selected the major habitat type as they are in most ants.

Although dulotic relationships are not known in the tropics an interesting alternative is perhaps more common than suspected. This is a condition in which two or more species use the same nest, share galleries and trails, tend the same bugs and are well disposed towards each other but keep their brood separate. In the known cases the species involved are not related and may belong to different subfamilies. Swain (1980) in Brazil has recently studied *Crematogaster limata* (Myrmicinae) which associates either with *Camponotus femoratus* (Formicinae) or with *Monacis debilis* (Dolichoderinae). *Crematogaster* (*Cl*) is only 2.3 mm long whereas *Camponotus* (*Cf*) is 5.0 mm and *Monacis* (*Md*) 4.0 mm. When living with *Md*, the *Cl* discovers sugar baits first but *Md*, which never discovers the bait themselves, take them over without violence. Although *Cl* extrudes its sting and elevates its abdomen, apparently releasing repellant, this does not drive *Md* away. *Md* thus uses *Cl* to find food. When baits of wasp larvae are used, *Cl* retain them and *Md* has to search independently.

The relationship between *Cl* and *Camponotus femoratus* (*Cf*) is clearer. *Cl* still find the recruit first, whatever the bait, but at sugar, *Cf* come and feed at the same time often standing over *Cl*; at wasp larvae they push in over the *Cl* and try to take the wasp grub away in one piece instead of cutting it up as the *Cl* do. *Cf* recruit and monopolize bait, keeping the *Cl* away by bending their abdomen under; they may actually lift a *Cl* worker up in their jaws, and take it away, but they do not crush it. *Cf* are thus completely in control and the *Cl* do not benefit even when the prey is taken back to the nest, as brood chambers are not shared. Swain points out that the *Cl*, though they find food for the others and are subordinate, may obtain protection and a nest since *Cf* constructs the nest that *Cl* comes to live in, not the other way round. Jeanne (1979a) in his latitudinal transect comments that *Cf* and *Cl* share trails and nests, and together account for 70% of understorey predation in Brazilian forest.

16.3.3 ANTS IN PLANTATIONS

The ant community of coconut and cocoa plantation has been given a lot of attention and has already been described by Brian (1965) but here some more recent research is considered and community structure analysed. In the Solomon Islands, four common species share the coconut palms with over 60 minor species (Brown, 1959; Greenslade, 1971). Two of these four are indigenous: *Iridomyrmex caudatus* (*Ic*, Dolichoderinae), *Oecophylla longinoda* (*Ol*, Formicinae) and two are invaders (*Pheidole megacephala* (*Pm*, Myrmicinae) and *Anoplolepis longipes* (*Al*, Formicinae). The minor species rarely share trees with the major four but there are important exceptions. Any attempt to identify a dominance order within the four is made difficult by unaccountable fluctuations in *Al* and *Pm*. In one plot with all but *Pm* present, monthly assessment of changes in occupancy showed 136 involving *Ol*, 96 involving *Al* and only 6 involving *Ic*. This suggests that the last was able to hold what it gained, and, after further observation the success order arrived at was *Ic* > *Al* = *Pm* > *Ol*; the last could eliminate all the minor species though it did not always do so. When *Pm* dies out for any reason, subdominants take over.

Fighting method is only a small part of the success pattern. In one oceanic coconut plantation it was thought that about six factors were important. (1) *Ol* foraged in hot sunshine midday, *Al* had a maximum forage period at dusk and *Ic* and *Pm* were nocturnal. (2) The immigrant species, *Al* and *Pm*, had unstable populations. (3) Individual size variation from small to large goes *Ic*, *Pm*, *Al*, *Ol*; the small ants, can hide in crevices and collectively can pin down a big one, but *Ol* workers can combine to pin down and kill prey too big for the others. (4) In this community only *Ol* is monogyne; if the queen dies it has to start all over again but can fly to locate new sites. *Ic*, a polygyne, spreads slowly and surely and occupies any available territory at an economic density. (5) Living entirely in a tree, as does *Ol*, is not always an advantage, for ground nests provide refuges from which to expand later. *Ic* can nest in the ground and is, in general, more adaptable than *Ol*. (6) Ability to live on honey-dew for a while is an advantage in cases of siege and in this the success order is *Ic*, *Ol*, *Pm* and *Al*; the last, a hunter, is dangerously deprived when cut off up a tree. The overall picture is of a fluctuating competition between four major species with many minor ones dodging in and out when and where opportunity allows. Perhaps before *Pm* and *Al* arrived there was a balance between the two distinctive arboreal forms: the slow-spreading polygyne *Ic* with small workers, active at night, and the flying, but vulnerable, monogyne *Ol* with large workers active by day.

Whilst the invasion in the Solomon Islands by *Pm* and *Al* has apparently been arrested by the two powerful residents, in other tropical islands this is not so (Haskins and Haskins, 1965). Hawaii is at present occupied by the 'Argentine' ant, *Iridomyrmex humilis* (*Ih*) which now lives between 30° and 36° either side of the equator and by *Pheidole megacephala* (*Pm*) from Central Africa. They have settled to a zonation in which the warm, lowland areas are used by *Pm* and the cooler, higher mountain areas over a 1000 m by *Ih*, indicating that geographical origin and adaptation govern the final settlement areas. Fluker and Beardsley (1970) have pointed out that *Pm* vanishes from the zone into which *Ih* is expanding but may move back later though it is often followed by other subordinate ants such as species of *Tetramorium* or *Solenopsis*. *Solenopsis geminata* is the only one to establish colonies and these are attacked directly by *Ih*, which may take two weeks to kill the *Solenopsis* and push 30 m or so into their territory. In the southern US, *Solenopsis invicta* has halted the northward advance of *Ih* (Brown, 1973). The technique of *Pm* is to go for appendages and hang on until their soldiers arrive and cut the foe up. Their workers are more aggressive but smaller than those of *Ih* which are only aggressive when attacked. *Anoplolepis longipes* (*Al*, Formicinae) is restricted in Hawaii to nesting under large rocks and in ditches; it is the least aggressive of the trio but has an effective spray of formic acid.

In Bermuda, however, a more complex situation is found (Lieberburg *et al.*, 1975). *Pm* arrived in the late 1800s and spread over the whole area, then *Ih* came in the mid 1900s and began to spread quickly in its turn and it was anticipated that (as had happened in Madeira in the 1800s) it would replace *Pm* within some 50 years, but the advance of *Ih* began to slow about 1959 (Crowell, 1968) and in certain areas *Pm* is now actually regaining land. *Ih* first spreads by fragmentation and fission of worker/queen groups; this happens periodically and during the expansion phase it is more aggressive than normal. Migratory columns of workers and queens may be seen spreading into *Pm* regions. *Pm* soldiers rise to the nest surface and fighting by mandibular grapples goes on for weeks. In addition to this common ant technique, Lieberberg and co-workers describe a chemical assault by *Ih*: they direct their gaster tips at the head of a *Pm* worker and stun it in a matter of seconds, no doubt with the famous terpenoid lactone, iridomyrmecin (Cavill and Locksley, 1957). Chemical interference of this sort is probably only a feature of the expansion phase of *Ih* and these 'soldiers' may be specially induced physiological phenotypes. In this way the *Ih* can displace *Pm* but the spread is limited by population growth which, in turn, depends on adaptation to climate and habitat. An interesting observation that may be relevant is that although *Pm* recruit to and guard large baits *Ih* does not.

On the east coast of Africa (Tanzania) some time ago, Way (1953, 1954) and later Vanderplank (1960, summarized in Brian, 1965) studied a coconut community similar to that in the Solomon Islands. *Ol* and a species of *Pheidole* (this time *P. punctulata*, *Pp*) were resident, and *Al* was an invader. Here *Ol* was under attack by the *Al* and the *Pp*, and another *Anoplolepis custodiens* (*Ac*) which was also hostile to the *Al*. *Ac* just picks up *Ol* one at a time and sprays them to death as they mass in defence at the bottom of their tree, but they appear unable to survive, much less attack *Ol* where ground vegetation shades the soil. *Pp* does not seem to be as adept at dealing with *Ol* as their Solomon Islands congener, *Pm*. They nest at the foot of the tree in soil and crevices and destroy *Ol* queens and small, incipient colonies. The workers just hang on whilst the soldiers cut the formicine up: but the weather must be dry! *Ol* workers are also the victims of some species of *Crematogaster*. Thus, in coconut *Ol* is well adapted but subject to much varied pressure from many other ants.

Room (1971) compared species distribution in various cocoa plantations in Ghana and found 108 species living in the cocoa plants alone. Although the trees were old, the canopy was not quite closed and patches of sunshine reached the herbs on the light loamy soil. Whereas most of the species nested in dead wood on the soil, 10 could dominate the canopy. Five of these are: *Oecophylla longinoda* (*Ol*), *Crematogaster striatula* (*Crs*), *Crematogaster africana* (*Cra*), *Crematogaster clariventris* (*Crc*) and *Crematogaster depressa* (*Crd*). All establish a patchwork of mutually exclusive colonies but the dominant species varies from place to place. One never entirely eliminates another as their feeding is slightly different though all collect honey-dew and they share nest resources (carton) except for *Ol* which has its silk-bound leaf nests. Soil surface insolation does not enable soil nesting species to build up a strong enough population to challenge these arboreal ants.

Each dominant has a set of associated species only two of which are actually shared. The most striking of these sub-communities is that dominated by *Ol* with its associate *Crc*, which are 90% associated. They are truly parabiotic in that a territory of *Crc* is included within one of *Ol* and the individuals are so linked to the colony they live in that the *Ol* do not attack them, though they will attack those of another colony. It is thought that the *Crc* are protected from other ants by the presence of *Ol* but what they give in return is problematical. Do they find food?

In an area of great heterogeneity, canopy cover varied from very thin, letting in *Camponotus acvapimensis* (*Ca*), to very thick, sheltering *Tetramorium aculeatum* (*Ta*). After removing the nests of dominant species, Majer

(1976a, b, c) found that *T. aculeatum*, a monogyne though polydomous species, spread into thinner canopy implying that it is normally constrained by pressures from surrounding ants. Similarly, *Crs*, a polygyne species, only spreads into the territory of *Ol* if its nests are first cut away. *Crd* moves into most cocoas whose dominants have been removed and are thus not very specific in their habitat requirements. *Ol* is not able to move, nor are the subsidiary species. These experiments demonstrate that each species is part of a dominance system that is flexible and sensitive to the state of the habitat.

Aggression is very rarely seen when food is plentiful for confrontation is normally avoided. Fighting follows food shortage and is followed by re-ingestion. *Ol* once invaded a tree used by a *Crematogaster* and licked a variety of coccid that the *Crematogaster* did not use: the latter withdrew into the foliage and *Ol* gained a source of honey-dew. This is not competition for food; merely an adjustment of territories that improves the total utilization of resources, and probably helps to stabilize the community.

Again, *Crc* spread into a *Crd* territory and started a fight on the trunk of a shade tree, not a cocoa, at least 14 m from the *Crd* and 9 m from the *Crc* nest. This went on for 2 or 3 months until the *Crd* had routed the invaders, pushed into their territory and taken over their nest. Majer also describes a revealing event when a tree used by *Crd* fell into a territory of *Ol* and was engulfed. He suggests that these communities are metastable. The colony territories of *Oecophylla longinoda* are adjustable like those of *Formica polyctena* and *Tetramorium caespitum* and this enables minor species to establish. A good example of adjustment to reduce friction is when *Tetramorium aculeatum* avoids the activity period of *Crd* by shifting its own; it does this where territories are in contact. The same *Tetramorium* has been reported to avoid *Crc* in this way (Aryeetey in Leston, 1973); it is evidently a flexible species.

Majer studied another community on cocoa in Ghana (6° N) and used radiophosphorus to determine the shape and size of the colonies of *Cremato-gaster*. Exclusive spacing between species of this genus was the rule and sharing with *Ol* rare. If *Ol* and *T. aculeatum* used the same tree they had distinct territories. In this area *Ol* was the most abundant ant and contributed 71% of total ant biomass though using only 12% of trees (it is, of course, a big individual). Second in order (numbers) came *Crs*, though with only 14% of the total ant biomass (wet). In another place their relative frequencies were reversed and *Crs* had 73% of trees and *Ol* only 3%.

In Nigeria a similar set of dominants occurs (Fig. 16.8). Of 3 800 trees surveyed 33% had *Ol*, 14% had *Ta*, 13% had *Pheidole megacephala* (*Pm*), and 12% had *Crematogaster africana* or *C. depressa* with relatively little localization (Taylor and Adedoyin, 1978). *Ol* was commonest in cocoa and coffee but absent from trees with unsuitable leaves like plantain whose leaves are too big, oil palm whose leaves are too narrow, or cashew whose leaves are too small and leathery. This species is very vulnerable to rain and wind and of

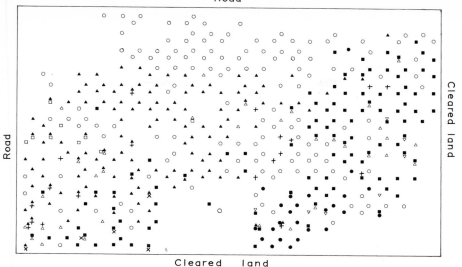

Fig. 16.8 The ants on trees in a cocoa plantation near Ibadan, Nigeria. (After Taylor, 1977.) Each tree is represented by o if no ants were seen but another symbol or group of symbols otherwise. ▲ = *Oecophylla longinoda*; △ = *Pheidole megacephala*; ● = *Crematogaster clariventris*; ■ = *Acantholepis capensis*; □ = *Crematogaster* sp. A; ∇ = *Crematogaster acvapimensis*. For minor species see Taylor's paper.

course leaf-fall in the dry season. *C. striatula* was common in coffee and cola but rare in cocoa, and it is possible that *Pm* may have kept it out. The leaf axils of oil palms made excellent nest sites for *Pm* and *Acantholepis capensis*, though normally both of these are ground nesters. The seral stages of a plantation were again shown to be very important. Some species are extremely limited, others indifferent to the stage of the canopy closure. Stenotypic species are: *T. aculeatum* and *C. clariventris* in dense canopy; *C. gabonensis* in old trees with flaky bark; *Acantholepis capensis* and *Camponotus acvapimensis* in open canopy where sunlight reaches the ground. *P. megacephala* is eurytypic and ranges widely on the ground and up trees. Taylor (1977) found that the predominant ant species on 100 indigenous wild forest trees were *Oecophylla longinoda* and *Crematogaster africana* (46%) *Camponotus acvapimensis*, *C. vividus* and *Cataglyphis guineensis* (44%) others (10%). *O. longinoda* and *Crematogaster* clean up the corpses and debris after dorylines have passed by (Leroux, 1979b).

The fauna associated with each dominant ant species on cocoa in Ghana comprised 85 species of bugs, beetles, flies, moths and crickets of which 54 were common. The canopy density, shade, edge distance, forest distance, leaf production season and dominant ant were recorded and used as attributes in a species ordination. The ant species with most associated insects were (in diminishing order): *C. striatula*, assorted species, *T. aculeatum* and, *O. longinoda*. This means that from the human viewpoint the last is the species with the 'cleanest' trees and *C. striatula* with the 'dirtiest' trees. *O. longinoda*, which cultures scale insects as well as being a predator, is independent of other insects and able to exterminate them, whereas *T. aculeatum*, which is largely predacious, cannot do so. Majer (1978) points out that in spraying to control capsid bugs, trees carrying *O. longinoda* should be left. The interests of this ant are quite compatible with those of man.

In a Camerounian cocoa plantation the same set of dominant species form a mosaic in the tree foliage but they do not extend on to the floor. On the floor about 80 species live and have a dominance system affecting nest site and food but they cannot form a mosaic. No doubt the ability to do this depends on the tree structure as well as the organization of the ant society. Moreover the ground herbs and grasses are constantly abused by man with herbicides and machetes (Jackson, 1982).

The complexity of ant communities is revealed by Samways (1981) in a study of S. African citrus orchards under control for a scale insect. When parathion applications stopped, ant diversity rose and a mosaic of dominants established. This culminated after several years in one ant, *Pheidole megacephala* (*Pm*) which though a honey-dew feeder (unlike some of the ousted subdominants) confined its interests to grass and herbs between trees and was not a pest. Samways concludes: 'Clearly no matter what the external influences, the ant species hierarchy is relatively constant and robust . . . (and) . . . seems to be a fundamental feature of ant community biology'.

16.4 Conclusions

In conclusion, ask the question: do communities of social insects really exist? They do. Species use the habitat resources differently and after a phase of colonization where segregation is imperfect establish a system in which friction is minimized. That comes partly from differences in habitat selection and partly from competitive adjustments which end in compressed, tightly interlocking niches. Long-lasting communities go on to evolve polygyny, social parasitism, dulosis and many degrees of co-operation from sharing nests to territorial parabiosis.

CHAPTER 17

Two themes

This is the place to speculate. I have selected two topics: (1) mutualistic relations with plants, and (2) social organization.

17.1 Plant mutualism

Clearly many mutualistic relationships can be very loose and long-term, like that of the wood-eating termites and the trees that 'wait' for the decomposition products. One feature that social insects have evolved that is absent from the non-social ones, is that they gather their food into nests inside which a large proportion of their population is packed. These are defended, sanitized and atmospherically conditioned for this intense population concentration runs counter to the usual strategy of dispersion; only the nests themselves show signs of over dispersal. Plants have played a part in this assortment into nests and forage fields and have contributed suitable pieces of food and nest building material benefiting in various ways in return.

They have come in on the issue of food transport. Some have baited their seeds with oily arils and in exchange have managed to get their seeds dispersed, not simply to a distant place but even to a suitable place such as an ant's nest in a sunny spot on the forest floor. Then the seeds stand a good chance of being planted out of reach of foraging birds and mice either by being left in the ant's nest where they were or by being covered in rubbish: organic rubbish. The bait of an oily aril is not necessary for the group of ants that have 'learnt' that seeds have food value and can be stored in dry airy chambers (the *Messor* group). Evidently, they transport and bury many more seeds than they eat: certainly those that germinate are taken away from the nest and thrown in the midden. This of course is not a tight symbiosis but can be regarded as a mutualism however loose. In this way a plant such as one of the grasses can be helped to establish in a more arid habitat than it could invade alone. Many other seeds may be collected as nesting material and not used as food at all. So the ant habit of collecting and carrying and burying helps plant dispersal.

Another interesting case of the use of social insects for transport concerns pollen and its use as food by bees whilst being carried from one flower to another: ideally on a different plant of the same species. This does not necessitate that the bee be a social one but the question: have social bees advanced further than non-social ones in this arrangement is worth asking. Unfortunately the answer is uncertain. Some solitary bees have life cycles that mesh closely with one plant group and may be ideal pollinators. Social bees seem to have the opportunity to visit many different plants though they may specialize briefly: bumblebees collect from several and make mixed loads of pollen but honeybees (*Apis* sp.) are more 'single minded'. Has *Apis* evolved to suit the plants? This is difficult to believe for they are great opportunists and will collect nectar and pollen by biting through plant tissue and they will go for honey-dew when it is easily available. In fact, plants which only supply pollen may have an advantage today as pollen collecting honeybees do not brush it off their hairs and discard it like nectar collectors can be seen to do. Even though honeybees have narrowed the species of plants used at any moment they still tend to collect locally and not to travel over the plant's field of distribution in the bee range from its hive. This must reduce their effectiveness as pollinators of bushes and trees. Perhaps then their ability to 'wait for' a plant's season and to recruit many bees simultaneously enables them to be better pollination instruments than less social bees. Of course we are trying to visualize that natural state before man transported and protected *Apis* for use as a collector of honey and as a pollinator of crops, which is always difficult and uncertain.

The most notable group of plants to invade the steady microclimate of the nest are the fungi. Yeasts may have been the first to enter, on fruits and grass and caterpillar faeces. Many filamentous fungi could easily have started in the rubbish left in chambers no longer needed for brood. Then mutual co-evolution must have taken a long time to reach its present state of perfection, where one species of fungus gives special fruiting bodies to the insect in exchange for cutting, collecting, delivering and preparing its substrate. In both the ant and termite groups this tightly knit symbiosis has enabled colonization of new habitats. Termites take dead plant material before other decomposers get it whereas the ants acquire an entirely new food niche and now coexist in tropical forest with the formidable doryline predators.

The defensive behaviour of social wasps, bees and ants has been used by many different organisms including small vertebrates. Green plants protect themselves from leaf-eating, sap-sucking and gall-forming insects as well as herbivorous vertebrates and even other plants and fire. Though a neat way of doing this is to have an extrafloral nectary that supplies attractive juices in the right place at the right time, the best results come if the ants are given a nest as they defend their nests savagely. Many plants come ready made with pith cavities in their stems but new features like swollen thorns and nodes with soft patches for entry (recognizable in some way to founding queens) and food bodies all help.

Some of these plants actually harbour scale insects inside their stems as an additional attraction for ants. If plants use this and other groups of sap-sucking bugs to attract ants to zones in need of protection then many of the ant/bug intimate interrelationships may be trios rather than duos. Presumably the plants have evolved a way of restricting the population of bugs to certain parts at certain times. In fact the original function of the ants may have been to clear up these insects. Then the ant/bug relationship could become a balanced predator/prey relationship as with *Oecophylla* and scale insects or *Lasius flavus* and root-feeding aphids. Whereas the gain to the tree from the former system has been known for millenia the gain to grasses in the latter system is less obvious and needs more research. I suggest that it comes through the soil structure improvement that ant excavation of deep layers brings about for these mineral rich layers are well known to be beneficial to surface plant growth. The mounds then attract rabbits which keep the grasses trim, subdue their scrubby competitors and add fresh dung pellets. The warmer microclimate that all this brings then enables plants that flourish in the Mediterranean area to live patchily in eastern England. Does *Lasius* plant their seeds? Thus a microsystem or ecological node of high energy exchange forms and endures around the ant.

17.2 Social organization

All insect societies are machines that manufacture their own sort of material out of food whatever it may be. They need to produce enough to replace their own losses, to grow to an economically apt size and then to reproduce without detriment to their social engine. They are, in the words of systems analysts: linear second order systems. They start small, grow faster and faster and then slow down either to a perpetual wobble or to a wobble that damps down to a steady state. Badly controlled ones wobble to extinction, though they are likely to reproduce in the process. The wobble frequency depends on internal structure such as the development time from egg to adult and later on the amount of effort in energy and material that is put into reproduction. This reproduction is, of course, the real aim of the society as it is of all living matter. This, since it detracts from the maintenance effort, should start slowly and increase to a maximum slowly. This maximum should come when the society is at its most powerful and has most surplus material available, for if this is not exported as sexuals or as clusters of workers around a sexual the systems will cease to operate efficiently. In addition to blowing off surplus it is important from the genetic point of view to devote a certain ratio of effort to each sex. How this is properly allocated will depend on several factors like whether the male eggs are laid by workers, how many queens there are, whether the society reproduces by fission or by individual sexual emission and whether it is well placed ecologically.

Even if export of sexuals keeps the society lean and effective it is still important for it to avoid gross inefficiency. There are again several causes for this: an obvious one is overpopulation or congestion that leads to collision and direct interference in activities by the individual workers. Equally important is the need to avoid overlap either in search of food (a well-known parasite/host search problem in ecology) or overlap of effort as when feeding larvae inside the nest. Communication can solve this last problem but only the most advanced societies seem to have got anywhere near an efficient food delivery system. Of course, as the system grows food has to be sought and carried in from further afield. Then a number of aids can be used: scouts, leaders, guides, permanent ways, depots for storage and, not least, team work. In spite of all these aids as the society grows food must 'cost' more; this is the inevitable consequence of concentrating the machinery of growth into nest centres.

Given that food is collected and stored for preparation there is still the delicate question of priority. Distribution can't be a mad scramble but how is the system organized? Should the queen get first pick so that she can provide eggs? Can she match her egg laying to food intake in any way? Do the biggest larvae come next as they have least to eat before they are ready to meta-morphose and feedback into the society (i.e. work for it). In general it seems that even in social insects with a cellular arrangement of nest space more eggs are laid than mature into larvae. Much of this mortality may be due to defective embryogenesis but some could well be adjustment to food supply, measured in general terms as the number of nurses and the food they can offer. This may well arise from a greater care with larvae taken by nurses that are well-fed, for each larva must have a chemical veneer that protects it against the predatory 'hunger' of the workers. Interestingly there is evidence that the moult from one instar to another is the stage when adjustment takes place: the older a larva gets the more secure it is from destruction.

The trick of making small workers out of female larvae does of course use less food and hasten the return on outlay. Many of the stored winter larvae in temperate ant species are on their way to become workers. In fact female larvae all have to pass a series of tests when they are subjected to worker determination pressure by assessing their development stability. Then those under-nourished and those intrinsically weak are shunted off to worker formation. There is also evidence that workers are able to recognize the state and age of a larva and manipulate it, e.g. by diluting the food and giving them plenty. At an early stage this enables them to go straight for metamorphosis; later, (after diapause) it prevents them building up, under JH influence, a complete, well stocked gyne-producing individual. Massage helps too; it is a treatment intermediate between nursing and predatory destruction and is usually undertaken by foragers strongly subordinate to queens.

The queens in fact are the moderating influence that prevent workers going all out for male and gyne production as they do when queenless. Queens prevent the society maturing, have in other words a juvenilizing effect analogous to the CA in larvae which keep the post-embryo in check. In both cases the ultimate change of state (from worker producing to sexual producing) is better; more successful in both a qualitative and a quantitative sense than it would have been if undertaken prematurely. So, queens in the society and the CA in the individual larva are merely restraining an endogenous trend, forcing as it were, a population pressure to build up so that when restraint is no longer possible the outcome has a greater chance of success. Queen power declines from several different causes. (1) A programmed aging that fades out the maternally implanted worker bias of eggs and leaves unbiassed or gyne biassed eggs. (2) Increase in the number of workers reduces their contact with the queens so that queen power to change their treatment of larvae fades either through switching fewer completely or through changing all to a lesser degree. This dilution effect applies even if the workers can spread the queen pheromones themselves. (3) After a great burst of population growth young workers preponderate and they are less responsive to queens either intrinsically or because the inevitably high worker to queen ratio prevents their sensitization during the short period after emergence when they are susceptible to imprinting.

Thus the insect society is an organism. It establishes, grows, matures and reproduces. It is as well defined and regulated as any living system.

References

Abbott, A. (1978) Nutrient dynamics of ants, in *Production ecology of ants and termites* (ed. M.V. Brian) Cambridge University Press, Cambridge, pp. 233–44.

Abe, T. (1979) Studies on the distribution and ecological role of termites in a lowland rain forest of West Malaysia. 2. Food and feeding habits of termites in Pasoh Forest Reserve. *Japanese Journal of Ecology*, **29**, 121–35.

Abe, T. (1980) Studies on the distribution and ecological role of termites in a lowland rain forest of West Malaysia. 4. The role of termites in the process of wood decomposition in Pasoh Forest Reserve. *Revue d'ecologie et de biologie de sol*, **17** (1) 23–40.

Abe, T. and Matsumoto, T. (1979) Studies on the distribution and ecological role of termites in a lowland rain forest of West Malaysia. 3. Distribution and abundance of termites in Pasoh Forest Reserve. *Japanese Journal of Ecology*, **29**, 337–51.

Abraham, M. and Pasteels, J.M. (1980) Social behaviour during nest-moving in the ant *Myrmica rubra* L. (Hym. Form.). *Insectes Sociaux*, **27** (2), 127–47.

Abrams, J. and Eickwort, G.C. (1981) Nest switching and guarding by the communal sweat bee *Agapostemon virescens* (Hymenoptera, Halictidae). *Insectes Sociaux*, **28** (2), 105–16.

Alexander, R.D. and Sherman, P.W. (1977) Local mate competition and parental investment in social insects. *Science*, **196**, 494–500.

Alford, D.V. (1975) *Bumblebees*, Davis-Poynter, London, 352 pp.

Alibert-Berthot, J. (1969) La trophallaxie chez le termite à cou jaune, *Calotermes flavicollis* (Fabr.), étudiée à l'aide de radio-éléments. I. Participation des différentes castes à la trophallaxie 'globale'. Les catégories d'aliments transmis. *Annales des sciences naturelles (Zoologie), Sér. 12*, **11**, 235–325.

Alloway, T.M. (1979) Raiding behaviour of two species of slave-making ants, *Harpagoxenus americanus* (Emery) and *Leptothorax duloticus* (Wesson) (Hymenoptera: Formicidae). *Animal Behaviour*, **27**, 202–10.

Ambrose, J.T., Morse, R.A. and Boch, R. (1979) Queen discrimination by honeybee swarms. *Annals of the Entomological Society of America*, **72** (5), 673–5.

Archer, M.E. (1981a) A simulation model for the colonial development of *Paravespula vulgaris* (Linnaeus) and *Dolichovespula sylvestris* (Scopoli) (Hymenoptera: Vespidae). *Melanderia*, **36**, 59 pp.

Archer, M.E. 1981b. Successful and unsuccessful development of colonies of *Vespula vulgaris* (Linn.). (Hymenoptera: Vespidae). *Ecological Entomology*, **6** (1), 1–10.

Arnold, G. and Delage-Darchen, B. (1978) Nouvelles données sur l'equipement enzymatique des glandes salivaires de l'ouvrière d'*Apis mellifica* (Hymenoptère Apidé). *Annales des sciences naturelles, zoologie, Sér. 12*, **20** (4), 401–22.

Auclair, J.L. (1963) Aphid feeding and nutrition. *Annual Review of Entomology*, **8**, 439–490.

Autuori, M. (1974) Der Staat der Blattschneider-ameisen, in *Sozialpolymorphismus bei Insekten* (ed. G.H. Schmidt), Wissenschaftliche verlagsgesellschaft, Stuttgart, pp. 631–56.

Avitabile, A. (1978) Brood rearing in honeybee colonies from late autumn to early spring. *Journal of Apicultural Research,* 17 (2), 69–73.

Avitabile, A., Stafstrom, D. and Donovan, K.J. (1978) Natural nest sites of honeybee colonies in trees in Connecticut, U.S.A. *Journal of Apicultural Research,* 17 (4), 222–6.

Bailey, L. (1952) The action of the proventriculus of the worker honeybee, *Apis mellifera* L. *Journal of Experimental Biology,* 29 (2), 310–27.

Baldridge, R.S., Rettenmeyer, C.W. and Watkins, J.F. II. (1980) Seasonal, nocturnal and diurnal flight periodicities of Nearctic army ant males (Hymenoptera: Formicidae). *Journal of the Kansas Entomological Society,* 53 (1), 189–204.

Barlin, M.R., Blum, M.S. and Brand, J.M. (1976) Fire ant trail pheromones: analysis of species specificity after gas chromatographic fractionation. *Journal of Insect Physiology,* 22, 839–44.

Baroni Urbani, C. (1968) Monogyny in ant societies. *Sonderdruck aus Zoologischer Anzeiger,* 181 (3/4), 269–77.

Baroni Urbani, C. (1969) Ant communities of the high-altitude Appennine grasslands. *Ecology,* 50, 488–92.

Baroni Urbani, C. (1974) Polymorphismus in der Ameisen-gattung *Camponotus* aus morphologischer Sicht, in *Sozialpolymorphismus bei Insekten* (ed. G.H. Schmidt), Wissenschaftliche verlagsgesellschaft, Stuttgart, pp. 543–64.

Baroni Urbani, C. (1979) Territoriality in social insects, in *Social Insects* (ed. H.R. Hermann), Academic Press, London and New York, pp. 91–121.

Baroni Urbani, C. and Kannowski, P.B. (1974) Patterns of the red imported fire ant settlement of a Louisiana pasture: some demographic parameters, interspecific competition and food sharing. *Environmental Entomology,* 3, 755–60.

Barrows, E.M. (1975) Individually distinctive odors in an invertebrate. *Behavioral Biology,* 15, 57–64.

Barrows, E.M., Bell, W.J. and Michener, C.D. (1975) Individual odor differences and their social functions in insects. *Proceedings of the National Academy of Science, U.S.A., Zoology,* 72 (7), 2824–8.

Batra, L.R. and Batra, S.W.T. (1966) Fungus-growing termites of tropical India and associated fungi. *Journal of the Kansas Entomological Society,* 39, 725–38.

Beetsma, J. (1979) The process of queen-worker differentiation in the honeybee. *Bee World,* 60 (1) 24–39.

Beig, D. (1972) The production of males in queenright colonies of *Trigona (Scaptotrigona) postica. Journal of Apicultural Research,* 11, 33–9.

Beig, D. (1977) Endocrine control in Meliponinae: 'corpus allatum' activity and ovary growth in stingless bees, Proceedings, VIIIth International Congress of the International Union for the Study of Social Insects, Wageningen, pp. 228–31.

Beig, D., Bueno, O.C., da Cunha, R.A. and de Moraes, H.J. (1982) Differences in quantity of food in worker and male brood cells of *Scaptotrigona postica* (Latr. 1807) (Hymenoptera, Apidae). *Insectes Sociaux,* 29 (2), 189–94.

Benois, A. (1969) Étude morphologique, biologique et éthologique de *Camponotus vagus* Scop. *(pubescens* Fab.) (Hyménoptère, Formicoides, Formicidae) Doctoral Thesis, University of Toulouse.

Benois, A. (1972) Etude écologìque de *Camponotus vagus* Scop. (= *pubescens* Fab.) (Hymenoptera, Formicidae) dans la région d'Antibes: nidification et architecture des nids. *Insectes Sociaux,* 19 (2) 111–29.

Bentley, B.L. (1977) The protective function of ants visiting the extrafloral nectaries of *Bixia orellana* (Bixaceae). *Journal of Ecology,* 65, 27–38.

Bequaert, J. (1922) Ants in the diverse relations to the plant world, in *Ants of the Belgian Congo*

W.M. Wheeler, Bulletin, American Museum of Natural History, **45**, 333–583.

Berg, R.Y. (1975) Myrmecochorous plants in Australia and their dispersal by ants. *Australian Journal of Botany*, **23**, 475–508.

Bergström, G. (1975) Development of an integrated system for the analyses of volatile communication substances in social Hymenoptera, in *Pheromones and defensive secretions in social insects*, Proceedings, Symposium of the International Union for the Study of Social Insects, Dijon, pp. 173–87.

Bergström, G. and Löfquist, J. (1972) Similarities between the Dufour's gland secretions of the ants *Camponotus ligniperda* (Latvi) and *Camponotus herculeanus* (L). *Entomologica Scandinavica*, **3**, 225–38.

Bergström, G., Svensson, B.G., Appelgren, M. and Groth, I. (1981) Complexity of bumble bee marking pheromones: biochemical ecological and systematical interpretation. Systematics Association Special Volume 19, *Biosystematics of Social Insects* (eds. P.E. Howse and J.–L. Clement), Academic Press, London and New York, pp. 175–83.

Bernardi, C., Cardani, D., Ghiringhelli, D., Selva, A., Baggini, A. and Pavan, M. (1967) On the components of secretion of mandibular glands of the ant *Lasius (Dendrolasius) fulginosus*. *Tetrahedron Letters*, **40**, 3893–6.

Berndt, K.P. (1977) The physiology of reproduction in the Pharoah's ant (*Monomorium pharaonis* L.) l. Pheromone mediated cyclic production of sexuals. *Wiadomosci Parazytologcizne*, **23**, 163–6.

Berndt, K.P. and Nitschmann, J. (1979) The physiology of reproduction in the Pharoah's ant (*Monomorium pharaonis* L.) 2. The unmated queens. *Insectes Sociaux*, **26** (2), 137–45.

Bernstein, R.A. (1975) Foraging strategies of ants in response to variable food density (Hym. Formicidae). *Ecology*, **56** (1), 213–19.

Bernstein, R.A. (1979a) Schedules of foraging activity in species of ants. *Journal of Animal Ecology*, **48** (3), 921–30.

Bernstein, R.A. (1979b) Relations between species diversity and diet in communities of ants. *Insectes Sociaux*, **26** (4), 313–21.

Bernstein, R.A. and Gobbel, M. (1979) Partitioning of space in communities of ants. *Journal of Animal Ecology*, **48** (3), 931–42.

Benton, A.W. (1967) Esterases and phosphatases of honeybee venom. *Journal of Apicultural Research*, **6** (2), 91–4.

Bier, K.H. (1953) Beziehungen zwischen Nährzellkerngrösse und Ausbildung ribonukleinsäurehaltiger Strukturen in den Oocyten von *Formica rufa rufopratensis minor* Gösswald. *Zoologische Anzeiger Supplement*, **16**, 369–74.

Bigley, W.S. and Vinson, S.B. (1975) Characterization of a brood pheromone isolated from the sexual brood of the imported fire ant, *Solenopsis invicta*. *Annals of the Entomological Society of America*, **68** (2), 301–4.

Bignell, D.E., Oskarsson, H. and Anderson, J.M. (1980) Distribution and abundance of bacteria in the gut of a soil feeding termite *Procubitermes aburiensis*. *Journal of General Microbiology*, **117**, 393–403.

Bignell, D.E., Oskarsson, H. and Anderson, J.M. (1981) Association of *Actinomycetes* with soil-feeding termites: a novel symbiotic relationship?, in *Actinomycetes*, (eds. V. Schaal and G. Pulverer) Zentralblatt für Bakteriologie Supplementheft, Vol. 11, pp. 201–6.

Blum, M.S. (1974) Pheromonal sociality in the hymenoptera, in *Pheromones* (ed. M.C. Birch) North Holland, Amsterdam, pp. 223–49.

Blum, M.S. and Brand, J.M. (1972) Social insect pheromones: their chemistry and function. *American Zoologist*, **12**, 553–76.

Blum, M.S. and Portocarrero, C.A. (1964) Chemical releasers of social behaviour. IV. The hindgut as the source of the odor trail pheromone in the neotropical army ant genus, *Eciton*. *Annals of the Entomological Society of America*, **57** (6), 793–4.

Blum, M.S., Jones, T.H., Hölldobler, B., Fales, H.M. and Jaouni, T. (1980) Alkaloidal venom mace: offensive use by a thief ant. *Naturwissenschaften,* **67,** 144.

Boch, R. (1979) Queen substance pheromone produced by immature queen honeybees. *Journal of Apicultural Research,* **18** (1), 12–15.

Boch, R. and Morse, R.A. (1979) Individual recognition of queens by honey bee swarms. *Annals of the Entomological Society of America,* **72** (1), 51–3.

Bodot, P. (1966) Études écologique et biologique des Termites des savanes de Basse Côte-d'Ivoire. Thèse, Université d'Aix Marseille.

Bodot, P. (1966) Observations sur l'essaimage et les premières étages du développment de la colonie d'*Allodontotermes giffardii* Silv. (Isoptera, Termitidae). *Insectes Sociaux,* **14** (4), 351–8.

Bolton, B. (1980) The ant tribe Tetramoriini (Hymenoptera: Formicidae) The genus *Tetramorium* Mayr in the Ethiopian zoogeographical region. *Bulletin of the British Museum (Natural History), Entomology series,* **40** (3), 193–384.

Boomsma, J.J., van der Lee, G.A., and van der Have, T.M. (1982) On the production ecology of *Lasius niger* L. (Hymenoptera, Formicidae) in successive coastal dune valleys. *Journal of Animal Ecology,* **51,** 975–92.

Boomsma, J.J., Leusink, A. (1981) Weather conditions during nuptial flights of four european ant species. *Oecologia,* **50** (2), 236–41.

Boomsma, J.J. and van Loon, A.J. (1982) Structure and diversity of ant communities in successive coastal dune valleys. *Journal of Animal Ecology,* **51,** 957–74.

Boomsma, J.J. and DeVries, A. (1980) Ant species distribution in a sandy coastal plain. *Ecological Entomology,* **5** (3), 189–204.

Bouillon, A. (1970) Termites of the Ethiopian region, in *Biology of the termites,* Vol. II, (eds. K. Krishna and F.M. Weesner), Academic Press, New York, pp. 154–280.

Boven J.K.A. Van (1970) Le polymorphisme des ouvrières de *Megaponera foetens* Fabr. *Publicatiës van het Natuurhistorisch genootschap in Limburg,* **20,** 5–9.

Boyd, N.D. and Martin, M.M. (1975) Faecal proteinases of the fungus growing ant, *Atta texana:* their fungal origin of ecological significance. *Journal of Insect Physiology,* **21** (11), 1815–20.

Bradshaw, J.W.S., Baker, R. and Howse, P.E. (1979a) Multicomponent alarm pheromones in the mandibular glands of major workers of the African weaver ant, *Oecophylla longinoda. Physiological Entomology,* **4,** 15–25.

Bradshaw, J.W.S., Baker, R. and Howse, P.E. (1979b) Caste and colony variations in the chemical composition of the cephalic secretions of the African weaver ant, *Oecophylla longinoda. Physiological Entomology,* **4** (1), 27–38.

Bradshaw, J.W.S., Baker, R. and Howse, P.E. (1979c) Chemical composition of the poison apparatus secretions of the African weaver ant, *Oecophylla longinoda,* and their role in behaviour. *Physiological Entomology,* **4** (1), 39–46.

Brandt, D. CH. (1980) The thermal diffusivity of the organic material of a mound of *Formica polyctena* Foerst in relation to the thermoregulation of the brood (Hymenoptera, Formicidae). *Netherlands Journal of Zoology,* **30** (2), 326–44.

Breed, M.D. and Gamboa, G.J. (1977) Control of worker activities by queen behavior in a primitively eusocial bee. *Science,* **195,** 694–6.

Breed, M.D., Silverman, J.M. and Bell, W.J. (1978) Agonistic behaviour, social interactions and behavioural specialization in a primitively eusocial bee. *Insectes Sociaux,* **25** (4), 351–64.

Breen, J. (1977) Patterns of distribution of *Formica lugubris* Zett. nests in Irish plantation woods (Hymenoptera, Formicidae). Proceedings, VIIIth International Congress of the International Union for the Study of Social Insects, Wageningen, p. 156.

Breen, J. (1979) Worker populations of *Formica lugubris* Zett. nests in Irish plantation woods. *Ecological Entomology,* **4,** 1–7.

Brian, A.D. (1951) Brood development in *Bombus agrorum*. *Entomologists Monthly Magazine*, 87, 207–12.

Brian, A.D. (1952) Division of labour and foraging in *Bombus agrorum* Fabricus. *Journal of Animal Ecology*, 21, 223–40.

Brian, A.D. (1954) The foraging of bumble-bees. *Bee World*, 35, 61–7, 81–91.

Brian, A.D. (1957) Differences in the flowers visited by four species of bumble-bees and their causes. *Journal of Animal Ecology*, 26, 71–98.

Brian, M.V. (1951) Summer population changes in colonies of the ant *Myrmica*. *Physiologia Comparata et Oecologia*, 2 (3), 248–62.

Brian, M.V. (1953) Brood-rearing in relation to worker number in the ant *Myrmica*. *Physiological Zoology*, 26 (4), 355–66.

Brian, M.V. (1955a) Food collection by a Scottish ant community. *Journal of Animal Ecology*, 24, 336–51.

Brian, M.V. (1955b) Studies of caste differentiation in *Myrmica rubra* L. 2. The growth of workers and intercastes. *Insectes Sociaux*, 2, 1–34.

Brian, M.V. (1964) Ant distribution in a southern English heath. *Journal of Animal Ecology*, 33, 451–61.

Brian, M.V. (1965) *Social Insect Populations*, Academic Press, New York and London, 135 pp.

Brian, M.V. (1967) Production by social insects, in *Secondary productivity of Terrestrial ecosystems* (ed. K. Petrusewicz) Warszawa, pp. 835–9.

Brian, M.V. (1969) Male production in the ant *Myrmica rubra* L. *Insectes sociaux*, 16, 249–68.

Brian, M.V. (1970) Communication between queens and larvae in the ant *Myrmica*. *Animal Behaviour*, 18, 467–72.

Brian, M.V. (1972) Population turnover in wild colonies of the ant *Myrmica*. *Ekologia Polska*, 20, 43–53.

Brian, M.V. (1973a) Feeding and growth in the ant *Myrmica*. *Journal of Animal Ecology*, 42, 37–53.

Brian, M.V. (1973b) Temperature choice and its relevance to brood survival and caste determination in the ant *Myrmica rubra* L. *Physiological Zoology*, 46 (4) 245–52.

Brian, M.V. (1973c) Queen recognition by brood-rearing workers of the ant *Myrmica rubra* L. *Animal Behaviour*, 21, 691–8.

Brian, M.V. (1974a) Kastendetermination bei *Myrmica rubra* L., in *Sozialpolymorphismus bei Insekten* (ed. G.H. Schmidt), Wissenschaftliche Verlagsgesellschaft, Stuttgart., pp. 565–89.

Brian, M.V. (1974b) Caste differentiation in *Myrmica rubra*: the role of hormones. *Journal of Insect Physiology*, 20, 1351–65.

Brian, M.V. (1975) Larval recognition by workers of the ant *Myrmica*. *Animal Behaviour*, 23, 745–56.

Brian, M.V. (1977) The synchronisation of colony and climatic cycles. Proceedings, VIIIth International Congress of the International Union for the Study of Social Insects, Wageningen, pp. 202–6.

Brian, M.V. (1978) *Production ecology of ants and termites*. International Biological Programme 13, (ed. M.V. Brian) Cambridge University Press, 409 pp.

Brian, M.V. (1979a) Caste differentiation and division of labour, in *Social Insects I* (Ed. H. Hermann), Academic Press, New York, pp. 121–222.

Brian, M.V. (1979b) Habitat differences in sexual production by two co-existent ants. *Journal of Animal Ecology*, 48, 943–53.

Brian, M.V. (1980) Social control over sex and caste in bees, wasps and ants. *Biological Review*, 55, 379–415.

Brian, M.V. (1981) Treatment of male larvae in ants of the genus *Myrmica*. *Insectes Sociaux*, 28, (2), 161–6.

Brian, M.V. and Abbott, A. (1977) The control of food flow in a society of the ant *Myrmica rubra*

L. *Animal Behaviour,* **25,** 1047–55.

Brian, M.V. and Brian, A.D. (1949) Observations on the taxonomy of the ants *Myrmica rubra* L. and *Myrmica laevinodis* (Nylander) (Hym., Formicidae). *Transactions of the Royal Entomological Society of London,* **100,** 393–409.

Brian, M.V. and Brian, A.D. (1951) Insolation and ant population in the west of Scotland. *Transactions of the Royal Entomological Society of London,* **102,** 303–30.

Brian, M.V. and Brian, A.D. (1952) The wasp *Vespula sylvestris* (Scop.): Feeding, foraging and colony development. *Transactions of the Royal Entomological Society of London,* **103,** 1–26.

Brian, M.V. and Brian, A.D. (1955) On the two forms macrogyna and microgyna of the ant *Myrmica rubra* (L). *Evolution,* **9,** 280–90.

Brian, M.V. and Carr, C.A.H. (1960) The influence of the queen on brood rearing in ants of the genus *Myrmica. Journal of Insect Physiology,* **5,** 81–94.

Brian, M.V. and Downing, B.M. (1958) The nests of some British ants. Proceedings Xth International Congress of Entomology, Montreal, 1956, Vol. 2, pp. 539–40.

Brian, M.V. and Elmes, G.W. (1974) Production by the ant *Tetramorium caespitum* in a southern English heath. *Journal of Animal Ecology,* **43,** 889–903.

Brian, M.V. and Evesham, E.J.M. (1982) The role of young workers in *Myrmica* colony development. Proceeding IXth International Congress of the International Union for the Study of Social Insects, Colorado.

Brian, M.V. and Hibble, J. (1964) Studies of caste differentiation in *Myrmica rubra* L. 7. Caste bias queen age and influence. *Insectes Sociaux,* **11,** 223–38.

Brian, M.V. and Jones, R.M. (1980) Worker population structure and gyne production in the ant *Myrmica. Behavioral Ecology and Sociobiology,* **7,** 281–6.

Brian, M.V. and Rigby, C. (1978) The trophic eggs of *Myrmica rubra* L. *Insectes Sociaux,* **25,** 89–110.

Brian, M.V., Hibble, J. and Stradling, D.J. (1965) Ant pattern and density in a southern English heath. *Journal of Animal Ecology,* **34,** 545–55.

Brian, M.V., Hibble, J. and Kelly, A.F. (1966) The dispersion of ant species in a southern English heath. *Journal of Animal Ecology,* **35,** 281–90.

Brian, M.V., Elmes, G. and Kelly, A.F. (1967) Populations of the ant *Tetramorium caespitum* Latreille. *Journal of Animal Ecology,* **36,** 337–42.

Brian, M.V., Mountford, M.D., Abbott, A. and Vincent, S. (1976) The changes in ant species distribution during ten years post-fire regeneration of a heath. *Journal of Animal Ecology,* **45,** 115–33.

Brian, M.V., Clarke, R.T. and Jones, R.M. (1981a) A numerical model of an ant society. *Journal of Animal Ecology,* **50,** 387–405.

Brian, M.V., Jones, R.M. and Wardlaw, J.C. (1981b) Quantitative aspects of queen control over reproduction in the ant *Myrmica. Insectes Sociaux,* **28** (2), 161–6.

Breed, M.D. (1977) Interactions among individuals and queen replacement in an eusocial halictine bee. Proceedings VIIIth Congress of the International Union for the Study of Social Insects, Wageningen, pp. 228–31.

Breed, M.D. and Gamboa, G.J. (1977) Control of worker activities by queen behaviour in a primitively eusocial bee. *Science,* **195,** 694–6.

Brown, C.A., Watkins, J.F. and Eldridge, D.W. (1979) Repression of bacteria and fungi by the army ant secretion: skatole. *Journal of the Kansas Entomological Society,* **52** (1), 119–22.

Brown, E.S. (1959) Immature nutfall of coconuts in the Solomon Islands. II Changes in ant populations and their relation to vegetation. *Bulletin of Entomological Research,* **50,** 523–58.

Brown, W.L., Jr (1973) A comparison of the hylaean and Congo-West African rain forest ant Faunas, in *Tropical forest ecosystems in Africa and South America: a comparative review,* (eds. B.J. Meggers, E.S. Ayensu and W.D. Duckworth), Smithsonian Institute Press, Washington D.C., pp. 161–85.

Brückner, D. (1980) Hoarding behaviour and life span of inbred, non-inbred and hybrid honeybees. *Journal of Apicultural Research,* **19** (1), 35–41.

Bruinsma, O. and Leuthold, R.H. (1977) Pheromones involved in the building behaviour of *Macrotermes subhylinus* (Rambur). Proceedings, VIIIth International Congress of the International Union for the Study of Social Insects, Wageningen, pp. 257–8.

Buchli, H. (1950) Recherche sur la fondation et le développment des nouvelles colonies chez le termite lucifuge (*Reticulitermes lucifugus* (Rossi)). *Physiologia Comparata et Oecologia,* **2**, 145–60.

de Bruyn, G.D. and Mabelis, A.A. (1972) Food territories in *Formica polyctena* Först. Proceedings, XIIIth International Congress of Entomology, Vol. 3, pp. 358–60.

Buckle, G.R. (1982) Queen–worker behavior and nestmate interactions in young colonies of *Lasioglossum zephyrum. Insectes Sociaux,* **29** (2), 125–37.

Büdel, A. (1968) Le microclimat de la ruche, in *Traité de Biologie de l'Abeille,* Vol. 4, (ed. R. Chauvin) Masson, Paris, p. 379.

Büdel, A. and Herold, E. (1960) *Biene und Bienenzucht,* Ehrenwirth, München.

Burgett, D.M. (1974) Glucose oxidase: A food protective mechanism in social Hymenoptera. *Annals of the Entomological Society of America,* **67** (4), 545–6.

Büschinger, A. (1974) Polymorphismus und Kastendetermination in Ameisentribus *Leptothoracini,* in *Sozialpolymorphismus bei Insekten* (ed. G.H. Schmidt), Wissenschaftliche Verlagsgesellschaft, Stuttgart, pp. 604–23.

Büschinger, A. (1975) Sexual pheromones in ants, in *Pheromones and defensive secretions in ants,* symposium of the International Union for the Study of Social Insects, Dijon, pp. 225–33.

Büschinger, A. (1978) Genetisch bedingte Entstehung geflügelter Weibchen bei der Sklavenhattenden Ameise *Harpagoxenus sublaenis* (Nyl.) (Hym. Form.) *Insectes Sociaux,* **25**, 163–72.

Büschinger, A. (1979) Functional monogyny in the american guest ant *Formicoxenus hirticornis* (Emery) (= *Leptothorax hirticornis), (Hym. Form.). Insectes Sociaux,* **26** (1), 61–8.

Büschinger, A., Ehrhardt, W. and Winter, V. (1980) The organization of slave raids in dulotic ants – a comparative study (Hymenoptera: Formicidae). *Zeitschrift für Tierpsychologie,* **53**, 245–64.

Butler, C.G. (1940) The ages of the bees in a swarm. *Bee World,* **21**, 9–10.

Butler, C.G. (1957) The process of queen supersedure in colonies of honeybees (*Apis mellifera* Linn.). *Insectes Sociaux,* **4** (3), 211–23.

Butler, C.G., Jeffree, E.P. and Kalmus, H. (1943) The behaviour of a population of honeybees on an artifical and on a natural crop. *Journal of Experimental Biology,* **20** (1), 65–73.

Butler, C.G., Fletcher, D.J.C. and Watler, D. (1969) Nest-entrance marking with pheromones by the honeybee *Apis mellifera* L., and a wasp, *Vespula vulgaris* L. *Animal Behaviour,* **17**, 142–7.

Byron, P.A., Byron, E.R. and Bernstein, R.A. (1980) Evidence of competition between two species of desert ants. *Insectes Sociaux,* **27** (4), 351–60.

Cagniant, G. (1979) La parthogénèse thélytoque et arrhénotoque chez la fourmi *Cataglyphis cursor* Fonsc (Hym. Form.) cycle biologique en élevage des colonies avec reine et des colonies sans reine. *Insectes Sociaux,* **26** (1), 51–60.

Calaby, J.H. and Gay, F.J. (1956) The distribution and biology of the genus *Coptotermes* (Isoptera) in Western Australia. *Australian Journal of Zoology,* **4**, 19–39.

Calam, D.H. (1969) Species and sex-specific compounds from the heads of male bumblebees (*Bombus* spp). *Nature,* **221**, 856–7.

Cammaerts, M.C. (1978) Recruitment to food in *Myrmica rubra* L. *Biology of Behaviour,* **4**, 159–72.

Cammaerts, M.C., Morgan, E.D. and Tyler, R.C. (1977) Territorial marking in the ant *Myrmica rubra* L. *(Formicidae). Biology of Behaviour*, **2**, 263–72.

Cammaerts, M.C., Inwood, M.R., Morgan, E.D., Parry, K. and Tyler, R.C. (1978) Comparative study of the pheromones emitted by workers of the ants *Myrmica rubra* and *Myrmica scabrinodis* (Hym. Formicidae). *Journal of Insect Physiology*, **24** (3), 207–14.

Cammaerts, M.C., Evershed, R.P. and Morgan, E.D. (1981a) Comparative study of the mandibular gland secretion of four species of *Myrmica* ants. *Journal of Insect Physiology*, **27** (4), 225–31.

Cammaerts, M.C., Evershed, R.P. and Morgan, E.D. (1981b) Comparative study of the dufour gland secretions of workers of four species of *Myrmica* ants. *Journal of Insect Physiology*, **27**, 59–65.

Cammaerts–Tricot, M.C. (1975) Ontogenesis of the defence reactions in the workers of *Myrmica rubra* L. (Hymenoptera, Formicidae). *Animal Behaviour*, **23**, 124–30.

Cammaerts–Tricot, M.C., Morgan, E.D. and Tyler, R.C. (1977) Isolation of the trail pheromone of *Myrmica rubra* (Hym., Formicidae). *Journal of Insect Physiology*, **23** (3), 421–7.

Carmargo, C.A. de (1979) Sex determination in bees. XI production of diploid males and sex determination in *Melipona quadrifasciata. Journal of Apicultural Research*, **18**, 77–84.

Carr, C.A.H. (1962) Further studies on the influence of the queen in ants of the genus *Myrmica. Insectes Sociaux*, **9**, 197–211.

Cavill, G.W.K. and Locksley, H.D. (1957) The chemistry of ants. II. Structure and configuration of irodolactone (isoiridomyrmecin). *Australian Journal of Chemistry*, **10** (3), 352–8.

Cavill, G.W.K. and Robertson, P.L. (1965) Ant venoms, attractants and repellents. *Science*, **149**, 1337–45.

Cederberg, B. (1977) Chemical basis for defense in bumblebees. Proceedings, VIIIth International Congress of the International Union for the Study of Social Insects, Wageningen, p. 77.

Chadab, R. and Rettenmeyer, C.W. (1975) Mass recruitment by army ants. *Science*, **188**, 1124–5.

Chapman, S.B., Hibble, J. and Rafarel, C.R. (1975) Net aerial production by *Calluna vulgaris* on lowland heath in Britain. *Journal of Ecology*, **63**, 233–58.

Charlesworth, B. (1978) Some models of the evolution of altruistic behaviour between siblings. *Journal of Theoretical Biology*, **72**, 297–319.

Charnov, E.L. (1978) Sex-ratio selection in eusocial Hymenoptera. *The American Naturalist*, **112**, 317–26.

Chauvin, R. (1970) The world of ants (Transl. George Ordish), Victor Gollancz, London, 216 pp.

Chen, S.C. (1937) The leaders and followers among the ants in nest-building. *Physiological Zoology*, **10** (4), 437–55.

Cherix, D. (1980) Note préliminaire sur la structure, la phénologie et le régime alimentaire d'une super-colonie de *Formica Lugubris* Zett. *Insectes Sociaux*, **27** (3), 226–36.

Cherix, D. and Gris, G.Ch. (1977) The giant colonies of the red wood ant in the Swiss Jura *(Formica lugubris* Zett). Proceedings, VIIIth International Congress of the International Union for the Study of Social Insects, Wageningen, p. 296.

Cherrett, J.M. (1968) The foraging behaviour of *Atta cephalotes* (L.) (Hymenoptera, Formicidae) I. Foraging pattern and plant species attacked in tropical rain forest. *Journal of Animal Ecology*, **37**, 387–403.

Cherrett, J.M. (1972) Some factors involved in the selection of vegetable substrate by *Atta cephalotes* (L.) (Hymenoptera, Formicidae) in tropical rain forest. *Journal of Animal Ecology*, **41**, 647–60.

Coenen-Stass, D., Schaarschmidt, B. and Lamprecht, I. (1980) Temperature distribution and calorimetric determination of heat production in the nest of the wood ant, *Formica* polyctena (Hymenoptera, Formicidae). *Ecology,* **61** (2), 238–44.

Cole, B.J. (1980) Repertoire convergence in two Mangrove ants, *Zacryptocerus varians* and *Camponotus (Colobopsis)* sp. *Insectes Sociaux,* **27** (3), 265–75.

Collingwood, C.A. (1979) The Formicidae (Hymenoptera) of Fennoscandia and Denmark. *Fauna Entomologica Scandinavica,* Vol. 8, Scandinavian Science Press, Klampenborg, Denmark, 174 pp.

Collins, N.M. (1979) The nest of *Macrotermes bellicosus* (Sneathman) from Mokwa, Nigeria. *Insectes Sociaux,* **26** (3), 240–6.

Collins, N.M. (1981a) Populations; age structure and survivorship of colonies of *Macrotermes bellicosus* (Isoptera, Macrotermitinae). *Journal of Animal Ecology,* **50**, 293–311.

Collins, N.M. (1981b) The role of termites in the decomposition of wood and leaf litter in the Southern Guinea Savanna of Nigeria. *Oecologia,* **51**, 389–99.

Colombel, P. (1974) L'élevage artificial du couvain d'*Odontomachus heamatodes* L. et la différenciation trophogènique des ouvrières et des reines. *Compte rendu de l'Académie des Sciences, Paris, Sér. D,* **279**, 489–91.

Colombel, P. (1978) Biologie d'*Odontomachus heamatodes*, déterminisme de la caste femelle. *Insectes Sociaux,* **25**, 141–51.

Corbet, S.A., Unwin, D.M. and Prŷs–Jones, O.E. (1979) Humidity, nectar and insect visits to flowers, with special reference to *Crataegus, Tilia* and *Echium. Ecological Entomology,* **4** (1), 9–22.

Craig, R. (1979) Parental manipulation, kin selection, and the evolution of altruism. *Evolution,* **33** (1), 319–34.

Craig, R. and Crozier, R.H. (1979) Relatedness in the polygynous ant *Myrmecia pilosula. Evolution,* **33** (1), 335–41.

Crowell, K.L. (1968) Rates of competitive exclusion by the Argentine ant in Bermuda. *Ecology,* **49** (3), 551–5.

Crozier, R.H. (1968) Interpopulation karyotype differences in Australian *Iridomyrmex* of the '*Detectus*' group. (Hymenoptera: Formicidae: Dolichoderinae). *Journal of the Australian Entomological Society,* **7**, 25–7.

Crozier, R.H. (1971) Heterozygosity and sex-determination in haplodiploidy. *American Naturalist,* **105**, 399–412.

Crozier, R.H. (1977) Genetic differentiation between populations of the ant *Aphaenogaster 'rudis'* in the south eastern United States. *Genetica,* **47** (1), 17–36.

Crozier, R.H. (1979) Genetics of sociality, in *Social Insects,* Vol. I (ed. G.H. Hermann), Academic Press, New York and London, pp. 223–286.

Crozier, R.H. (1980) Genetical structure of social insect populations, in *Evolution of social behaviour: Hypotheses and Empirical Tests* (ed. H. Markl), Dahlem Konferenzen 1980, Verlag Chemie GmbH, Weinheim, pp. 129–46.

Crozier, R.H. and Brückner, D. (1981) Sperm clumping and the population genetics of Hymenoptera. *American Naturalist,* **117**, 561–3.

Crozier, R.H. and Dix, M.W. (1979) Analysis of two genetic models for the innate components of colony odor in social Hymenoptera. *Behavioral Ecology and Sociobiology,* **4**, 217–24.

Culver, D.C. and Beattie, A.J. (1978) Myrmecochory in *Viola*: Dynamics of seed–ant interactions in some West Virginia species. *Journal of Ecology,* **66**, 53–72.

Cumber, R.A. (1949) The biology of bumble-bees, with special reference to the production of the worker caste. *Transactions of The Royal Entomological Society of London,* **24**, 119–27.

Cunha, M.A.S. da (1979) Ovarian development in *Scaptotrigona postica* Latr. 1807 (Hym. Apidae) II A quantitative study. *Insectes Sociaux,* **26** (3), 196–203.

Czechowski, W. (1977) Pdikaliczne kolonie mrowek. *Przegląd Zoologiczny,* **11** (4), 284–98.

Darchen, R. (1954) Quelques régulations sociales dans la construction chez les Abeilles. *Insectes Sociaux,* **1** (3), 219–28.

Darchen, R. (1958) Construction et reconstruction de la cellule de rayons d'*Apis mellifica*. *Insectes Sociaux*, **5** (4), 357–72.

Darchen, R. (1971) *Trigona (Axestotrigona) oyani* Darchen (Apidae, Trigoninae), une nouvelle espece d'abeille africaine. Description du nid inclus dans une fourmilière. *Biologia Gabonica*, **7** (4), 407–21.

Darchen, R. (1972) Éthologie comparative de l'économie des matériaux de construction chez divers apides sociaux. *Revue du Comportement Animal*, **6** (3), 201–6.

Darchen, R. (1973) La thermorégulation et l'écologie de quelques espèces d'abeilles sociales d'Afrique (*Apidae, Trigonini*) et *Apis mellifica* var. *Adansonii) Apidologie*, **4** (4), 341–70.

Darchen, R. (1976) *Ropalida cincta*, guêpe sociale de la savane de Lamto (Cote-d'Ivoire) (Hym. Vespidae). *Annales de la Société Entomologique de France (NS)* **12** (4), 579–601.

Darchen, R. (1977) L'essaimage chez les hypotrigones au Gabon: dynamique de quelques populations. *Apidologie*, **8** (1), 33–59.

Darchen, R. (1980) La cire, son recyclage et son rôle probable a l'interieur d'une colonie d'*Apis mellifica*. *Apidologie*, **11** (3), 193–202.

Darchen, R. and Delage-Darchen, B. (1974) Les stades larvaires de *Melipona beechii*. *Compte Rendu de l'Académie des Sciences, Paris*, **278**, 3115–18.

Darchen, R. and Delage-Darchen, B. (1975) Contribution a l'étude d'une abeille du Mexique *Melipona beecheii* B. (Hymenoptère: Apidae). *Apidologie*, **6** (4), 295–339.

Davidson, D.W. (1977) Foraging ecology and community organization in desert seed-eating ants. *Ecology*, **58**, 725–37.

Dawkins, R. (1976) *The selfish gene*. Oxford University Press, Oxford, p. 224.

Dejean, A. (1974) Etude du cycle biologique et de la reproduction de *Temnothorax recedens* (Nyl.) (Insecta, Formicidae). Doctoral thesis, University of Toulouse.

Dejean, A., Passera, L. (1974) Ponte des ouvrières et inhibition royale chez la fourmi *Temnothorax recedens* (Nyl.) (Formicidae, Myrmicinae). *Insects Sociaux*, **21** (4), 343–56.

Delage, B. (1968) Recherches sur les fourmis moissoneuses du Bassin Aquitain: éthologie, physiologie de l'alimentation. *Annales des Sciences Naturelles (Zoologie), Sér. 12*, **10** (2), 197–265.

Delage-Darchen, B. (1971) Contribution a l'étude écologique d'une savane de Côte d'Ivoire (Lamto): les fourmis des strates herbacées et arborées. *Biologia Gabonica*, **7** (4), 461–96.

Delage-Darchen, B. (1974) Écologie et biologie de *Crematogaster impressa* Émery, Fourmi savanicole d'Afrique. *Insectes Sociaux*, **21** (1), 13–34.

Delage-Darchen, B., Talec, S. and Darchen, R. (1979) Enzymatic studies on the salivary and midgut glands of the stingless bee, *Apotrigona nebulata* (Sm.), (Hymenoptera, Apoidea, Meliponidae). *Annales des Sciences Naturelles, (Zoologie), Paris, Sér. 13*, **1**, 261–7.

Délye, G. (1968) Recherches sur l'écologie, la physiologie et l'éthologie des Fourmis du Sahara. Thèses, Faculté des sciences, Université d'Aix-Marseille.

Délye, G. (1971) Observations sur le nid et le comportement constructeur de *Messor arenarius* (Hyménoptères: Formicidae). *Insectes Sociaux*, **18**, 15–20.

Dew, H.E. and Michener, C.D. (1978) Foraging flights of two species of *Polistes* wasps (Hymenoptera: Vespidae). *Journal of the Kansas Entomological Society*, **51** (3), 380–5.

Dlussky, G.M. (1967) *Muravi roda formika*, Moskva: Izdatel'stvo 'Nauka'. Moskva, pp. 236.

Dobrzańska, J. (1966) The control of the territory by *Lasius fuliginosus* Latr. *Acta Biologiae Experimentalis, Warsaw*, **26** (2), 193–213.

Dobrzańska, J. (1978) Problem of behavioural plasticity in slave-making Amazon-ant *Polyergus rufescens* Latr. and in its slave-ants *Formica fusca* L. and *Formica cinerea* Mayr. *Acta Neurobiologiae Experimentalis*, **38** (2–3), 113–32.

Dobrzański, J. (1971) Manipulatory learning in ants. *Acta Neurobiologiae Experimentalis*, **31**, 111–40.

Doncaster, C.P. (1981) The spatial distribution of ants on Ramsey Island, South Wales. *Journal of Animal Ecology,* **50** (1), 195–218.

Eastlake, O., Chew, A. and Chew, R.M. (1980) Body size as a determinant of small-scale distribution of ants in evergreen woodland, south-eastern Arizona. *Insectes Sociaux,* **27** (3), 189–202.

Edwards, J.P. (1982) Control of *Monomorium pharaonis* (L.) with methoprene baits: implications for the control of other pest species, in *The Biology of Social Insects,* Proceedings of the Ninth Congress of the International Union for the Study of Social Insects, Boulder, Colorado (eds. M.D. Breed, C.D. Michener and H.E. Evans), Westview Press, Boulder, Colorado, pp. 119–23.

Edwards, R. (1980) *Social wasps, their biology and control,* Rentokil Ltd, East Grinstead, 398 pp.

Eickwort, G.C. and Ginsberg, H.S. (1980) Foraging and mating behaviour in *Apoidea. Annual Review of Entomology,* **25,** 421–6.

Elmes, G.W. (1971) An experimental study on the distribution of heathland ants. *Journal of Animal Ecology,* **40,** 495–9.

Elmes, G.W. (1973) Observations on the density of queens in natural colonies of *Myrmica rubra* L. (Hymenoptera Formicidae). *Journal of Animal Ecology,* **42,** 761–71.

Elmes, G.W. (1974) The spatial distribution of a population of two ant species living in limestone grassland. *Pedobiologia,* **14,** 412–18.

Elmes, G.W. (1975) PhD Thesis, University of London.

Elmes, G.W. (1976) Some observations on the microgyne form of *Myrmica rubra* L. (Hymenoptera, Formicidae). *Insectes Sociaux,* **23,** 3–22.

Elmes, G.W. (1978a) A morphometric comparison of three closely related species of *Myrmica* (Formicidae), including a new species from England. *Systematic Entomology,* **3,** 131–45.

Elmes, G.W. (1978b) Populations of *Myrmica* (Formicidae) living on different types of *Calluna* moorland – a semi-natural habitat of southern England. *Memorabilia Zoologica,* **29,** 41–60.

Elmes, G.W. (1980a) Queen numbers in colonies of ants of the genus *Myrmica. Insectes Sociaux,* **27,** 43–60.

Elmes, G.W. (1980b) The comparative morphology of *Myrmica. Annual Report of the Institute of Terrestrial Ecology,* NERC, Swindon, pp. 51–4.

Elmes, G.W. and Wardlaw, J.C. (1981) The quantity and quality of overwintered larvae in five species of *Myrmica* (Hymenoptera: Formicidae). *Journal of Zoology, London,* **193,** 429–46.

Elmes, G.W. and Wardlaw, J.C. (1982a) A population study of the ants *Myrmica sabuleti* and *Myrmica scabrinodis,* living in two sites in the south of England. I A comparison of colony populations. *Journal of Animal Ecology,* **51** (2), 651–64.

Elmes, G.W., Wardlaw, J.C. (1982b) A population study of the ants *Myrmica sabuleti* and *Myrmica scabrinodis* living at two sites in the south of England. II Effect of above-nest vegetation. *Journal of Animal Ecology,* **51** (2), 665–80.

Elmes, G.W., Wardlaw, J.C. (1982c) A comparison of the effect of a queen upon the development of large hibernated larvae of four species of *Myrmica* (Hym. Formicidae). *Insectes Insectes Sociaux,* (in press).

Elmes, G.W., Wardlaw, J.C. (1982d) A comparison of the effect of temperature on the development of large hibernated larvae of four species of *Myrmica* (Hum. Formicidae). *Insectes Sociaux,* (in press).

Elton, E.T.G. (1958) The artificial establishment of wood ant colonies for biological control in the Netherlands. Proceedings, 10th International congress of entomology, pp. 573–8.

Emmert, W. (1968) Die Postembryonalentwicklung sekretorischer Kopfdrüsen von *Formica pratensis Retz. und Apis mellifica* L. (Ins., Hym). *Zeitschrift für Morphologie und Ökologie der Tiere,* **63,** 1–62.

Engels, W. and Engels, E. (1977) Vittelogenin and Fertilität bei Stachellösen Bienen. *Insectes Sociaux*, **24**, 71–94.

Esch, H. (1967) The sounds produced by swarming honeybees. *Zeitschrift für vergleichende Physiologie*, **56**, 408–11.

Esch, H., Esch, I. and Kerr, W.E. (1965) Sound: An element common to communication of stingless bees and to dances of the honey bee. *Science*, **149**, 320–1.

Ettershank, G. (1968) The three dimensional gallery structure of the nest of the meat ant *Iridomyrmex purpureus* (Sm.) (Hymenoptera: Formicidae). *Australian Journal of Zoology*, **16**, 715–23.

Ettershank, G., Ettershank, J.A. and Whitford, W.G. (1980) Location of food sources by subterranean termites. *Environmental Entomology*, **9** (5), 645–8.

Evans, H.E. and West-Eberhard, M.J. (1973) *The Wasps*. David and Charles, Newton Abbot, 265 pp.

Evesham, E.J.M. (1982) Regulation of the production of female sexual morphs in the ant *Myrmica rubra* L. Doctoral thesis, University of Southampton.

Fell, D.R., Ambrose, J.T., Burgett, D.M. *et al* (1977) The seasonal cycle of swarming in honeybees. Journal of Apicultural Research **16** (4), 170–3.

Ferguson, A.W. and Free, J.B. (1979) Production of a forage-marking pheromone by the honeybee *Journal of Apicultural Research*, **18** (2), 128–35.

Ferguson, A.W. and Free, J.B. (1980) Queen pheromone transfer within honeybee colonies. *Physiological Entomology*, **5** 359–66.

Ferguson, A.W., Free, J.B., Pickett, J.A. and Winder, M. (1979) Techniques for studying honeybee pheromones involved in clustering and experiments on the effect of Nasonov and queen pheromones. *Physiological Entomology*, **4**, 339–44.

Fisher, R.A. (1930) *The genetical theory of natural selection*, Oxford University Press, 272 pp.

Flanders, S.E. (1962) Physiological prerequisites of social reproduction in the Hymenoptera. *Insectes Sociaux*, **9**, 375–88.

Flanders, S. (1969) Social aspects of facultative gravidity and agravidity in social Hymenoptera. Proceedings, VIth International Congress of the International Union for the Study of Social Insects, Bern, pp. 47–53.

Fletcher, D.J.C. (1973) 'Army Ant' behaviour in the Ponerinae: a reassessment. Proceedings, VIIth International Congress of the International Union for the Study of Social Insects, London, pp. 116–21.

Fletcher, D.J.C. (1975–1976) New perspectives in the causes of absconding in the African bee (*Apis mellifera adansonii* L.). *South African Bee Journal I, II*, **47** (6), 11–14; **48** (1), 6–9.

Fletcher, D.J.C. (1978a) The influence of vibratory dances by worker honeybees on the activity of virgin queens. Vibration of queen cells by worker honeybees and its relation to the issue of swarms with virgin queens. *Journal of Apicultural Research*, **17** (1), 3–13, 14–26.

Fletcher, D.J.C. (1978b) The African bee, *Apis mellifera adansonii*, in Africa. *Annual Review of Entomology*, **23**, 151–71.

Fletcher, D.J.C. and Blum, M.S. (1981a) Pheromonal control of deälation and oogenesis in virgin queen fire ants. *Science*, **212**, 73–5.

Fletcher, D.J.C. and Blum, M.S. (1981b) A bioassay technique for an inhibitory primer pheromone of the fire ant, *Solenopsis invicta* Buren. *Journal of the Georgia Entomological Society*, **16** (3), 352–6.

Fletcher, D.J.C. and Brand, J.M. (1968) Source of the trail pheromone and method of trail laying in the ant *Crematogaster peringueyi*. *Journal of Insect Physiology*, **14** (6), 783–8.

Fletcher, D.J.C. Blum, M.S., Whitt, T.V. and Temple, N. (1980) Monogyny and polygyny in the Fire Ant, *Solenopsis invicta*. *Annals of the Entomological Society of America*, **73**, 658–61.

Fluker, S.S. and Beardsley, J.W. (1970) Sympatric associations of three ants: *Iridomyrmex humilis*, *Pheidole megacephala* and *Anoplolepis longipes* in Hawaii. *Annals of the*

Entomological Society of America, **63** (5), 1290–6.

Fowler, H.G. and Robinson, S.W. (1979) Foraging by *Atta sexdens* (Formicidae: Attini): seasonal patterns, caste and efficiency. *Ecological Entomology*, **4** (3), 239–47.

Free, J.B. (1959) The transfer of food between the adult members of a honeybee community. *Bee World*, **40**, 193–201.

Free, J.B. (1965) The allocation of duties among worker honey bees. *Symposium of the Zoological Society of London*, **14**, 39–59.

Free, J.B. (1967) Factors determining the collection of pollen by honeybee foragers. *Animal Behaviour*, **15** (1), 134–44.

Free, J.B. (1968) The conditions under which foraging honeybees expose their Nasonov gland. *Journal of Apicultural Research*, **7** (3), 139–45.

Free, J.B. (1970) *Insect Pollination of crops*, Academic Press, London and New York, 544 pp.

Free, J.B. (1971) Effect of flower shapes and nectar guides on the behaviour of foraging honeybees. *Behaviour*, **24**, 269–85.

Free, J.B. (1977) The seasonal regulation of drone brood and drone adults in a honeybee colony. Proceedings, VIIIth Congress of the International Union for the Study of Social Insects, Wageningen, pp. 207–10.

Free, J.B. and Spencer-Booth, Y. (1959) The longevity of worker honeybees (*Apis mellifera*). *Proceedings of the Royal Entomological Society of London (A)*, **34**, 141–50.

Free, J.B. and Williams, I. (1971) The effect of giving pollen and pollen supplement to honeybee colonies on the amount of pollen collected. *Journal of Apicultural Research*, **10** (2), 87–90.

Free, J.G. and Williams, I.H. (1972) The transport of pollen on the body hairs of honeybees (*Apis mellifera* L.) and bumblebees (*Bombus* spp. L.). *Journal of Applied Ecology*, **9**, 609–15.

Free, J.B. and Williams, I.H. (1973) Genetic determination of honeybee (*Apis mellifera* L.) foraging preferences. *Annals of Applied Biology*, **73**, 137–41.

Free, J.B. and Williams, I.H. (1975) Factors determining the rearing and rejection of drones by the honey-bee colony. *Animal Behaviour*, **23**, 650–75.

Free, J.B. and Williams, I.H. (1979) Communication by pheromones and other means in *Apis florea* colonies. *Journal of Apicultural Research*, **18** (1), 16–25.

Freeland, J. (1958) Biological and social patterns in the Australian bulldog ants of the genus *Myrmecia*. *Australian Journal of Zoology*, **6**, 1–18.

Fresneau, D. (1979) Étude du rôle sensoriel de l'antenne dans l'ethogenèse des soins aux cocons chez *Formica polyctena* Forst (Hym: Form.) *Insectes Sociaux*, **26** (3), 170–95.

Friesen, L.J. (1973) The search dynamics of recruited honeybees. *Biological Bulletin*, **144**, 107–31.

Frith, G. (1979) Digging and related behaviour in *Myrmica ruginodis*. Ph.D. Thesis, University of Hull.

Frogner, K.J. (1980) Variable developmental period: intraspecific competition models with conditional age-specific maturity and mortality schedules. *Ecology*, **61** (5), 1099–106.

Fukuda, H. and Sakagami, S.F. (1968) Worker brood survival in honeybees. *Researches on Population Ecology*, **10**, 31–9.

Gallé, L. Jr (1980) Dispersion of high density ant populations in sandy soil grassland ecosystems. *Acta Biologica Szeged*, **26** (1–4), 129–35.

Gamboa, G.J. (1975) Foraging and leaf-cutting of the desert gardening ant *Acromyrmex versicolor versicolor* (Pergande) (Hymenoptera: Formicidae). *Oecologia*, **20**, 103–10.

Gamboa, G.J. (1976) Effects of temperature on the surface activity of the desert leaf-cutter ant, *Acromyrmex versicolor versicolor* (Pergande) (Hymenoptera: Formicidae). *The American Midland Naturalist*, **95** (2), 485–91.

Gamboa, G.J. (1978) Intraspecific defense: advantage of social cooperation among paper wasp foundresses. *Science*, **199**, 1463–5.

Gamboa, G.J. (1980) Comparative timing of brood development between multiple- and single-foundress colonies of the paper wasps, *Polistes metricus*. *Ecological Entomology*, **5** (3), 221–5.

Gamboa, G.J., Bradley, D.H. and Wiltjer, S.L. (1978) Division of labor and subordinate longevity in foundress associations of the paper wasp, *Polistes metricus* (Hymenoptera: Vespidae). *Journal of the Kansas Entomological society*, **51** (3), 343–52.

Garofalo, C.A. (1978) Bionomics of *Bombus (Fervidobombus) morio* 2. Bodysize, length of life of workers. *Journal of Apicultural Research*, **17** (3), 130–6.

Gibo, D.L. (1974) A laboratory study on the selective advantage of foundress associations in *Polistes fuscatus* (Hymenoptera: Vespidae). *Canadian Entomologist*, **106**, 101–106.

Glancey, B.M., Stringer, C.E., Jr, Craig, C.H., Bishop, P.M. and Martin, B.B. (1973) Evidence of a replete caste in the fire ant *Solenopsis invicta*. *Annals of the Entomological Society of America*, **66**, 233–4.

Glancey, B.M., Stringer, C.E., Craig, C.H. and Bishop, P.M. (1975) An extraordinary case of polygyny in the red imported fire ants. *Annals of the Entomological Society of America*, **68**, 922.

Glancey, B.M., St Romain, M.K. and Crozier, R.H. (1976) Chromosome numbers of the red and black imported fire ants, *Solenopsis invicta* and *S. richteri*. *Annals of the Entomological Society of America*, **69** (3), 469–70.

Glover, P.E., Trump, E.C. and Waterridge, L.E.D. (1964) Termitaria and vegetation patterns on the Loita plains of Kenya. *Journal of Ecology*, **52**, 367–77.

Glunn, F.J., Howard, D.F. and Tschinkel, W.R. (1981) Food preference in colonies of the fire ant *Solenopsis invicta*. *Insectes Sociaux*, **28** (2), 217–22.

Göetsch, W. and Eisner, H. (1930) Beiträge zur Biologie Körnersammelnder Ameisen, II. Teil. *Zeitschrift für Morphologie und Ökologie der Tiere*, **16** (3–4), 371–452.

Gontarski, H. (1949) Mikrochemische Futtersaftuntersuchungen und die Frage der Königinentstehung. *Hessische Biene*, **85**, 89–92.

Gorton, R.E. (1978) Observations on the nesting behaviour of *Mischocyttarus immarginatus* (Rich.) (Vespidae: Hymenoptera) in a dry forest in Costa Rica. *Insectes Sociaux*, **25** (3), 197–204.

Gösswald, K. and Bier, K. (1953) Untersuchungen zur Kastendetermination in der Gattung *Formica*. 2. Die Aufzucht von Geschlechtstieren bei *Formica rufa pratensis* (Retz). *Zoologische Anzeiger*, **151**, 126–34.

Gould, J.L. (1976) The dance-language controversy. *The Quarterly Review of Biology*, **51** (2), 211–44.

Grassé, P.P. (1944) Recherches sur la biologie des termites champignonnistes (Macrotermitinae). *Annales des Science Naturelles (Zoologie)*, **6**, 97–171.

Grassé, P.P. (1945) Recherches sur la biologie des termites champignonnistes (Macrotermitinae). *Annales des Sciences Naturelles. (Zoologie)*, *Sér. 11*, **7**, 115–46.

Grassé, P.P. (1949) Ordre des Isoptères ou termites, in *Traité de zoologie* (ed. P.P. Grassé), Vol. IX, Masson et Cie, Paris, pp. 408–544.

Grassé, P.P. (ed.) (1951) *Traité de Zoologie*, Tome X, Masson et Cie, Paris.

Grassé, P.P. and Noirot, C. (1955) La fondation de nouvelles sociétés par *Bellicositermes natalensis* Hav. *Insectes Sociaux*, **2** (3), 213–20.

Grassé, P.P. and Noirot, C. (1961) Nouvelles recherches sur la systematique et l'éthologie des termites champignonnistes du genre *Bellicositermes*. Emerson. *Insectes Sociaux*, **8**, 311–59.

Gray, B. (1973) A morphometric study of worker variation in three *Myrmecia* species (Hymenoptera: Formicidae). *Insectes Sociaux*, **20**, 323–31.

Greenberg, L. (1979) Genetic component of bee odor in kin recognition. *Science*, **206**, 1095–7.

Greenslade, P.J.M. (1971) Interspecific competition and frequency changes among ants in Solomon Islands coconut plantations. *Journal of Applied Ecology*, **8**, 323–49.

Greenslade, P.J.M. (1975) Dispersion and history of a population of the meat ant. *Iridomyrmex*

References 335

purpureus (Hymenoptera: Formicidae). *Australian Journal of Zoology,* **23**, 495–510.

Greenslade, P.J.M. (1979) *A guide to the Ants of South Australia.* Special Educational Bulletin series, South Australian Museum, Adelaide.

Grogan, D.E. and Hunt, J.H. (1977) Digestive proteases of two species of wasps of the genus *Vespula. Insect Biochemistry,* **7**, 191–6.

Haas, A. (1949) Arttypische Flugbahnen von Hummelmännchen. *Zeitschrift für vergleichende Physiologie,* **31**, 281–307.

Haines, I.H. and Haines, J.B. (1978) Colony structure, seasonality and food requirements of the crazy ant, *Anoplolepis longipes* (Jerd.), in the Seychelles. *Ecological Entomology,* **3** (2), 109–18.

Halberstadt, K. (1980) Elektrophoretische Untersuchungen zur Sekretionstätigkeit der Hypopharynxdrüse der Honigbiene (*Apis mellifera* L.). *Insectes Sociaux,* **27** (1), 61–78.

Hamilton, W.D. (1964) The genetical evolution of social behaviour. Parts I and II. *Journal of Theoretical Biology,* **7** (1), 1–52.

Handel, S.N. (1976) Dispersal ecology of *Carex pedunculata* (Cyperaceae), a new North American myrmecochore. *American Journal of Botany,* **63** (8), 1071–9.

Handel, S.N. (1978) The competitive relationship of three woodland sedges and its bearing on the evolution of ant-dispersal of *Carex pedunculata. Evolution,* **32**, 151–63.

Handel, S.N., Fisch, S.B. and Schatz, G.E. (1981) Ants disperse a majority of herbs in a mesic forest community in New York State. *Bulletin of the Torrey Botanical Club,* **108** (4), 430–7.

Hangartner, W., Reichson, J.M. and Wilson, E.O. (1970) Orientation to nest material by the ant *Pogonomyrmex badius* (Latreille). *Amimal Behaviour,* **18**, 331–4.

Hansell, M.H. (1982) Brood development in the subsocial wasp *Parischnogaster mellyi* (Saussure) (Stenogastrinae, Hymenoptera). *Insectes Sociaux,* **29** (1), 3–14.

Harbo, J.R. (1979) The rate of depletion of spermatozoa in the queen honeybee spermatheca. *Journal of Apicultural Research,* **18** (3), 204–7.

Harper, J.L. (1977) *Population biology of plants,* Academic Press, London, 892 pp.

Harris, W.V. (1955) Termites and the soil, in *Soil Zoology,* (ed. D.K. Mc E. Kevan), Butterworths, London, pp. 62–72.

Harris, W.V. (1961) *Termites, their recognition and control,* Longmans, Green and Co., London, p. 187.

Harrison, J.S. and Gentry, J.B. (1981) Foraging pattern, colony distribution and foraging range of the Florida harvest ant, *Pogonomyrmex badius. Ecology,* **62** (6), 1467–73.

Haskins, C.P. and Haskins, E.F. (1950) Notes on the biology and social behaviour of the archaic ponerine ants of the genera *Myrmecia* and *Promyrmecia. Annals of the Entomological Society of America,* **43** (4), 461–91.

Haskins, C.P. and Haskins, E.F. (1964) Notes on the biology and social behaviour of *Myrmecia inquilina.* The only known myrmeciine social parasite. *Insectes Sociaux,* **11** (3), 267–82.

Haskins, C.P. and Haskins, E.F. (1965) *Pheidole megacephala* and *Iridomyrmex humilis* in Bermuda – equilibrium or slow replacement? *Ecology,* **46** (5), 736–40.

Haskins, C.P. and Haskins, E.F. (1980) Notes on female worker survivorship in the archaic ant genus *Myrmecia. Insectes Sociaux,* **27** (4), 345–50.

Hassell, M.P. and Southwood, T.R.E. (1978) Foraging strategies of insects. *Annual Review of Ecology and Systematics,* **9**, 75–95.

Havarty, M.I. (1979) Soldier production and maintenance of soldier proportions in laboratory experimental groups of *Coptotermes formosanus* Shiraki. *Insectes Sociaux,* **26** (1), 69–84.

Havarty, M.I. and Nutting, W.L. (1975) A simulation of wood consumption by the subterranean Termite *Heterotermes aureus* (Snyder), in an Arizona desert grassland. *Insectes Sociaux,* **22** (1), 93–102.

Havarty, M.L., Nutting, W.L. and LaFage, J.P. (1975) Density of colonies and spatial distribution of foraging territories of the desert subterranean termite, *Heterotermes aureus*

(Snyder). *Environmental Entomology,* **4,** 105–9.

Heinrich, B. (1974) Thermoregulation in bumblebees, I. Brood incubation by *Bombus vosmesenskii* queens. *Journal of Comparative Physiology,* **88,** 129–40.

Heinrich, B. (1979) *Bumblebee Economics,* Harvard University Press, Cambridge, Masachusetts and London, England, 245 pp.

Heinrich, B. and Raven, P.H. (1972) Energetics and Pollination Ecology. *Science,* **176,** 597–602.

Heinrich, B., Mudge, P.R. and Deringis, P.G. (1977) Laboratory analysis of flower constancy in foraging bumblebees: *Bombus ternarius* and *B. terricola. Behavioral Ecology and Sociobiology,* **2,** 247–65.

Hemmingsen, A.M. (1973) Nocturnal weaving on nest surface and division of labour in weaver ants (*Oecophylla smaragdina* Fabricius, 1775) (Hym., Formicidae). *Videnskabelige Meddelelser fra Dansk naturhistorik Forening: Kjøbenhavn,* **136,** 49–56.

Hepworth, D., Pickard, R.S. and Overshott, K.J. (1980) Effects of the periodically intermittent application of a constant magnetic field on the mobility in darkness of worker honeybees. *Journal of Apicultural Research,* **19,** 179–86.

Herbers, J. (1978) Trends in sex ratios of the reproductive broods of *Formica obscuripes.* (Hymenoptera: Formicidae). *Annals of the Entomological Society of America,* **71,** 791–3.

Herbers, J.M. (1979) The evolution of sex-ratio strategies in hymenopteran societies. *The American Naturalist,* **114** (6), 818–34.

Herbert, E.W., Svoboda, J.A., Thompson, M.J. and Shimanuki, H. (1980) Sterol utilization in honeybees fed on synthetic diet: effects on brood rearing. *Journal of Insect Physiology,* **26,** 287–9.

Hermann, H.R. (1971) Sting autotomy, a defensive mechanism in certain social Hymenoptera. *Insectes Sociaux,* **18** (2), 111–20.

Hermann, H. (ed) (1979–1982) *Social Insects,* Vol I, 437pp; Vol II, 491pp; Vol III, 459pp, Academic Press, New York.

Hermann, H.R. and Blum, M.S. (1967) The morphology and histology of the hymenopterous poison apparatus III *Eciton hamatum* (Formicidae). *Annals of the Entomological Society of America,* **60** (6), 1282–91.

Hermann, H.R. and Dirks, T.F. (1975) Biology of *Polistes annularis* 1. Spring behaviour. *Psyche,* **82,** 97–108.

Hermann, F.J. and Leese, B.M. (1956) A grass (*Munroa squarrosa*) apparently cultivated by ants. *The American Midland Naturalist,* **56** (2), 506–7.

Hewitt, P.H., Nel, J.J.C. and Conradie, S. (1969a) The role of chemicals in communication in the harvester termites *Hodotermes mossambicus* (Hagen) and *Trinervitermes trinervoides* (Sjöstedt). *Insectes Sociaux,* **16,** 79–86.

Hewitt, P.H., Nel, J.J.C. and Conradie, S. (1969b) Preliminary studies on the control of caste formation in the harvester termite *Hodotermes mossambicus* (Hagen). *Insectes Sociaux,* **16,** 159–72.

Higashi, S. (1976) Nest proliferation by budding and nest growth in *Formica (Formica) yessensis* in Ishikari Shore. *Journal of the Faculty of Science, Hokkaido Univeristy, Series VI, Zoology,* **20** (3), 359–89.

Higashi, S. (1978) Analysis of internest drifting in a supercolonial ant *Formica (Formica) yessensis* by individually marked workers. *Kontyû,* **46** (2), 176–91.

Higashi, S. (1979) Polygyny, nest budding and internest mixture of individuals in *Formica (Serviformica) japonica* Motschulsky at Ishikari Shore. *Kontyû,* **47** (3), 381–9.

Higashi, S. and Yamauchi, K. (1979) Influence of a supercolonial ant *Formica (Formica) yessensis* Forel on the distribution of other ants in Ishikari coast. *Japanese Journal of Ecology,* **29,** 257–64.

Hodges, C.M. (1981) Optimal foraging in bumblebees: hunting by expectation. *Animal Behaviour,* **29** (4), 1166–71.

References   337

Holdaway, F.G. and Gay, F.J. (1948) Temperature studies of the habitat of *Eutermes exitiosus* with special reference to the temperatures within the mound. *Australian Journal of Scientific Research, B,* **1**, 464–93.

Hölldobler, B. (1962) Zur Frage der Oligogynie bei *Camponotus ligniperda* (Latr.) und *Camponotus herculeanus* (L.) (*Hym. Formicidae*). *Sonderdruck aus Zoologischer angewandte Entomologie,* **49**, 337–52.

Hölldobler, B. (1971) Communication between ants and their guests. *Scientific American,* **224** (3), 86–93.

Hölldobler, B. (1973) Chemische Strategie beim Nahrungserwerb der Diebsameise (*Solenopsis fugax* Latr.) und der pharaoameise (*Monomorium pharaonis* L.). *Oecologia* **11**, 371–80.

Hölldobler, B. (1976a) Recruitment behaviour, home range orientation and territoriality in harvester ants, *Pogonomyrmex*. *Behavioral Ecology and Sociobiology* **1**, 3–44.

Hölldobler, B. (1976b) The behavioral ecology of mating in harvester ants (Hymenoptera: Formicidae: (*Pogonomyrmex*). *Behavioral Ecology and Sociobiology,* **1**, 405–23.

Hölldobler, B. (1977) Communication in social Hymenoptera, in *How animals communicate,* (ed. T.A. Sebeok), Indiana University Press, Bloomington, pp 418–471.

Hölldobler, B. (1979) Territories of the African Weaver ant (*Oecophylla longinoda*): a field study. *Zeitschrift für Tierpsychologie,* **51**, 201–13.

Hölldobler, B. (1980) Canopy orientation: A new kind of orientation in ants. *Science,* **210**, 86–8.

Hölldobler, B., and Engel, H. (1978) Tergal and sternal glands in ants. *Psyche,* **85** (4), 285–330.

Hölldobler, B. and Haskins, C.P. (1977) Sexual calling behaviour in primitive ants. *Science,* **195**, 793–4.

Hölldobler, B. and Michener, C.D. (1980) Mechanisms of identification and discrimination in social hymenoptera, in *Evolution of social behavior: Hypotheses and Empirical Tests,* (ed. H. Markl), Dahlem Konferenzen 1980, Verlag Chemie GmBH, Weinheim, pp. 35–8.

Hölldobler, B. and Möglich, M. (1980) The foraging system of *Pheidole militicida* (Hymenoptera: Formicidae). *Insectes Sociaux,* **27** (3), 237–64.

Hölldobler, B. and Traniello, J. (1980a) Tandem running pheromone in ants. *Naturwissenschaften,* **67**, 360.

Hölldobler, B. and Traniello, J.F.A. (1980b) The pygidial gland and chemical recruitment communication in *Pachycondyla* (= *Termitopone*) *laevigata*. *Journal of Chemical Ecology,* **6** (5), 883–93.

Hölldobler, B. and Wilson, E.O. (1977a) Weaver ants. *Scientific American,* **237**, 146–54.

Hölldobler, B. and Wilson, E.O. (1977b) The number of queens: an important trait in ant evolution. *Naturwissenschaften,* **64**, 8–15.

Hölldobler, B. and Wilson, E.O. (1978) The multiple recruitment systems of the African weaver ant *Oecophylla longinoda*. *Behavioral Ecology and Sociobiology,* **3**, 19–60.

Hölldobler, B. and Wüst, M. (1973) Ein Sexualpheromon bei der Pharaoameise *Monomorium pharaonis* (L.). *Zeitschrift für Tierpsychologie,* **32**, 1–9.

Hölldobler, B., Möglich, M. and Maschwitz, U. (1974) Communication by tandem running in the ant *Camponotus sericeus*. *Journal of Comparative Physiology,* **90**, 105–27.

Hölldobler, B., Stanton, R.C. and Markl. H. (1978) Recruitment and food-retrieving behaviour in *Novomessor* (Formicidae, Hymenoptera) I. Chemical signals. *Behavioral Ecology and Sociobiology,* **4**, 163–81.

Honk, C.G.J., Velthuis, H.H.W., Röseler, P.F. and Malotaux, M.E. (1980) The mandibular glands of *Bombus terrestris* queens as a source of queen pheromones. *Entomologia Experimentalis Applicata,* **28**, 191–8.

Howard, D.F. and Tschinkel, W.R. (1980) The effect of colony size and starvation on food flow and the fire ant, *Solenopsis invicta* (Hymenoptera: Formicidae). *Behavioral Ecology and Sociobiology,* **7**, 293–300.

Howse, P.E. (1968) On the division of labour in the primitive termite *Zootermopsis nevadensis*

(Hagen). *Insectes Sociaux,* 15, 45–56.

Howse, P.E. (1970) *Termites, a study in Social Behaviour,* Hutchinson, London.

Hubbard, M.D. (1974) Influence of nest material and colony odour on digging in the ant *Solenopsis invicta* (Hymenoptera, Formicidae). *Journal of the Georgia Entomological Society,* 9, 127–32.

Hubbard, M.D. and Cunningham, W.G. (1977) Orientation of mounds in the ant *Solenopsis invicta* (Hymenoptera, Formicidae, Myrmicinae). *Insectes Sociaux,* 24 (1), 3–7.

Hubbell, S.P. and Johnson, L.K. (1978) Comparative foraging behavior of six stingless bee species exploiting a standardized resource. *Ecology,* 59 (6), 1123–36.

Hughes, I.G. (1975) Changing altitude and habitat preferences of two species of wood-ant (*Formica rufa* and *F. lugubris*) in North Wales and Salop. *Transactions of the Royal Entomological Society of London,* 127 (3), 227–39.

Hung, A.C.F. and Vinson, S.B. (1977) Interspecific hybridization and caste specificity of protein in Fire Ant. *Science,* 196, 1458–60.

Hussain, A., Forrest, J.M.S. and Dixon, A.F.G. (1974) Sugar, organic acid, phenolic acid and plant growth regulator content of extracts of honeydew of the aphid *Myzus persicae* (Hom. Aphididae) and of its host plant *Raphanus sativus. Annals of Applied Biology,* 78 (1), 65–73.

Huwyler, S., Grob, K. and Viscontini, M. (1975) The trail pheromone of the ant *Lasius fuliginosus* identification of six components. *Journal of Insect Physiology,* 21, 299–304.

Imai, H.T., Crozier, R.H. and Taylor, R.W. (1977) Karyotype evolution in Australian ants. *Chromosoma,* 59, 341–93.

Imperatriz-Fonseca, V.L. (1977) Queen supersedure in *Paratrigona subnuda* Proceedings, VIIIth International Congress of the International Union for the Study of Social Insects, Wageningen, p. 78.

Inouye, D.W. (1978) Resource partitioning in bumblebees: experimental studies of foraging behavior. *Ecology,* 59 (4), 672–77.

Inouye, D.W. and Taylor, O.R. (1979) A temperate region plant–ant–seed predator system: consequences of extra-floral nectar secretion by *Helianthella quinqueneruis. Ecology,* 60 (1), 1–7.

Inozemtsev, A.A. (1974) Dinamika troficheskikh svyazey ryzikh lesnykh murav'ev i ikh rol'v regulyatsii chislennosti vrednykh bespozvonochnykn v dubravakh Tul'skov oblasti. *Ekologiya,* 3, 63–71.

Ishay, J. (1972) Thermoregulatory pheromones in wasps. *Experientia,* 28, 1185–7.

Ishay, J. (1975) Caste determination by social wasps: cell size and building behaviour. *Animal Behaviour,* 23, 425–31.

Ishay, J. (1976) Comb building by the oriental hornet (*Vespa orientalis*). *Animal Behaviour,* 24, 72–83.

Ishay, J. and Ikan, R. (1968) Food exchange between adults and larvae in *Vespa orientalis* F. *Animal Behaviour,* 16, 298–303.

Ishay, J. and Schwartz, A. (1973) Acoustical communication between the members of the oriental hornet (*Vespa orientalis*) colony. *Journal of the Acoustical Society of America,* 63, 640–9.

Jackson, D.A. (1982) The structure and organisation of a tropical ant community: studies on a Camerounian Cocoa Plantation. PhD Thesis, University of Oxford.

Jaffé, K. and Howse, P.E. (1979) The mass recruitment system of the leaf cutting ant, *Atta cephalotes* L. *Animal Behaviour,* 27 (2), 930–9.

Jaffé K., Bazire-Benazét, M. and Howse, P.E. (1979) An integumentary pheromone-secreting gland in *Atta* sp: territorial marking with a colony-specific pheromone in *Atta cephalotes. Journal of Insect Physiology,* 25, 833–9.

Jaisson, P. (1975) L'imprègnation dans l'ontogenèse comportements de soins aux cocons chez la jeune Fourmi rousse (*Formica polyctena* Först.). *Behaviour*, **52**, 1–37.

Jaisson, P. and Fresneau, D. (1978) The sensitivity and responsiveness of ants to their cocoons in relation to age and methods of measurement. *Animal Behaviour*, **26**, 1064–71.

Jander, R. (1976) Grooming and pollen manipulation in bees (Apoidea): the nature and evolution of movements involving the foreleg. *Physiological Entomology*, **1**, 179–94.

Janzen, D.H. (1966) Coevolution of mutualism between ants and acacias in Central America. *Evolution*, **20** (3), 249–75.

Janzen, D.H. (1969) Allelopathy by myrmecophytes: the ant *Azteca* as an allelopathic agent of *Cecropia*. *Ecology*, **50** (1), 147–53.

Janzen, D.H. (1972) Protection of *Barteria* (Passifloraceae) by *Pachysima* ants (Pseudomyrmecinae) in a Nigerian rain forest. *Ecology*, **53** (5), 885–92.

Janzen, D.H. (1973) Evolution of polygnous obligate acacia ants in western Mexico. *Journal of Animal Ecology*, **42**, 727–50.

Jeanne, R.L. (1972) Social biology of the neotropical wasp *Mischocyttarus drewseni*. *Bulletin of the Museum of Comparative Zoology at Harvard College*, **144** (3), 63–150.

Jeanne, R.L. (1975) The adaptiveness of social wasp nest architecture. *Quarterly Review of Biology*, **50**, 267–87.

Jeanne, R.L. (1979a) A latitudinal gradient in rates of ant predation. *Ecology*, **60** (6), 1211–24.

Jeanne, R.L. (1979b) Construction and utilization of multiple combs in *Polistes canadensis* in relation to the biology of a predaceous moth. *Behavioral Ecology and Sociobiology*, **4**, 293–310.

Jeanne, R.L. (1980) Evolution of social behaviour in the Vespidae. *Annual Review of Entomology*, **25**, 371–96.

Jeanne, R.L. (1981) Chemical communication during swarm emigration in the social wasp, *Polybia sericea* (Olivier). *Animal Behaviour*, **29**, 102–13.

Johansson, T.S.K. and Johansson, M.P. (1977) Feeding honey-bees pollen and pollen substitutes. *Bee World*, **58** (3), 105–18.

Johnson, L.K. and Hubbell, S.P. (1974) Aggression and competition among stingless bees: field studies. *Ecology*, **55**, 120–7.

Johnson, L.K. and Hubbell, S.P. (1975) Contrasting foraging strategies and coexistence of two bee species on a single resource. *Ecology*, **56**, 1398–406.

Johnson, R.A. (1981) Colony development and establishment of fungus comb in *Microtermes* sp. nr. *usambaricus* (Sjöstedt) (Isoptera: Macrotermitinae) from Nigeria. *Insectes Sociaux*, **28** (1), 3–12.

Johnson, R.A. and Wood, T.G. (1979) Termites of the acid zones of Africa and the Arabian peninsula. *Sociobiology*, **5** (3), 279–93.

Jones, R.J. (1979) Expansion of the nest of *Nasutitermes costalis*. *Insectes Sociaux*, **26** (4), 322–42.

Jones R.J. (1980) Gallery construction by *Nasutitermes costalis*: Polyethism and the behaviour of individuals. *Insectes Sociaux*, **27** (1), 5–24.

Josens, G. (1971a) Le renouvellement des meules à champignons contruites par quatre Macrotermitinae (Isoptères) des savanes de Lamto-Pacobo (Côte d'Ivoire) *Compte Rendu Hebdomadaire des Séances de l'Académie des Sciences, Paris, Sér. D*, **272**, 3329–32.

Josens, G. (1971b) Variations thermiques dans les nids de *Trinervitermes geminatus* Wasmann, en relation avec le milieu extérieur dans la savane de Lamto (Côte d'Ivoire). *Insectes Sociaux*, **18** (1), 1–14.

Josens, G. (1972) Etudes biologiques et écologiques des termites (Isoptera) de la savane de Lamto-Pakobo (Côte d'Ivoire). Doctoral thesis, University of Brussels.

Jutsum, A.R. (1979) Interspecific aggression in leaf-cutting ants. *Animal Behaviour*, **27**, 833–8.

Jutsum, A.R. and Fisher, M. (1979) Reserves in sexual forms of *Acromyrmex octospinosus* (Reich) (Formicidae, Attini). *Insectes Sociaux*, **26** (2), 111–20.

Jutsum, A.R. Saunders, T.S. and Cherret, J.M. (1979) Intraspecific aggression in the leaf-cutting ant *Acromyrmex octospinosus. Animal Behaviour,* **27** (3), 839–44.

Kalmus, H. and Ribbands, C.R. (1952) The origin of the odours by which honeybees distinguish their companions. *Proceedings of the Royal Society, (B)* **140**, 50–9.

Kalshoven, L.G.E. (1959) Observations on the nests of initial colonies of *Neotermes tectonae* (Damm.) in teak trees. *Insectes Sociaux,* **6**, 231–242.

Kefuss, J.A. (1978) Influence of photoperiod on the behaviour and brood rearing activities of the honeybee. *Journal of Apicultural Research,* **17** (3), 137–51.

Kerr, W.E. and Hebling, N.J. (1964) Influence of the weight of worker bees on the division of labor. *Evolution,* **18**, 267–70.

Kerr, W.E., Sakagami, S.F., Zucchi, R., Portugal-Araújo, V. and de Camargo, J.M.F. (1967) Observações sobre a arquitetura dos ninhos e comportamento de algumas espécies de abelhas sem ferrão das vizinhanças de Manaus, Amazonas (Hymenoptera, Apoidea). *Atlas do Simpósio sôbre a Biota Amazônica, Conselho Nacional de Pesquisas, Rio de Janeiro,* **5** (*Zool.*), 255–309.

King, T.J. (1977) The plant ecology of ant hills in calcareous grasslands. *Journal of Ecology,* **65**, 235–315.

Klahn, J.E. (1979) Philopatric and nonphilopatric foundress associations in the social wasp *Polistes fuscatus. Behavioral Ecology and Sociobiology,* **5**, 412–24.

Kleinfeldt, S.E. (1978) Ant-gardens: the interaction of *Codonanthe crassifolia* (Gesneriaceae) and *Crematogaster longispina* (Formicidae). *Ecology,* **59** (3), 449–56.

Klimetzek, D. (1981) Population studies on hill building wood-ants of the *Formica rufa* group. *Oecologia,* **48**, 418–21.

Knerer, G. (1977) Caste and male production in halictine bees. Proceedings, VIIIth Congress of the International Union for the Study of Social Insects, Wageningen, pp. 132–3.

Knuth, P. (1906–1909) *Handbook of flower pollination,* (Translator J.R. Ainsworth Davis) Vol. I, 1906; Vol. II, 1908; Vol. III, 1909, Oxford University Press, Oxford.

Koeniger, N. (1975) Observations on alarm behaviour and colony defence of *Apis dorsata.* Proceedings, International Union for the Study of Social Insects, Dijon, pp. 153–172.

Koeniger, N. and Fuchs, S. (1973) Sound production as colony defense in *Apis cerana* Fabr. Proceedings of the VIIth International Congress of the International Union for the Study of Social Insects, London, pp. 199–204.

Koeniger, N. and Vorwohl, G. (1979) Competition for food among four sympatric species of Apini in Sri Lanka (*Apis dorsata, Apis cerana, Apis florea, Trigona iridipennis*). *Journal of Apicultural Research,* **18** (2), 95–109.

Koeniger, N. and Wijayagunasekera, H.N.P. (1976) Time of drone flight in the three Asiatic honeybee species (*Apis cerana, Apis florea, Apis dorsata. Journal of Apicultural Research,* **15** (2), 67–71.

Köhler, F. (1955) Wache und Volksduft im Bienenstaat. *Zeitschrift für Bienenforschung,* **3**, 57–63.

Kondoh, M. (1968) Bioeconomic studies on the colony of an Ant, species *Formica japonica* Motschulsky. I Nest structure and seasonal change of the colony members. *Japanese Journal of Ecology,* **18**, 124–33.

Korst, P.J.A.M. and Velthuis, H.H.W. (1982) The nature of trophallaxis in honey bees. *Insectes Sociaux,* **29** (2), 209–21.

Krishna, K. and Weesner, F.M., (eds) (1969–70) *Biology of the termites,* Vols. I and II, Academic Press, New York and London.

Kugler, C. (1979a) Evolution of the sting apparatus in the Myrmicine ants. *Evolution,* **33** (1), 117–130.

Kugler, C. (1979b) Alarm and defense: a function for the pygidial gland of the myrmicine ant, *Pheidole biconstricta. Annals of the Entomological Society of America,* **72** 532–6.

Kugler, J., Orion, T. and Ishay, J. (1976) The number of ovarioles in the Vespinae. *Insectes Sociaux*, **23**, 525–33.

Kugler, J., Motro, M. and Ishay, J.S. (1979) Comb building abilities of *Vespa orientalis* L. queenless workers. *Insectes Sociaux*, **26** (2), 143–9.

Kukuk, P.F., Breed, M.D., Sobti, A. and Bell, W.S. (1977) The contributions of kinship and conditioning to nest recognition and colony member recognition in a primitively eusocial bee, *Lasioglossum zephyrum* (Hymenoptera: Halictidae). *Behavioral Ecology and Sociobiology*, **2**, 319–27.

Kutter, H. (1977) Hymenoptera, Formicidae. *Insectes Helvetica*, **6**, 1–298.

La Fage, J.P. and Nutting, W.L. (1978) Nutrient dynamics of termites, in *Production ecology of ants and termites*, (ed. M.V. Brian), Cambridge University Press, pp. 165–232.

Laine, K.J. and Niemelä, P. (1980) The influence of ants on the survival of mountain birches during an *Oporinia autumnata* (Lep., Geometridae) outbreak. *Oecologia*, **47**, 39–42.

Landolt, P.J. and Akre, R.D. (1979) Occurrence and location of exocrine glands in some social Vespidae (Hymenoptera). *Annals of the Entomological Society of America*, **72**, 141–8.

Landolt, P.J., Akre, R.D. and Green, A. (1977) Effects of colony division on *Vespula atropilosa* (Sladen) (Hymenoptera: Vespidae). *Journal of the Kansas Entomological Society*, **50**, 135–47.

Lavigne, R.J. (1969) Bionomics and nest structure of *Pogonomyrmex occidentalis*. *Annals of The Entomological Society of America*, **62**, 1166–75.

Ledoux, A. (1950) Recherche sur la biologie de la Fourmi Fileuse (*Oecophylla longinoda* Latr.). *Annales des Sciences Naturelles: Zoologie et Biologie Animale*, **12**, 313–461.

Lee, K.E. and Wood T.G. (1971) *Termites and Soils*, Academic Press, London and New York, 251 pp.

LeMasne, G. and Bonavita, A. (1969) La fondation des sociétés selon le 'type *Myrmecia*' chez la Fourmi *Manica rubida* Latr. Proceedings, VIth International Congress of the International Union for the Study of Social Insects, Bern, pp. 137–47.

Lenoir, A. (1979a) Le comportement alimentaire et la division du travail chez la Fourmi *Lasius niger*. *Bulletin biologique de la France et de la Belgique. Tome CXIII*, (2–3), 79–314.

Lenoir, A. (1979b) Feeding behaviour in young societies of the ant *Tapinoma erraticum* L. trophallaxis and polyethism. *Insectes Sociaux*, **26** (1) 19–37.

Lensky, Y. (1963) Études sur la physiologie et l'écologie de l'Abeille (*Apis mellifica* L. var *ligustica*) en Israël. Thèse, Université Hébraïque de Jerusalem.

Lensky, Y. and Slabezki, Y. (1981) The inhibiting effect of the queen bee (*Apis mellifera* L.) foot-print pheromone on the construction of swarming queen cups. *Journal of Insect Physiology*, **27** (5), 313–24.

Lenz, M., McMahan, E.A. and Williams, E.R. (1982) Neotenic production in *Cryptotermes brevis* (Walker): influence of geographical origin, group composition, and maintenance conditions (Isoptera: Kalotermitidae). *Insectes Sociaux*, **29** (2), 148–63.

Lepage, M.G. (1981a) Etude de la prédation de *Megaponera foetens* (F.) sur les populations récoltantes de Macrotermitinae dans un écosystème semi-aride (Kajiado–Kenya). *Insectes Sociaux*, **28** (3), 247–62.

Lepage, M.G. (1981b) L'impact des populations récoltantes de *Macrotermes michaelseni* (Sjostedt) (Isoptera, Macrotermitinae) dans un écosystème semi-aride (Kajiado–Kenya). *Insectes Sociaux*, **28** (3), 297–308.

Leprun, J.C. and Roy-Noël, J. (1977) Les caractères analytiques distinctifs des matériaux des nids du genre *Macrotermes* au Sénégal Occidental. Leurs rapports avec les sols. *Pedobiologia*, **17**, 361–8.

Leroux, J.M. (1979a) Possibilities de scissions multiples pour des colonies de dorylines *Annoma nigricans* Illiger, Hymenoptères: Formicidae, en Côte d'Ivoire. *Insects Sociaux,* **26** (1), 13–17.

Leroux, J.M. (1979b) Sur quelques modalités de disparitions des colonies *d'Annoma nigricans* Illiger (Formicidae: Dorylinae) dans la region de Lamto (Cote d'Ivoire). *Insectes Sociaux,* **26** (2), 92–9.

Leston, D. (1973) The ant mosaic – tropical tree crops and the limiting of pests and diseases. *Pest Articles and News Summaries* **19**, 311–41.

Letendre, M. and Huot, L. (1972) Considérations préliminaires en vue de la revision taxonomique des fourmis du groupe *microgyna* genre *Formica* (Hymenoptera: Formicidae). *Annales de la Société entomologique de Québec,* **17** (3), 117–32.

Leuthold, R.H. (1968) A tibial gland scent-trail and trail-laying behaviour in the ant *Cremato-gaster ashmeadi* Mayr. *Psyche, Cambridge,* **75** (3), 233–48.

Leuthold, R.H. (1977) Post flight communication in two termite species, *Trinervitermes bettonianus* and *Hodotermes mossambicus.* Proceedings VIIIth International Congress of the International Union for the Study of Social Insects, Wageningen, pp. 62–4.

Leuthold, R.H., Bruinsma, O. and van Huis, A. (1976) Optical and pheromonal orientation and memory for homing distance in the harvester termite *Hodotermes mossambicus* (Hagen) (Isopt., Hodotermitidae) *Behavioral Ecology and Sociobiology,* **1** (2), 127–39.

Lévieux, J. (1971) Mise en évidence de la structure des nids et de l'implantation des zones de chasse de deux espèces de *Camponotus* (Hym., Form.) à l'aide de radio-isotopes. *Insectes Sociaux,* **18** (1), 29–48.

Lévieux, J. (1975) La nutrition des fourmis tropicales. *Insectes Sociaux,* **22** (4), 381–90.

Lévieux, J. (1976) Deux aspects de l'action des fourmis (Hymenoptera, Formicidae) sur le sol d'une savane préforestière de Côte-d'Ivoire. *Bulletin d'Ecologie,* **7** (3), 283–95.

Lévieux, J. (1977) The nutrition of tropical ants: V. Modes of exploitation of the resources of the community (Hym., Formicidae). *Insectes Sociaux,* **24** (3), 235–60.

Lévieux, J. (1979) La nutrition des fourmis granivore – IV Cycle d'activité et régime alimentaire de *Messor galla* et de *Messor* (= *Cratomyrmex*) *regalis* en saison des pluies fluctuations anuelles. Discussion. *Insectes Sociaux,* **26** (4), 279–94.

Lévieux, J. and Diomande, T. (1978a) La nutrition des fourmis granivores I. Cycle d'activité et régime alimentaire de *Messor galla* et de *Messor* (= *Cratomyrmex*) *regalis* (Hymenoptera, Formicidae). *Insectes Sociaux,* **25** (2), 127–39.

Lévieux, J. and Diomande, T. (1978b) La nutrition des fourmis granivores. II. Cycle d'activité et régime alimentaire de *Brachyponera senaarensis* (Mayr) (Hymenoptera Formicidae). *Insectes Sociaux,* **25** (3), 187–96.

Lévieux, J. and Louis, D. (1975) La nutrition des fourmis tropicales. II. Comportement alimentaire et régime de *Camponotus vividus* (Smith) (Hym. Form.) comparaison intragenérique. *Insectes Sociaux,* **22** (4), 391–404.

Levin, B.R. and Kilmer, W.L. (1974) Interdemic selection and the evolution of altruism: a computer simulation study. *Evolution,* **28**, 527–45.

Lewis, T., Pollard, G.V. and Dibley, G.C. (1974) Micro-environmental factors affecting diel patterns of foraging in the leaf-cutting ant *Atta cephalotes* (L.) (Formicidae: Attini). *Journal of Animal Ecology,* **43**, 143–53.

Lieberburg, I., Kranz, P.M. and Seip, A. (1975) Bermudan ants revisited: the status and interaction of *Pheidole megacephala* and *Iridomyrmex humilis*. *Ecology,* **56**, 473–8.

Light, S.F. and Weesner, F.M. (1955) The incipient colony of *Tenuirostritermes tenuirostris* (Desneux). *Insectes Sociaux,* **2**, 135–46.

Lindauer, M. (1955) Schwarmbienen aus Wohnungssuche. *Zeitschrift für vergleichende Physiologie,* **37**, 263–324.

Lindauer, M. and Kerr, W.E. (1958) Die gegenseitige Verständigung bei den stachellosen Bienen. *Zietschrift für vergleichende Physiologie,* **41**, (4), 405–34.

Lindauer, M. and Kerr, W.E. (1960) Communication between the workers of stingless bees. *Bee World*, **41**, 29–41, 65–71.

Litte, M. (1977) Behavioral ecology of the social wasp *Mischocyttarus mexicanus*. *Behavioral Ecology and Sociobiology*, **2**, 229–46.

Lofgren, C.S., Banks W.A. and Glancey, B.M. (1975) Biology and control of imported fire ants. *Annual Review of Entomology*, **20**, 1–30.

Longhurst, C. and Howse, P.E. (1978) The use of kairomones by *Megaponera foetens* (Fab.) (Hymenoptera: Formicidae) in the detection of its termite prey. *Animal Behaviour*, **26**, 1213–18.

Longhurst, C. and Howse, P.E. (1979a) Foraging recruitment and emigration in *Megaponera foetens* (Fab.) (*Hym. For.*) from the Nigerian Guinea savanna. *Insectes Sociaux*, **26** (3), 204–15.

Longhurst, C. and Howse, P.E. (1979b) Some aspects of the biology of the males of *Megaponera foetens* (Fab.) (Hymenoptera: Formicidae). *Insectes Sociaux*, **26** (2), 85–91.

Longhurst, C., Baker, R., Howse, P.E. and Speed, W. (1978a) Alkylpyrazines in ponerine ants: their presence in three genera, and caste specific behavioural responses to them in *Odontomachus troglodytes* (Hym., Formicidae). *Journal of Insect Physiology*, **24** (12), 833–7.

Longhurst, C., Johnson, R.A. and Wood, T.G. (1978b) Predation by *Megaponera foetens* (Fabr.) (Hymenoptera: Formicidae) on termites in the Nigerian Southern Guinea Savanna. *Oecologia*, **32**, 101–7.

Longhurst, C., Baker, R. and Howse, P.E. (1979) Termite predation by *Megaponera foetans* (Fabr.) (Hymenoptera: Formicidae): coordination of raids by glandular secretions. *Journal of Chemical Ecology*, **5**, 703–19.

Longhurst, C., Baker, R. and Howse, P.E. (1980) A multicomponent mandibular gland secretion in the ponerine ant *Bothroponera soror* (Emery). *Journal of Insect Physiology*, **26**, 551–5.

Lövgren, B. (1958) A mathematical treatment of the development of colonies of different kinds of social wasps. *Bulletin of Mathematical Biophysics*. **20**, 119–48.

Lüscher, M. (1961) Social control of polymorphism in termites. *Symposium of The Royal Entomological Society of London*, **1**, 57–67.

Lüscher, M. (1964) Die spezifische Wirkung männlicher und weiblicher Ersatzgeschlechtstiere auf die Entstehung von Ersatzg schlechtstieren bei der Termite *Kalotermes flavicollis* (Fabr.). *Insectes Sociaux*, **11**, 79–90.

Lüscher, M. (1974) Kasten und Kastendifferenzierung bei niederen Termiten, in *Sozialpolymorphismus bei Insecten*, (ed. G.H. Schmidt), Wissenschaftliche Verlagsgesellschaft, Stuttgart, pp. 694–762.

Lüscher, M. (1976) Evidence for an endocrine control of caste determination in higher termites, in *Phase and caste determination in insects*, Proceedings of the Section physiology and biochemistry of the XVth International Congress of Entomology, Washington D.C. (ed. M. Lüscher), Pergamon Press, Oxford pp. 91–104.

Lynch, J.F., Balinsky, E.C. and Vail, S.G. (1980) Foraging patterns in three sympatric forest ant species, *Prenolepis imparis*, *Paratrechina melanderi* and *Aphaenogaster rudis* (Hymenoptera: Formicidae). *Ecological Entomology*, **5** (4), 353–71.

Mabelis, A.A. (1979a) Wood ant wars: the relationship between aggression and predation in the red wood ant (*Formica polyctena* Först). *Netherlands Journal of Zoology*, **29** (4), 451–620.

Mabelis, A.A. (1979b) Nest splitting by the red wood ant (*Formica polyctena* Foerster). *Netherlands Journal of Zoology*, **29** (1), 109–25.

MacKay, W.P. (1981) A comparison of the nest phenologies of three species of *Pogonomyrmex* harvester ants (Hymenoptera: Formicidae). *Psyche*, **88** (1–2), 25–74.

Mackensen, O. and Nye, W.P. (1969) Selective breeding of honeybees for alf-alfa pollen collection; sixth generation and outcrosses. *Journal of Apicultural Research*, **8**, 9–12.

McGurk, D.J., Frost, J., Eisenbraun, E.J., Vick, K., Drew, W.A. and Young, J. (1966) Volatile compounds in ants: Identification of 4-methyl-3-heptanone from *Pogonomyrmex* ants. *Journal of Insect Physiology,* 12, 1435–41.

McLellan, A.R., Rowland, C.W. and Fawcett, R.H. (1980) A monogynous eusocial insect worker population model with particular reference to honey bees. *Insectes Sociaux,* 27 (4), 305–11.

Majer, J.D. (1976a) The maintenance of the ant mosaic in Ghana cocoa farms. *Journal of Applied Ecology,* 13, 123–44.

Majer, J.D. (1976b) The ant mosaic in Ghana cocoa farms: further structural considerations. *Journal of Applied Ecology,* 13, 145–55.

Majer, J.D. (1976c) The influence of ants and ant manipulation on the cocoa farm fauna. *Journal of Applied Ecology,* 13, 157–75.

Majer, J.D. (1978) The influence of blanket and selective spraying on ant distribution in a west African cocoa farm. *Revista Theobroma (Brasil),* 8, 87–93.

Malyshev, S.I. (1966) *Genesis of the Hymenoptera* Translation by National Lending Library for Science and Technology, 1968 (eds. O.W. Richards and B. Uvarov) Methuen, London, 319 pp.

Marikovsky, P.I. (1963) The ants *Formica sanguinea* (Latr.) as pillagers of *Formica rufa* (Lin.) nests. *Insectes Sociaux,* 10, 119–28.

Markin, G.P. (1970) Food distribution within laboratory colonies of the Argentine ant, *Iridomyrmex humilis* (Mayr). *Insectes Sociaux,* 17 (2), 127–57.

Markin, G.P., Dillier, J.H., Hill, S.O., Blum, M.S. and Hermann, H.R. (1971) Nuptial flight and flight ranges of the imported fire ant, *Solenopsis invicta. Annals of the Entomological Society of America,* 66, 803–8.

Markin, G.P., Collins, H.L. and Dillier, J.H. (1972) Colony founding by queens of the red imported fire ant *Solenopsis invicta. Annals of the Entomological Society of America,* 65, 1053–8.

Markin, G.P., Dillier, J.H. and Collins, H.L. (1973) Growth and development of colonies of the red imported fire ant, *Solenopsis invicta. Annals of the Entomological Society of America,* 66 (4), 803–8.

Markl, H. (1964) Geomenotaktische fehlorientierung bei *Formica polyctena* Förster. *Zeitschrift für vergleichende Physiologie,* 48, 552–86.

Markl, H. (1973) The evolution of stridulatory communication in ants. Proceedings, VIIth International Congress of the International Union for the Study of Social Insects, pp. 258–65.

Markl, H. and Hölldobler, B. (1978) Recruitment and food-retrieving behaviour in *Novomessor* (Formicidae, Hymenoptera) II. Vibration signals. *Behavioral Ecology and Sociobiology,* 4 183–216.

Markl, H., Hölldobler, B. and Hölldobler, T. (1977) Mating behavior and sound production in harvester ants (*Pogonomyrmex,* Formicidae). *Insectes Sociaux,* 24 (2), 191–212.

Martin, H. and Lindauer, M. (1966) Sinnesphysiologische Leistungen beim Wabenbau der Honigbiene. *Zeitscrift für vergleichende Physiologie,* 53 (3), 372–404.

Martin, M.M., Carman, R.M. and MacConnell, J.G. (1969) Nutrients derived from the fungus cultured by the fungus-growing ant *Atta colombica tonsipes. Annals of the Entomological Society of America,* 62 (1), 11–13.

Martin, M.M., Boyd, N.D., Gieselmann, M.J. and Silver, R.G. (1975) Activity of faecal fluid of a leaf-cutting ant toward plant cell wall polysaccharides. *Journal of Insect Physiology,* 21, 1887–92.

Maschwitz, U. (1966) Das Speichelsekret der Wespenlarven und seine biologische Bedeutung. *Zeitschrift für vergleichende Physiologie,* 53 (3), 228–52.

Maschwitz, U. (1974) Vergleichende Untersuchungen zur Funktion der Ameisenmetathora-kaldrüse. *Oecologia (Berl.),* 16, 303–10.

Maschwitz, U. (1975a) Old and new trends in the investigation of chemical recruitment in ants. Pheromones and defensive secretions in social insects, in *Pheromones and defensive secretions in social insects*. Symposium of International Union for the Study of Social Insects, Dijon, pp. 47–57.

Maschwitz, U. (1975b) Old and new chemical weapons in ants, in *Pheromones and defensive secretions in social insects*. Symposium of the International Union for the Study of Social Insects, Dijon, pp. 41–6.

Maschwitz, U. and Hölldobler, B. (1970) Der Kartonnestbau bei *Lasius fuliginosus* Latr. (Hym. Formicidae). *Zeitschrift für vergleichende Physiologie*, **66**, 176–89.

Maschwitz, U. and Muhlenberg, M. (1972) *Camponotus rufoglaucus*, eine wegelagernde Ameise. *Zoologischer Anzeiger Leipzig*, **191** (5/6), 364–8.

Maschwitz, U. and Schöernegge, P. (1977) Recruitment gland of *Leptagenys chinensis*. A new type of pheromone gland in ants. *Naturwissenshaften*, **64** (11), 589–90.

Maschwitz, U. and Tho, Y.P. (1974) Chinone als wehrsubstanzen bei einigen orientalische Macrotermitinen. *Insectes Sociaux*, **21** (3), 231–4.

Maschwitz, U., Koob, K. and Schildrnecht, H. (1970) Ein Beitrag zur Funktion der metathoracaldrüse der ameisen. *Journal of Insect Physiology*, **16**, 387–404.

Maschwitz, U., Jander, R. and Burkhardt, D. (1972) Wehrsubstanzen und wehrerhalten der Termite *Macrotermes carbonarius*. *Journal of Insect Physiology*, **18** 1715–20.

Maschwitz, U., Hölldobler, B. and Möglich, M. (1974) Tandemlaufen als Rekrutierungsverhalten bei *Bothroponera tesserinoda* Forel. *Zeitschrift für Tierpsychologie*, **35**, 113–23.

Masevicz, S. (1979) Some consequences of Fisher's sex ratio principle for social hymenoptera that reproduce by colony fission. *The American Naturalist*, **113** (3), 363–71.

Masne, G. Le (1953) Observations sur la biologie de la fourmi *Ponera eduardi* Forel. *Insectes Sociaux*, **3**, 239–59.

Matsumoto, T. and Abe, T. (1979) The role of termites in an equatorial rain forest ecosystem of west Malaysia. II. Leaf litter consumption on the forest floor. *Oecologia*, **38**, 261–74.

Maynard Smith, J. (1974) *Models in Ecology*, Cambridge University Press, 146 pp.

Maynard Smith, J. (1978a) *The evolution of sex*, Cambridge University Press, 209 pp.

Maynard Smith, J. (1978b) Optimization theory in evolution. *Annual Review of Ecology and Systematics*, **9**, 31–56.

Messina, F.J. (1981) Plant protection as a consequence of an ant-membracid mutualism: interactions on goldenrod (*Solidago* sp.). *Ecology*, **62** (6), 1433–40.

Metcalf, R.A. and Whitt, G.S. (1977a) Intra-nest relatedness in the social wasp *Polistes metricus*. A genetic analysis. *Behavioral Ecology and Sociobiology*, **2** (4), 339–51.

Metcalf, R.A. and Whitt, G.S. (1977b) Relative inclusive fitness in the social wasp *Polistes metricus*. *Behavioral Ecology and Sociobiology*, **2** (4), 353–60.

Meudec, M. (1977) Le comportement de transport du couvain lors d'une perturbation du nid chez *Tapinoma erraticum* (Dolichoderinae). Rôle de l'individu. *Insectes Sociaux*, **24** (4), 345–52.

Michener, C.D. (1974) *The Social Behaviour of the Bees*, Harvard University Press, Cambridge, Massachussetts, 404 pp.

Michener, C.D. (1975) The Brazilian bee problem. *Annual Review of Entomology*, **20**, 399–416.

Michener, C.D. and Brothers, D.J. (1974) Were workers of eusocial hymenoptera initially altruistic or oppressed? *Proceedings of the National Academy of Science, USA*, **71** (3), 671–4.

Mintzer, A. (1979) Foraging activity of the Mexican leaf cutting ant *Atta mexicana* (F. Smith), in a sonoran desert habitat (Hymenoptera, Formicidae). *Insectes Sociaux*, **26** (4), 365–72.

Möglich, M. (1975) Recruitment of *Leptothorax*. Proceedings, International Union for the Study of Social Insects, Dijon, pp. 235–42.

Möglich, M.H.J. and Alpert, G.D. (1979) Stone dropping by *Conomyrma bicolor* (Hymenoptera: Formicidae): A new technique of interference competition. *Behavioral Ecology and Sociobiology*, **6** (2), 105–14.

Möglich, M. and Hölldobler, B. (1975) Communication and orientation during foraging and emigration in the ant *Formica fusca*. *Journal of Comparative Physiology*, **101**, 275–88.

Moli, F. Le (1978) Social influence on the acquisition of behavioural patterns in the ant *Formica rufa* L. *Bollettino di Zoologia, Pubblicato dall' Unione Zoologica Italiana*, **45**, 399–404.

Moli, F. Le (1980) On the origin of slaves in dulotic ant societies. *Bollettino di Zoologia, Pubblicato dall' Unione Zoologica Italiana*, **47**, 207–12.

Moli, F. Le and Passetti, M. (1978) The effect of early learning on recognition, acceptance and care of cocoons in the ant *Formica rufa* L. *Atti della Società Italiana di Scienze Naturali, e del Museo Civile de Storia Naturale di Milano*, **118** (1), 49–64.

Montagner, H. (1966) Le mécanisme et les conséquences des comportements trophallactiques chez les guêpes du genre *Vespa*. Thèses, Faculté des Sciences de l'Université de Nancy, Nancy.

Moore, B.P. (1968) Studies on the chemical composition and function of the cephalic gland secretion in Australian termites. *Journal of Insect Physiology*, **14** (1), 33–9.

Morgan, E.D., Inwood, M.R. and Cammaerts, M.C. (1978) The mandibular gland secretion of the ant, *Myrmica scabrinodis*. *Physiological Entomology*, **3**, 107–14.

Morgan, E.D., Parry, D. and Tyler, R.C. (1979) The chemical composition of the Dufour gland secretion of the ant *Myrmica scabrinodis* (sic). *Insect Biochemistry*, **9** (1), 117–21.

Morse, D.H. (1978) Size-related foraging differences of bumblebee workers. *Ecological Entomology*, **3** (3), 189–92.

Morse, D.H. (1980) The effect of nectar abundance on foraging patterns of bumblebees. *Ecological Entomology*, **5** (1), 53–9.

Morse, R.A. and Laigo, F.M. (1969) *Apis dorsata* in the Philippines (including an annotated bibliography) *Monograph of the Philippine Association of Entomologists, Inc.* University of the Philippines, The College, Laguna, P.I. No. I, 96 pp.

Moser, J.E., Brownlee, R.G. and Silverstein, R.M. (1968) The alarm pheromones of *Atta texana*. *Journal of Insect Physiology*, **14**, 529–35.

Motro, M., Motro, U., Ishay, J.S. and Kugler, J. (1979) Some social and dietary prerequisites of oocyte development in *Vespa orientalis* L. workers. *Insectes Sociaux*, **26** (2), 150–64.

Muir, D.A. (1959) The ant–aphid relationship in west Dunbartonshire. *Journal of Animal Ecology*, **28**, 133–40.

Nagin, R. (1972) Caste determination in *Neotermes jouteli* (Banks). *Insectes Sociaux*, **19**, 39–61.

Naulleau, G. and Montagner, H. (1961) Construction de cellules irrégulières chez *Apis mellifica*. *Insectes Sociaux*, **8** (3), 203–11.

Naumann, M.G. (1970) The nesting behaviour of *Protopolybia pumila* in Panama (Hymenoptera, Vespidae). Ph.D. Dissertation, University of Kansas.

Nel, J.J.C. (1968) Aggressive behaviour of the harvest termites *Hodotermes messambicus* (Hagen.) and *Trinervitermes trinervoides* (Sjöstedt). *Insectes Sociaux*, **15**, 145–6.

Nielsen, M.G. (1972) An attempt to estimate energy flow through a population of workers of *Lasius alienus* (Först) (Hymenoptera: Formicidae). *Natura Jutlandica*, **16**, 99–107.

Nielsen, M.G. (1977) Nests of *Lasius flavus* F. on tidal meadow in Denmark. Proceedings VIIIth Congress of the International Union for the Study of Social Insects, Wageningen, pp. 140–1.

Nielsen, M.G. and Josens, G. (1978) Production by ants and termites, in *Production ecology of ants and termites* (ed. M.V. Brian), Cambridge University Press, Cambridge, pp. 45–54.

Nijhout, H.F. and Wheeler, D.E. (1982) Juvenile hormone and the physiological basis of insect polymorphisms. *Quarterly Review of Biology*, **57** (2), 109–33.

Noirot, C. (1959) Remarques sur l'écologie des termites. *Annales de la Société Royale Zoo-*

logique de Belgique, **89**, 151–69.

Noirot, C. (1969) Formation of castes in the higher Termites, in *Biology of Termites*, Vol. I, (ed. K. Krishna and F. Weesner), Academic Press, New York and London, pp. 311–350.

Noirot, C. (1970) The nests of termites in *Biology of Termites*, Vol. 2 (ed. K. Krishna and F.M. Weesner), Academic Press, New York and London, pp. 73–125.

Noirot, C. (1974) Polymorphismus bei höheren Termiten, in *Sozialpolymorphismus bei Insekten* (ed. G.H. Schmidt), Wissenschaftliche Verlagsgesellschaft, Stuttgart, pp. 740–65.

Noirot, C. and Bodot, P. (1964) L'essaimage *d'Allognathotermes hypogeus* Silv. (Isoptera, Termitidae). *Compte Rendu de l'Académie des Sciences, Paris,* **258**, 3357–9.

Noirot, C. and Noirot-Timothée, C. (1969) The digestive system, in *Biology of Termites*, Vol. I (ed. K. Krishna and F.M. Weesner), Academic Press, New York and London, pp. 49–88.

Noonan, K.M. (1978) Sex ratio of parental investment in colonies of the social wasp *Polistes fuscatus. Science* **199**, 1354–6.

Nutting, W.L. (1969) Flight and colony foundation, in *Biology of Termites*, Vol. 1 (ed. K. Krishna and F.M. Weesner), Academic Press, New York and London, pp. 233–82.

Ofer, J. (1970) *Polyrachis simplex* the weaver ant of Israel. *Insectes Sociaux,* **17** (1), 49–82.

Ofer, J., Shulov, A. and Noy-Meir, I. (1978) Associations of ant species in Israel: a multivariate analysis. *Israel Journal of Zoology,* **27**, 199–208.

Ohiagu, C.E. and Wood, T.G. (1976) A method for measuring rate of grass harvesting by *Trinervitermes geminatus* (Wasmann) (Isoptera, Nasutitermitinae) and observation on its foraging behaviour in southern Guinea savanna, Nigeria. *Journal of Applied Ecology,* **13**, 705–13.

Ohly-Wüst, M. (1973) Stomodeale und proctodeale sekrete von ameisenlarven und ihre biologische bedeutung. Proceedings, VIIth International Congress of the International Union for the Study of Social Insects, London, pp. 412–18.

Ohly-Wüst, M. (1977) Soziale wechselbeziehungen zwischen larven und arbeiterinnen im ameisenstaat, mit besonderer beachtung der trophallaxis. Inaugural-dissertation Johann-Wolfgan-Goethe-Universität zu Frankfurt am Main.

Okot-Kotber, B.M. (1980) Histological and size changes in corpora allata and prothoracic glands during development of *Macrotermes michaelseni* (Isoptera). *Insectes Sociaux,* **27** (4), 351–76.

Oloo, G.W. and Leuthold, R.H. (1979) The influence of food on trail-laying and recruitment behaviour in *Trinervitermes bettonianus* (Termitidae: Nasutitermitinae). *Entomologia Experimentalis et Applicata,* **26** (3), 267–78.

Olubajo, O., Duffield, R.M. and Wheeler, J.W. (1980) 4-Heptanone in the mandibular gland secretion of the nearctic ant, *Zacryptocerus varians* (Hymenoptera: Formicidae). *Annals of the Entomological Society of America,* **73**, 93–4.

Oster, G.F. and Wilson, E.O. (1978) *Caste and Ecology in the Social Insects*, Princeton University Press, Princeton, 352 pp.

Owen, R.E., Rodd, F.H. and Plowright, R.C. (1980) Sex ratios in bumble bee colonies: complications due to orphaning. *Behavioral Ecology and Sociobiology,* **7** (4), 287–92.

Pamilo, P. and Varvio-Aho, S.-L. (1979) Genetic structure of nests in the ant *Formica sanguinea. Behavioral Ecology and Sociobiology,* **6** (2), 91–8.

Pamilo, P., Rosengren, R., Vepsäläinen, K., Varvio-Aho, S.-L. and Pisarski, B. (1978) Population genetics of *Formica* ants. I Patterns of enzyme gene variation. *Hereditas,* **89**, 233–48.

Pamilo, P., Vepsäläinen, K., Rosengren, R., Varvio-Aho, S.-L. and Pisarski, B. (1979) Population genetics of *Formica* ants II. Genic differentiation between species. *Annales entomologici fennici,* **45** (3), 65–76.

Pardi, L. and Marino, M.T. (1970) Studi sulla biologia di *Belongaster* (Hymenoptera, Vespidae) 2. Differenziamento castale incipiente in *B. griseus* (Fab). *Monitore Zoologica Italiano, (n.s. Supplement III),* **11**, 235–65.

Parry, K. and Morgan, E.D. (1979) Pheromones of ants: a review. *Physiological Entomology,* **4,** 161–89.

Passera, L. (1969) Le rôle de la reine dans l'ovogenèse ouvrière chez la fourmi *Plagiolepis pygmaea* Latr. *Colloquium, Centre National de la Researche Scientifique,* **189,** 129–45.

Passera, L. (1974) Kastendetermination bei der Ameise *Plagiolepis pygmaea* Latr., in *Sozialpolymorphismus bei Insekten* (ed. G.H. Schmidt), Wissenschaftliche Verlagsgesellschaft, Stuttgart, pp. 513–32.

Passera, L. (1980a) La ponte d'oeufs préorientés chez la fourmi *Pheidole pallidula* (Nyl.) (Hymenoptera: Formicidae). *Insectes Sociaux,* **27** (1) 79–95.

Passera, L. (1980b) La fonction inhibitrice des reines de la fourmi *Plagiolepis pygmaea* Latr.: rôle des phéromones. *Insectes Sociaux,* **27** (3), 212–25.

Passera, L. and Suzzoni, J.P. (1979) Le rôle de la reine de *Pheidole pallidula* (Nyl.) (Hymenoptera, Formicidae) dans la sexualisation du couvain après traitement par l'hormone juvénile. *Insectes Sociaux,* **26** (4), 343–53.

Pasteels, J.M. (1965) Polyethisme chez les ouvrières de *Nasutitermes lujae* (Termitidae Isoptères). *Biologia Gabonica,* **1,** 191–205.

Pasteels, J.M., Verhaeghe, J.C., Braekman, J.C., Daloze, D. and Tursch, B. (1980) Caste-dependent pheromones in the head of the ant *Tetramorium caespitum. Journal of Chemical Ecology,* **6** (2), 467–72.

Pavan, M. and Ronchetti, G. (1955) Studi sulla morfologia esternae anatomia interna dell 'operaia di *Iridomyrmex humilis* Mayr. e ricerche chimiche e biologiche sulla iridomirmecina. *Atti della Società Italiana di Scienze Naturali, e del Museo Civile di Storia Naturale di Milano,* **94** (3–4), 379–477.

Peacock, A.D. and Baxter, A.T. (1950) Studies in Pharoah's ant *Monomorium pharaonis* (L.) 3. Life History. *Entomologist's Monthly Magazine,* **86,** 171–8.

Peacock, A.D., Smith, I.C., Hall, D.W. and Baxter, A.T. (1954) Studies in Pharoah's ant *Monomorium pharaonis.* 8. Male production by parthenogenesis. *Entomologist's Monthly Magazine,* **90,** 154–8.

Peakin, G.J. and Josens, G. (1978) Respiration and energy flow in *Production ecology of ants and termites* (ed. M.V. Brian), Cambridge University Press.

Pearson, B. (1980) The distribution of an esterase polymorphism in macrogynes and microgynes of *Myrmica rubra* Latreille. *Evolution,* **34** (1), 105–9.

Pearson, B. (1981) The electrophoretic determination of *Myrmica rubra* microgynes as a social parasite and the possible significance of this with regard to the evolution of ant social parasites, in *Systematics Association* Special Vol. 19, *Biosystematics of social insects* (eds. P.E. Howse and J.-L. Clement), Academic Press, London and New York, pp. 75–84.

Pearson, B. (1982a) The taxonomic status of morphologically anomalous ants in the *Lasius niger/Lasius alienus* taxon. *Insectes Sociaux,* **29** (1), 95–101.

Pearson, B. (1982b) Relatedness amongst normal queens (macrogynes) in nests of the polygynous ant *Myrmica rubra* Latreille. *Evolution,* **36** (1), 107–12.

Pearson, B. and Child, A.R. (1980) The distribution of an esterase polymorphism in macrogynes and microgynes of *Myrmica rubra* Latreille. *Evolution,* **34** (1), 105–9.

Pendrel, B.A. and Plowright, R.C. (1981) Larval feeding by adult bumblebee workers (Hymenoptera: Apidae). *Behavioral Ecology and Sociobiology,* **8,** 71–6.

Percival, M.S. (1965) *Floral Biology,* Pergamon Press, Oxford.

Peregrine, D.J. and Mudd, A. (1974) The effect of diet on the composition of the postpharyngeal glands of *Acromyrmex octospinosus* (Reich). *Insectes Sociaux,* **21** (4), 417–24.

Peregrine, D.J., Mudd, A. and Cherrett, J.M. (1973) Anatomy and preliminary chemical analysis of the post-pharyngeal glands of the leaf-cutting ant *Acromyrmex octospinosus* (Reich) Hymenoptera, Formicidae. *Insectes Sociaux,* **20** (4), 355–63.

Pętal, J. (1972) Methods of investigating the productivity of ants. *Ekologia polska,* **20,** 9–22.

Pętal, J. (1978) The role of ants in ecosystems, in *Production ecology of ants and termites* (ed. M.V. Brian), Cambridge University Press, pp. 293–325.

Pętal, J. (1980) Ant populations, their regulation and effect on soil in meadows. *Ekologia polska*, **28** (3), 297–326.

Peterson-Braun, M. (1977) Studies on the endogenous breeding cycle in *Monomorium pharaonis*. Proceedings, VIIIth International Congress of the International Union for the Study of Social Insects, Wageningen, pp. 211–12.

Philips, S.A. and Vinson, S.B. (1980) Source of the post-pharyngeal gland contents in the red imported fire ant, *Solenopsis invicta* Buren. *Annals of the Entomological Society of America*, **73** (3), 257–61.

Pickard, R.S. (1976) Bees, Magnetism and Electricity. Lecture to the Central Association of Bee-Keepers, 25 September 1976, Ilford, Essex, 12 pp.

Pisarski, B. (1973) *Struktura spoleczna Formica (C) exsecta Nyl. (Hymenoptera, Formicidae i jej wplyw na morfologie ekologie i etologie gatunku.* PAN, Instytut Zoologiczny, 134 pp.

Plateaux, L. (1970) Sur le polymorphisme social de la fourmi *Leptothorax nylanderi* (Förster) I Morphologie et biologie comparées des castes. *Annales des Sciences Naturelles (Zoologie)*, **12**, 373–478.

Plateaux, L. (1971) Sur le polymorphisme sociale de la fourmi *Leptothorax nylanderi* II Activité des ouvrières et déterminisme des castes. *Annales des Sciences Naturelles et Biologie Animale*, **13**, 1–90.

Plateaux, L. (1981) The *Pallens* morph of the ant *Leptothorax nylanderi*: description formal genetics and study of populations, in *Systematics Association* Special Vol. 19, *Biosystematics of Social Insects* (eds. P.E. Howse and J.-L. Clement), Academic Press, London and New York, pp. 63–74.

Plateaux-Quénu, C. (1962) Biology of *Halictus marginatus* Brullé. *Journal of Apicultural Research*, **1**, 41–51.

Plateaux-Quénu, C. (1972) La biologie des abeilles primitives. Collection 'Les grands problèmes de la biologie' monographie 11, Masson et Cie.

Plateaux-Quénu, C. (1978) Les sexués de remplacement chez *Evylaeus calceatus* (Scop.) (*Hym., Halicitinae*. *Insectes Sociaux*, **25** (3), 227–36.

Plateaux-Quénu, C. (1979) Qui remplace la reine dans un groupe d'ouvrières orphelines d'*Evylaeus calceatus* (Scop.) (Hym., Halicitinae)? *Annales de Sciences Naturelles (Zoologie), Paris, Sér. 13*, **1** 213–18.

Pleasants, J.M. (1981) Bumblebee response to variation in nectar availability. *Ecology*, **62** (6), 1648–61.

Plowman, K.P. (1981) Resource utilization by two New Guinea rainforest ants. *Journal of Animal Ecology*, **50** (3), 903–16.

Plowright, R.C. (1979) Social facilitation at the nest entrances of bumble bees and wasps. *Insectes Sociaux*, **26** (3), 223–31.

Plowright, R.C. and Jay, S.C. (1968) Caste differentiation in bumble-bees: The determination of female size. *Insectes Sociaux*, **15**, 171–92.

Poldi, B. (1963) Studi sulla fondazione dei nidi nei Formicidi. I. *Tetramorium caespitum* L. Proceedings, IVth International Congress of the International Union for the Study of Social Insects. *Symposia Genetica et Biologica Italica*, **12**, 132–199.

Pomeroy, D.E. (1976) Studies on a population of large termite mounds in Uganda. *Ecological Entomology* **1** (1), 49–61.

Pontin, A.J. (1960) Field experiments on colony foundation by *Lasius niger* (L.) and *L. flavus* (F.) (Hym., Formicidae). *Insectes Sociaux*, **7**, 227–30.

Pontin, A.J. (1961a) The prey of *Lasius niger* (L.) and *L. flavus* (F.) (Hym., Formicidae). *Entomologist's Monthly Magazine*, **97**, 135–7.

Pontin, A.J. (1961b) Population stabilization and competition between the ants *Lasius flavus* (F.)

and *L. niger* (L.). *Journal of Animal Ecology,* **32**, 565–74.

Pontin, A.J. (1978) The numbers and distribution of subterranean aphids and their exploitation by the ant *Lasius flavus* (Fabr.). *Ecological Entomology,* **3** (3), 203–7.

Porter, S.D. and Jorgensen, C.D. (1981) Foragers of the harvester Ant, *Pogonomyrmex owyheei*: A disposable caste? *Behavioral Ecology and Sociobiology,* **9**, 247–56.

Portugal-Araújo, V. de (1958) A contribution to the bionomics of *Lestrimelitta cubiceps* (Hymenoptera, Apidae). *Journal of the Kansas Entomological Society,* **31** (3), 203–11.

Post, D.C. and Jeanne, R.L. (1981) Colony defense against ants by *Polistes fuscatus* (Hymenoptera: Vespidae) in Wisconsin. *Journal of the Kansas Entomological Society,* **54** (3), 599–615.

Prestwich, G.D. (1975) Chemical analysis of soldier defensive secretions of several species of East African termites. Proceedings, Symposium of the International Union for the Study of Social Insects, Dijon, pp. 149–52.

Proctor, M. and Yeo, P. (1973) The new naturalist, *The pollination of flowers*, Collins, London, 418 pp.

Provost, E. (1979) Étude de la fermeture de la société de Fourmis chez diverses espèces de *Leptothorax* et chez *Camponotus lateralis* (Hyménoptères, Formicidae) *Compte Rendu de l'Académie des Sciences, Paris, Série D,* **288**, 429–32.

Provost, E. (1981) Aggressivity and the closure of ant societies. XVIIe Conference Internationale d'Ethologie, Oxford, 1–9 Septembre 1981. Séance: 'Species and individual recognition'.

Prŷs-Jones, O.E. (1982) Ecological studies of foraging and life history in bumble bees. Ph.D. Thesis, University of Cambridge.

Pudlo, R.J., Beattie, A.J. and Culver, D.C. (1980) Population consequences of changes in an ant-seed mutualism in *Sanguinaria canadensis. Oecologia,* **146**, 32–7.

Pyke, G.H. (1978) Optimal foraging in bumblebees and coevolution with their plants. *Oecologia,* **36**, 281–93.

Pyke, G.H. (1979) Optimal foraging in bumblebees: rule of movement between flowers within inflorescences. *Animal Behaviour,* **27**, 1167–81.

Quennedy, A. (1973) Observations cytologiques et chimique sur la glande frontale des termites. Proceedings, VIIth International Congress of the International Union for the Study of Social Insects, London, pp. 324–5.

Quinlan, R.J. and Cherrett, J.M. (1979) The role of fungus in the diet of the leaf-cutting ant *Atta cephalotes* (L.). *Ecological Entomology,* **4** (2), 151–60.

Raignier, A. (1948) L'économie thermique dune colonie polycalique de la Fourmi des bois. *La Cellule,* **51** (3), 281–368.

Raignier, A. (1972) Sur l'origine des nouvelles sociétés des fourmis voyageuses africaine. *Insectes Sociaux,* **19**, 153–70.

Raignier, A. and van Boven, J.K.A. (1955) Etude taxonomique, biologique et biométrique des *Dorylus* due sous-genre *Anomma* (Hymenoptera: Formicidae). *Annales du Musée Royale du Congo Belge, Sér. 4, Sciences Zoologiques,* **2**, 1–359.

Rajashekharappa, B.J., (1979) Queen recognition and the rearing of new queens by honeybee colonies (*Apis cerana*). *Journal of Apicultural Research,* **18** (3), 173–8.

Regnier, F.E. and Wilson, E.O. (1971) Chemical communication and 'propaganda' in slave maker ants. *Science,* **172**, 267–9.

Regnier, F.E., Neih, M. and Hölldobler, B. (1973) The volatile Dufour's gland components of the harvester ants *Pogonomyrmex rugosus* and *P. barbatus. Journal of Insect Physiology,* **19**, 981–92.

Rettenmeyer, C.W. and Watkins, J.F. II (1978) Polygyny and monogyny in army ants (Hymenoptera: Formicidae). *Journal of the Kansas Entomological Society,* **51** (4), 581–91.

Reznikova, Zh. I (1971) The interaction of ants of different species inhabiting the same area. *Ants and Forest Protection,* **4**, 52–65.

Reznikova, Zh. I. (1975) Non-antagonistic interrelationships between ants occupying similar ecological niches. *Zoologicheskii Zhurnal,* 54 (7), 1020–31.

Reznikova, Zh. I. (1979) Forms of spatial organization in *Formica pratensis* (in Russian). *Zoologicheskii Zhurnal,* 58 (10), 1490–500.

Reznikova, Zh. I. and Kulikov, A.V. (1978) Features of the feeding and interaction of steppe ant species (Hymenoptera, Formicidae). *Entomological Review,* 57 (1), 43–51.

Ribbands, C.R. (1949) The foraging method of individual honey-bees. *The Journal of Animal Ecology,* 18, 47–66.

Ribbands, C.R. (1953) *The Behaviour and Social Life of Honey Bees,* Bee Research Association, London.

Richards, O.W. (1969) The Biology of some W. African social wasps. *Memorie della Società Entomologica Italiana* 48, 79–93.

Richards, O.W. (1971) The biology of the social wasps (Hymenoptera, Vespidae). *Biological Review,* 46, 483–528.

Richards, O.W. (1978) *The social wasps of the Americas excluding the vespinae,* British Museum of natural history, London, pp. 580.

Richards, O.W. and Richards, M.J. (1951) Observations on the social wasps of South America. *Transactions of the Royal Entomological Society of London,* 102 (1), 1–170.

Ritter, F.J., Brüggemann-Rotgans, I.E.M., Verkuil, E. and Persoons, C.J. (1975) The trail pheromone of the Pharoah's ant *Monomorium pharaonis*: components of the odour trail and their origin. Pheromones and defensive secretions in social insects: Symposium International Union for the Study of Social Insects. Dijon, pp. 99–103.

Robertson, P.L., Dudzinski, M.L. and Orton, C.J. (1980) Exocrine gland involvement in trailing behaviour in the Argentine ant (Formicidae: Dolichoderinae). *Animal Behaviour,* 28, 1255–73.

Robinson, S.W., Moser, J.C., Blum, M.S. and Amante, E. (1974) Laboratory investigations of the trail-following responses of four species of leaf-cutting ants with notes on the specificity of a trail pheromone of *Atta texana* (Buckley). *Insectes Sociaux,* 21, 87–94.

Rockstein, M. (1978) *Biochemistry of Insects* (ed. M. Rockstein), Academic Press, New York, London, 649 pp.

Rockwood, L.L. (1976) Plant selection and foraging patterns in two species of leaf-cutting ants (*Atta*). *Ecology,* 57, 48–61.

Rohrmann, G.F. and Rossmann, A.Y. (1980) Nutrient strategies of *Macrotermes ukuzii* (Isoptera: Termitidae). *Pedobiologia,* 20, 61–73.

Roland, C. (1976) Approche éco-éthologique et biologiques des sociétés de *Paravespula vulgaris* et *germanica*. Thesis, University of Nancy.

Room, P.M. (1971) The relative distribution of ant species in Ghana's cocoa farms. *Journal of Animal Ecology,* 40, 735–51.

Röseler, P.-F. (1977) Juvenile hormone control of oogenesis in bumble bee workers, *Bombus terrestris. Journal of Insect Physiology,* 23, 985–92.

Röseler, P.-F. and Röseler, I. (1977) Dominance in bumblebees. Proceedings, VIIIth International Congress of the International Union of the Study of Social Insects, Wageningen, pp. 232–5.

Röseler, P.-F. and Röseler, I. (1978) Studies on the regulation of the juvenile hormone titre in bumble-bee workers *Bombus terristris. Journal of Insect Physiology,* 24, 707–13.

Röseler, P.-F., Röseler, I. and Strambi, A. (1980) The activity of corpora allata in dominant and subordinated females of the wasp *Polistes gallicus. Insectes Sociaux* 27 (2), 97–107.

Röseler, P.-F., Röseler, I. and van Honk, C.G.J. (1981) Evidence for inhibition of corpora allata activity in workers of *Bombus terrestris* by a pheromone from the queens mandibular glands. *Experientia,* 37, 348–51.

Rosengren, R. (1977) The significance of age polyethism in social foraging of wood ants,

(*Formica rufa* group). Proceedings, VIIIth International Congress of the International Union for the Study of Social Insects, Wageningen, pp. 99–100.

Ross, N.M. and Gamboa, G.J. (1981) Nestmate discrimination in social wasps (*Polistes metricus* Hymenoptera: Vespidae). *Behavioral Ecology and Sociobiology*, **9**, 163–5.

Roubik, D.W. (1978) Competitive interactions between neotropical pollinators and africanized honey bees. *Science*, **201**, 1030–2.

Roubik, D.W. (1979) Nest and colony characteristics of stingless bees from French Guiana (Hymenoptera: Apidae). *Journal of the Kansas Entomological Society*, **52** (3), 443–70.

Ruttner, F., Woyke, J. and Koeniger, N. (1972) Reproduction in *Apis cerana*, I. Mating behaviour *Journal of Apicultural Research*, **11** (3), 141–6.

Ruttner, F., Woyke, J. and Koeniger, N. (1973) Reproduction in *Apis cerana*. 2. Reproductive organs and natural insemination. *Journal of Apicultural Research* **12** (1), 21–34.

Rutz, W., Gerig, L., Willie, H. and Lüscher, M. (1976) On the function of juvenile hormone in worker honey-bees. *Journal of Insect Physiology*, **22**, 1485–92.

Sakagami, S.F. (1959) Some interspecific relations between Japanese and European honey-bees. *Journal of Animal Ecology*, **28**, 51–68.

Sakagami, S.F. (1960) Ethological peculiarities of the primitive social bees, *Allodape* Lepeltier and allied genera. *Insectes Sociaux*, **7**, 231–49.

Sakagami, S.F. and Fukushima, K. (1957) *Vespa dybewskii* André as a facultative temporary social parasite. *Insectes Sociaux*, **4**, 1–12.

Sakagami, S.F. and Hayashida, K. (1962) Work efficiency in heterospecific ant groups composed of hosts and their labour parasites. *Animal Behaviour*, **10**, 96–104.

Sakagami, S.F. and Michener, C.D. (1962) *The nest architecture of the sweat bees (Halictinae)*. University of Kansas Press, Lawrence.

Sakagami, S.F. and Zucchi, R. (1965) Winterverhalten einer neotropischen Hummel. *Bombus atratus*, innerhalb des Beobachtungskastens. Ein Beitrag zur Biologie der Hummeln. *Journal of the Faculty of Science, Hokkaido University, Zoology*, **15**, 712–62.

Sakagami, S.F., Akahira, Y. and Zucchi, R. (1967a) Nest architecture and brood development in a neotropical bumblebee, *Bombus atratus*. *Insectes Sociaux*, **14** (4), 389–414.

Sakagami, S.F., Laroca, S. and Moure, J.S. (1967b) Two Brazilian apid nests worth recording in reference to comparative bee sociology, with description of *Euglossa melanotricha* Moure sp. n. (Hymenoptera, Apidae). *Annotationes Zoologicae Japonenses*, **40** (1), 45–54.

Sands, W.A. (1960) The initiation of fungus comb construction in laboratory colonies of *Ancistrotermes guineensis* (Sylvestri). *Insectes Sociaux*, **7**, 251–63.

Sands, W.A. (1965) Alate development and colony foundation in five species of *Trinervitermes* (Isoptera, Nasutiterminae) in Nigeria, West Africa. *Insectes Sociaux*, **12** (2), 117–30.

Sands, W.A. (1969) The association of termites and fungi, in *Biology of termites* Vol. 1 (ed. K. Krishna and F.M. Weesner), Academic Press, New York and London, pp. 495–524.

Samways, M.J. (1981) Comparison of ant community structure (Hymenoptera: Formicidae) in citrus orchards under chemical and biological control of red scale, *Aonidiella aurantii* (Maskell) (Hemiptera: Diaspididae). *Bulletin of Entomological Research*, **71** 663–70.

Schemske, D.W. (1980) The evolutionary significance of extrafloral nectar production by *Costus woodsonii* (Zingiberaceae): An experimental analysis of ant protection. *Journal of Ecology*, **68**, 959–67.

Scherba, G. (1958) Reproduction, nest orientation and population structure of an aggregation of mound nests of *Formica ulkei* (Emery) (Formicidae). *Insectes Sociaux*, **5**, 201–13.

Scherba, G. (1959) Moisture regulation in mound nests of the ant *Formica ulkei*. *American Midland Naturalist*, **61**, 499–509.

Scherba, G. (1962) Mound temperatures of the ant *Formica ulkei* Emery. *American Midland Naturalist*, **67** (2), 373–85.

Scherba, G. (1964) Analysis of inter-nest movement by workers of the ant *Formica opaciventris*

Emery (Hymenoptera: Formicidae). *Animal Behaviour,* **12,** 508–12.

Schmidt, G.H. (1974) *Sozialpolymorphismus bei Insekten* (ed. G.H. Schmidt), Wissenschaft-liche verlagsgesellschaft MBH, Stuttgart, 974 pp.

Schmidt, G.H. (1982) Egg dimorphism and male production in *Formica polyctena* Foerster, in *The biology of social insects,* Proceedings of the Ninth Congress of the International Union for the Study of Social Insects, Boulder, Colorado (eds. M.D. Breed, C.D. Michener and H.E. Evans), Westview Press, Boulder, Colorado, pp. 243–7.

Schmidt, J.O. and Blum, M.S. (1978) A harvester ant venom: chemistry and pharmacology. *Science,* **200,** 1064–6.

Schneider, P. (1972) Akustische Signale bei Hummeln. *Naturwissenschaften,* **59,** 168.

Schneirla, T.C. (1971) *Army ants: a study in social organisation* (ed. H. Topoff), Freeman, San Francisco, 349 pp.

Schumacher, A.M. and Whitford, W.G. (1974) Spatial and temporal variation in Chihauhuan desert ant faunas. *Southwestern Naturalist,* **21,** 1–8.

Seeley, T.D. (1974) Atmospheric carbon dioxide regulation in honey-bee (*Apis mellifera*) colonies. *Journal of Insect Physiology,* **20,** 2301–5.

Seeley, T.D. (1977) Measurement of nest cavity volume by the honeybee (*Apis mellifera* L.). *Behavioral Ecology and Sociobiology,* **2,** 201–27.

Seeley, T.D. (1978) Life history strategy of the honey bee, *Apis mellifera. Oecologia,* **32,** 109–18.

Seeley, T.D. (1979) Queen substance dispersal by messenger workers in honey-bee colonies. *Behavioral Ecology and Sociobiology,* **5,** 391–415.

Seeley, T. and Heinrich, B. (1981) Regulation of temperature in the nests of social insects, in *Insect thermoregulation* (ed. B. Heinrich), John Wiley, New York, pp. 160–234.

Seeley, T.D. and Morse, R.A. (1976) The nest of the honeybee *Apis mellifera. Insectes Sociaux,* **23** (4), 495–512.

Seeley, T.D. and Morse, R.A. (1977) Dispersal behavior of honey bee swarms. *Psyche,* **83** (3–4), 199–209.

Seeley, T.D. and Morse, R.A. (1978) Nest site selection by the honey bee *Apis mellifera. Insectes Sociaux,* **25** (4), 323–37.

Seeley, T.D. Morse, R.A. and Visscher, P.K. (1979) The natural history of the flight of honey bee swarms. *Psyche,* **86** (2–3), 103–13.

Selander, R.K. (1976) Genic variation in natural populations, in *Molecular evolution* (ed. F.J. Ayala), Sinauer, Sunderland, Massachusetts.

Simpson, J. (1955) The significance of the presence of pollen in the food of worker larvae of the honey-bee. *Quarterly Journal of Microscopical Science,* **96** (1), 117–20.

Simpson, J. (1960) The functions of the salivary glands of *Apis mellifera. Journal of Insect Physiology,* **4,** 107–21.

Simpson, J. (1974) The reproductive behaviour of European honeybee colonies. Lecture to the Central Association of bee-keepers, 14 March, 1974.

Skaife, S.H. (1954) The black-mound termites of the Cape, *Amitermes atlanticus* Fuller. *Transactions of the Royal Society of South Africa,* **34,** 251–71.

Skinner, G.J. (1980a) Territory, trail structure and activity patterns in the wood-ant, *Formica rufa* (Hymenoptera: Formicidae), in limestone woodland in north-west England. *Journal of Animal Ecology,* **49** (2), 381–94.

Skinner, G.J. (1980b) The feeding habits of the wood-ant, *Formica rufa* (Hymenoptera: Formicidae), in limestone woodland in north-west England. *Journal of Animal Ecology,* **49** (2), 417–34.

Skinner, G.J. and Whittaker, J.B. (1981) An experimental investigation of interrelationships between the wood-ant (*Formica rufa*) and some tree-canopy herbivores. *Journal of Animal Ecology,* **50** (1), 313–26.

Smeeton, L. (1980) Male production in the ant *Myrmica rubra* L. Doctoral thesis, University of Southampton.

Smeeton, L. (1981) The source of males in *Myrmica rubra* L. (Hym. Form.). *Insectes Sociaux,* **28** (3), 263–78.

Smith, R.H. and Shaw, M.R. (1980) Haplodiploid sex ratios and the mutation rate. *Nature,* **287**, 728–9.

Sommeijer, M.J., Beuvens, F.T. and Verbeek, H.J. (1982) Distribution of labour among workers of *Melipona favosa* F. : construction and provisioning of brood cells. *Insectes Sociaux,* **29** (2), 222–37.

Soulié, J. (1961) Quelques notes ethologiques sur la vie dans le nid chez deux espèces méditer-ranéennes de *Cremastogaster* (Hymenoptera–Formicoidea). *Insectes Sociaux*, **8** (1), 95–8.

Spradbery, J.P. (1971) Seasonal changes in the population structure of wasp colonies (Hymen-optera: Vespide). *Journal of Animal Ecology,* **40**, 501–23.

Spradbery, J.P. (1973a) *Wasps*, Sidgwick and Jackson, London, 408 pp.

Spradbery, J.P. (1973b) The european social wasp, *Paravespula germanica* (F.) (Hymenoptera: Vespidae) in Tasmania, Australia. Proceedings, VIIth International Congress of the Inter-national Union for the Study of Social Insects, London, pp. 375–80.

Springhetti, A. (1972) I feromoni nella differenziazione della caste di *Kalotermes flavicollis* (Fabr.) (Isoptera). *Bulletino di Zoologia,* **39**, 83–7.

Stebaev, I.V. and Reznikova, J.I. (1972) Two interaction types of ants living in steppe ecosystem in South Siberia, USSR. *Ekologia polska,* **20** (11), 103–9.

Stebaev, I.V. and Reznikova, Zh. I (1974) The system of spatial and temporal interrelationships in a settlement of steppe ants containing several species. *Zoologicheskii Zhurnal,* **53** (8), 1200–12.

Stradling, D.J. (1978a) The influence of size on foraging in the ant *Atta cephalotes* and the effect of some plant defence mechanisms. *Journal of Animal Ecology,* **47** (1), 173–88.

Stradling, D.J. (1978b) Food and feeding habits of ants, in *Production ecology of ants and termites*, (ed. M.V. Brian), Cambridge University Press, Cambridge, pp. 81–106.

Stuart, A.M. (1969) Social behaviour and communication, in *Biology of termites*, Vol. 1 (ed. K. Krishna and F.M. Weesner), Academic Press, New York and London, pp. 193–232.

Sturtevant, A.H. (1938) Essays on evolution. II On the effects of selection on social insects. *Quarterly Review of Biology,* **13**, 74–6.

Sudd, J.H. (1960) The foraging method of pharaoh's ant, *Monomorium pharaonis* (L.) *Animal Behaviour,* **8**, 67–75.

Sudd, J.H. (1965) Transport of prey by ants. *Behaviour,* **25** (3–4), 234–71.

Sudd, J.H. (1967) *An introduction to the behaviour of ants*, Arnold, London, 200 pp.

Sudd, J.H. (1969) The excavation of soil by ants. *Zeitschrift für Tierpsychologie,* **26** (3), 257–76.

Sudd, J.H. (1970a) Specific patterns of excavation in isolated ants. *Insectes Sociaux,* **17** (4), 253–60.

Sudd, J.H. (1970b) The response of isolated digging worker ants (*Formica lemani* Bondroit and *Lasius niger* (L.) to tunnels. *Insectes Sociaux,* **17** (4), 261–72.

Suzzoni, J.P. and Cagniant, H. (1975) Études histologiques des voies genitales chez l'ouvrière et la reine de *Cataglyphis cursor*. Arguments en faveur d'une parthogènese thélytoque chez cette espêce. *Insectes Sociaux,* **22**, 83–92.

Suzzoni, J.P. and Grimal, A. (1980) Variations biométriques des corps allates pendant la différenciation des castes reine et ouvrière chez *Plagiolepis pygmaea* Latr. (*Hymenoptera: Formicidae*). *Insectes Sociaux,* **27** (4), 399–414.

Swain, R.B. (1980) Trophic competition among parabiotic ants. *Insectes Sociaux,* **27** (4), 377–90.

Synge, A.D. (1947) Pollen collection by honeybees (*Apis mellifera*). *Journal of Animal Ecology,* **16** (2), 122–38.

Szlep, R. and Jacobi, T. (1967) The mechanism of recruitment to mass foraging in colonies of

Monomorium venustum Smith, *M. subopacum* ssp. *phoenicium* Gm., *Tapinoma israelis* For. and *T. simothi* v. *phoenicium* Em. *Insectes Sociaux,* **14** (1), 25–40.

Szlep–Fessel, R. (1970) The regulatory mechanism in mass foraging and the recruitment of soldiers in *Pheidole. Insectes Sociaux,* **17** (4), 233–44.

Taki, A. (1976) Colony founding of *Messor acicultatum* (Fr. Smith) (Hymenoptera: Formicidae) by single and grouped queens. *Physiology and Ecology, Japan,* **17** (1/2), 503–12.

Talbot, M. (1946) Daily fluctuations in above ground activity of three species of ants. *Ecology,* **27** (1), 65–70.

Talbot, M. (1965) Populations of ants in a low field. *Insectes Sociaux,* **12** (1), 19–48.

Taylor, B. (1977) The ant mosaic on cocoa and other tree crops in western Nigeria. *Ecological Entomology,* **2**, 245–55.

Taylor, B. and Adedoyin, S.F. (1978) The abundance and inter-specific relations of common ant species (Hymenoptera: Formicidae) on cocoa farms in western Nigeria. *Bulletin of Entomological Research,* **68**, 105–21.

Terada, Y., Garofalo, C.A. and Sakagami, S.F. (1975) Age-survival curves for workers of two eusocial bees (*Apis mellifera* and *Plebeia droryana*) in a subtropical climate, with notes on worker polyethism in *P. droryana. Journal of Apicultural Research,* **14** (3/4), 161–70.

Tevis, L., Jr (1958) Interrelations between the harvester ant *Veromessor pergandei* (Mayr.) and some desert ephemerals. *Ecology,* **39**, 695–704.

Thorne, B.L. (1982) Polygyny in Termites: multiple primary queens in colonies of *Nasutitermes corniger* (Motschuis) (Isoptera: Termitidae). *Insectes Sociaux,* **29** (1), 102–17.

Tilman, D. (1978) Cherries, ants and tent caterpillars: timing of nectar production in relation to susceptibility of caterpillars to ant predation. *Ecology,* **59** (4), 686–92.

Tohmé, G. (1972) Le nid et le comportement de construction de la fourmi *Messor ebeninus* Forel (Hymenoptera, Formicoidea). *Insectes Sociaux,* **19** (2), 95–103.

Tomlinson, J., McGinty, S. and Kish, J. (1981) Magnets curtail honey bee dance. *Animal Behaviour,* **29** (1), 307.

Topoff, H. (1971) Polymorphism in army ants related to division of labor and cyclic colony behaviour. *American Naturalist,* **105**, 529–48.

Topoff, H. and Lawson, K. (1979) Orientation of the army ant *Neivamyrmex nigrescens*: integration of chemical and tactile information. *Animal Behaviour,* **27**, 429–33.

Topoff, H. and Mirenda, J. (1978) Precocial behaviour of callow workers of the army ant *Neivamyrmex nigrescens*: importance of stimulation by adults during mass recruitment. *Animal Behaviour,* **26**, 698–706.

Topoff, H. and Mirenda, J. (1980) Army ants do not eat and run: influence of food supply on emigration behaviour in *Neivamyrmex nigrescens. Animal Behaviour,* **28**, 1040–5.

Topoff, H., Mirenda, J., Droual, R. and Herrick, S. (1980a) Behavioural ecology of mass recruitment in the army ant *Neivamyrmex nigrescens. Animal Behaviour,* **28**, 779–89.

Topoff, H., Mirenda, J., Droual, R. and Herrick, S. (1980b) Onset of the nomadic phase in the army ant *Neivamyrmex nigrescens* (Cresson) (Hym. Form.): distinguishing between callow and larval excitation by brood substitution. *Insectes Sociaux,* **27** (2), 175–9.

Topoff, H., Rothstein, A., Pujdak, S. and Dahlstrom, T. (1981) Statary behaviour in nomadic colonies of army ants: the effect of overfeeding. *Psyche,* **88**, 151–61.

Torossian, C. (1979) Importance quantitive des oeufs abortifs d'ouvrières dans le bilan trophique de la colonie de la fourmi *Dolichoderus quadripunctatus. Insectes Sociaux,* **26** (4), 295–9.

Traniello, J.F.A. (1977) Recruitment behavior, orientation, and the organization of foraging in the carpenter ant *Camponotus pennsylvanicus* DeGeer (Hymenoptera: Formicidae). *Behavioral Ecology and Sociobiology,* **2**, 61–79.

Traniello, J.F.A. (1978) Caste in a primitive ant: absence of age polyethism in *Amblyopone. Science,* **202**, 770–2.

Trivers, R.L. and Hare, H. (1976) Haplodiploidy and the evolution of the social insects. *Science* **191**, 249–63.

Tschinkel, W.R. and Close, P.G. (1973) The trail pheromone of the termite *Trinervitermes trinervoides* (Isopt., Termitidae). *Journal of Insect Physiology,* **19** (3), 707–21.

Tulloch, A. (1970) The composition of beeswax and other waxes secreted by insects. *Lipids,* **5**, 247–58.

Tumlinson, J.H. and Silverstein, R.M. (1971) Identification of the trail pheromone of a leaf-cutting ant, *Atta texana. Nature,* **234**, 348–9.

Uchmański, J. and Pętal, J. (1982) Long-term stability of ant colonies – a simulation model. *Journal of Animal Ecology,* **51**, 349–62.

Vanderplank, F.L. (1960) The bionomics and ecology of the red tree ant, *Oecophylla* sp., and its relationship to the coconut bug *Pseudotheraptus wayi* (Brown) (Coreidae). *Journal of Animal Ecology,* **29**, 15–33.

Vinson, S.B., Phillips, S.A. Jr and Williams, H.J. (1980) The function of the post-pharyngeal glands of the red imported fire ant, *Solenopsis invicta* Buren. *Journal of Insect Physiology,* **26**, 645–50.

Vroey, C. De (1979) Relations interspecifiques chez les fourmis. Proceedings, International Union for the Study of Social Insects, French section, Lausanne, pp. 107–13.

Wallis, D.J. (1962) Aggressive behaviour in the ant *Formica fusca. Animal Behaviour,* **10** (3–4), 267–74.

Wallis, D.I. (1964) The foraging behaviour of the ant *Formica fusca. Behaviour,* **23**, 149–76.

Waloff, N. (1957) The effect of the number of queens of the ant *Lasius flavus* (Fab.) (Hym. Formicidae) on their survival and on the rate of development of the first brood. *Insectes Sociaux,* **4**, 391–408.

Waloff, N. and Blackith, R.E. (1962) The growth and distribution of the mounds of *Lasius flavus* (F.) (Hymenoptera: Formicidae) in Silwood Park, Berkshire. *Journal of Animal Ecology,* **31**, 421–37.

Watson, J.A.L. (1973) The worker caste of the hodotermitid harvester termites. *Insectes Sociaux,* **20**, 1–20.

Watson, J.A.L. (1974) The development of soldiers in incipient colonies of *Mastotermes darwiniensis* Froggatt (Isoptera). *Insectes Sociaux,* **21**, 181–90.

Watson, J.A.L. and McMahan, E.A. (1978) Polyethism in the Australian harvester termite *Drepanotermes* (Isoptera, Termitinae) (Termitidae). *Insectes Sociaux,* **25** (1), 53–62.

Way, M.J. (1953) The relationship between certain ant species with particular reference to biological control of the coreid, *Theraptus* sp. *Bulletin of Entomological Research,* **44**, 669–91.

Way, M.J. (1954) Studies of the life history and ecology of the ant *Oecophylla longinoda* (Latr.). *Bulletin of Entomological Research,* **45**, 93–112.

Way, M.J. (1963) Mutualism between ants and honeydew-producing Homoptera. *Annual Review of Entomology,* **8**, 307–44.

Weaver, N. (1957) Effects of larval age on dimorphic differentiation of the female honey-bee. *Annals of the Entomological Society of America,* **50**, 283–94.

Weber, N.A. (1972) *Gardening ants – The attines*. Memoirs of the American Philosophical Society, Philadelphia, Vol. 92, 146 pp.

Wehner, R. (1976) Polarised-light navigation by insects. *Scientific American,* **235** (1), 106–14.

Weir, J.S. (1959) Changes in the retro-cerebral endocrine system of larvae of *Myrmica* and their relation to larval growth and development. *Insectes Sociaux,* **6**, 375–86.

Weir, J.S. (1973) Air flow, evaporation and mineral accumulation in mounds of *Macrotermes subhyalinus* (Rambur). *Journal of Animal Ecology,* **42**, 509–20.

Welch, R.C. (1978) Changes in the distribution of the nests of *Formica rufa* L. (Hymenoptera: Formicidae) at Blean Woods National Nature Reserve, Kent, during the decade following

Wellenstein, G. (1980) Auswirkung hügelbauxender Waldameisen der *Formica rufa* – Gruppe auf forstschädliche Raupen und das Wachstum der Waldbäume. *Zeitschrift für Angewandte Entomologie,* **89,** 144–57.

Wellington, W.G. and Cmiralova, D. (1979) Communication of height by foraging honey bees *Apis mellifera ligustica* (Hymenoptera, Apidae). *Annals of the Entomological Society of America,* **72,** 167–70.

Wells, P.H. (1973) 'Honey bees', in *Invertebrate Learning,* Vol. 2, *Arthropods and Gastropod Mollusks* (eds. W.C. Corning, J.A. Dyal and A.O.D. Willows), Plenum Press, New York, pp. 173–85.

Wenner, A.M. (1962) Sound production during the waggle dance of the honeybee. *Animal Behaviour,* **10,** 79–95.

Wenner, A.M. (1974) Information transfer in honey-bees. A population approach, in *Advances in the study of communication and affect,* Vol. I *Non verbal communication* (eds. L. Krames, P. Pliner and T. Alloway), Plenum Press, New York and London, pp. 133–69.

Werff, P.A. van der (1981) Two mound types of *Macrotermes* near Kajiado (Kenya): intraspecific variation or interspecific divergence? Systematics Association Special Vol. 19, *Biosystematics of Social Insects* (eds. P.E. Howse and J.L. Clément) Academic Press, London and New York, pp. 231–247.

West-Eberhard, M.J. (1969) The social biology of Polistine wasps. *Miscellaneous publications of the Museum of Zoology, University of Michigan,* **140,** 1–101.

West-Eberhard, M.J. (1973) Monogomy in 'polygynous' social wasps. Proceedings, VIIth International Congress of the International Union for the Study of Social Insects, London, pp. 396–403.

West-Eberhard, M.J. (1977) The establishment of reproductive dominance in social wasp colonies. Proceedings VIIIth Congress of The International Union for the Study of Social Insects, Wageningen, pp. 223–7.

West-Eberhard, M.J. (1978) Temporary queens in *Metapolybia* wasps: non-reproductive helpers without altruism? *Science,* **200,** 441–3.

Wheeler, D.E. (1982) Soldier determination in *Pheidole bicarinata*: inhibition by workers, in *The biology of social insects,* Proceedings of the Ninth Congress of the International Union for the Study of Social Insects, Boulder, Colorado (eds. M.D. Breed, C.D. Michener and H.E. Evans), Westview Press, Boulder, Colorado, p. 257.

Wheeler, D.E. and Nijhout, H.F. (1981) Imaginal wing discs in larvae of the soldier caste of *Pheidole bicarinata vinelandica* Forel. (Hymenoptera: Formicidae). *International Journal of Insect Morphology and Embryology,* **10** (2), 131–9.

Wheeler, J.W. and Blum, M.S. (1973) Alkylpyrazine alarm pheromones in Ponerine ants. *Science,* **182,** 501–3.

Wheeler, J.W., Evans, S.L., Blum, M.S. and Torgenson, R.L. (1975) Cyclopentyl ketones: Identification and function in *Azteca* ants. *Science,* **187,** 254–5.

Wheeler, W.M. (1910) *Ants, their structure, development and behaviour.* Columbia biological series, Vol. 9, Columbia University Press.

Wheeler, W.M. (1922) Ants of the American Museum Congo Expedition. A contribution to the myrmecology of Africa. *Bulletin of The American Museum of Natural History,* **45,** 1–1139.

Wheeler, W.M. (1936) Ecological relations of ponerine and other ants to termites. *Proceedings of the American Academy of Arts and Sciences,* **71,** 159–243.

Wheeler, W.M. (1942) Studies of neotropical ant-plants and their ants. *Bulletin of the Museum of Comparative Zoology at Harvard College,* **90,** 1–262.

Whitford, W.G. (1978) Structure and seasonal activity of Chihauhua desert ant communities. *Insectes Sociaux,* **25** (1), 79–88.

Whitford, W.G., Johnson, P.L. and Ramirez, J. (1976) Comparative ecology of the harvester ants

Pogonomyrmex barbatus (F. Smith) and Pogonomyrmex rugosus (Emery). *Insectes Sociaux*, **23**, 117–32.

Whitford, W.G., Depree, E. and Johnson, P. (1980) Foraging ecology of two Chihauhaun desert ant species; *Novomessor cockerelli* and *Novomessor albisetosus*. *Insectes Sociaux*, **27** (2), 148–56.

Whitham, T.G. (1977) Coevolution of foraging in *Bombus* and nectar dispensing in *Chilopsis*: A last dreg theory. *Science*, **197**, 593–6.

Wielgolaski, F.E. (ed.) (1975) *Fennoscandian tundra ecosystems*, Part I, Springer-Verlag, Berlin, Heidelberg, New York.

Wille, A. and Michener, C.D. (1973) The nest architecture of stingless bees with special reference to those of Costa Rica (Hymenoptera: Apidae). *Revista de Biologia Tropical, Universidad de Costa Rica*, **21** (1), 279 pp.

Williams, G.C. and Williams, D.C. (1957) Natural selection of individually harmful social adaptations among sibs with special reference to social insects. *Evolution*, **11**, 32–9.

Williams, R.M.C. (1959) Colony development in *Cubitermes ugandensis* Fuller (Isoptera: Termitidae). *Insectes Sociaux*, **6** (3), 291–304.

Willis, E.O. and Oniki, Y. (1978) Birds and army ants. *Annual Review of Ecology and Systematics*, **9**, 243–63.

Wilson, D.S. (1980) *The natural selection of populations and communities*, Benjamin/Cummings, Menlo Park, California, pp. 186.

Wilson, E.O. (1953) The ecology of some North American Dacetine ants. *Annals of the Entomological Society of America*, **46** (4), 479–95.

Wilson, E.O. (1955) A monographic revision of the ant genus *Lasius*. *Bulletin of the Museum of Comparative Zoology at Harvard College*, **113** (1), 1–199.

Wilson, E.O. (1962) Chemical communication among workers of the fire ant *Solenopsis saevissima* (Fr. Smith). 1. The organisation of mass-foraging. 2. An information analysis of the odor trail. 3. The experimental induction of social responses. *Animal Behaviour*, **10** (1–2), 134–64.

Wilson, E.O. (1964) The ants of the Florida keys. *Breviora, Museum of Comparative Zoology, Cambridge, Mass*, **210**, 1–14.

Wilson, E.O. (1971) *The Insect Societies*, The Belknap Press of Harvard University Press Cambridge, Massachusetts, 548 pp.

Wilson, E.O. (1974a) The population consequences of polygyny in the ant *Leptothorax curvispinosus*. *Annals of the Entomological Society of America*, **67** (5), 781–6.

Wilson, E.O. (1974b) Aversive behavior and competition with colonies of the ant *Leptothorax curvispinosus*. *Annals of the Entomological Society of America*, **67** (5), 777–80.

Wilson, E.O. (1974c) The soldier of the ant *Camponotus (Colobopsis) fraxinicola*, as a trophic caste. *Psyche*, **81**, 182–8.

Wilson, E.O. (1975) *Leptothorax duloticus* and the beginnings of slavery in ants. *Evolution*, **29**, 108–19.

Wilson, E.O. (1976a) The organization of colony defense in the ant *Pheidole dentata* Mayr. (Hymenoptera: Formicidae). *Behavioral Ecology and Sociobiology* **1**, 63–81.

Wilson, E.O. (1976b) A social ethogram of the neotropical arboreal ant *Zacryptocerus varians*. *Animal Behaviour*, **24**, 354–63.

Wilson, E.O. (1976c) The first workerless parasite in the ant genus *Formica* (Hymenoptera: Formicidae). *Psyche*, **83** (3–4), 277–81.

Wilson, E.O. (1976d) Which are the most prevalent ant genera? *Studia entomologica*, **19** (1–4), 187–200.

Wilson, E.O. (1978) Division of labor in fire ants based on physical castes (Hymenoptera: Formicidae: Solenopsis). *Journal of the Kansas Entomological Society*, **51** (4), 615–36.

Wilson, E.O. (1979) The evolution of caste systems in social insects. *Proceedings of the American*

Philosophical Society, **123** (4), 204–10.

Wilson, E.O. (1980a) Caste and division of labor in leaf-cutter ants (Hymenoptera: Formicidae: *Atta*). I. The overall pattern in *A. sexdens. Behavioral Ecology and Sociobiology,* 1, 143–56.

Wilson, E.O. (1980b) Caste and division of labor in leaf-cutter ants (Hymenoptera: Formicidae: *Atta*) II The ergonomic optimization of leaf cutting. *Behaviorial Ecology and Sociobiology,* 7, 157–65.

Wilson E.O. (1981) Communal silk-spinning by larvae of *Dendromyrmex* tree-ants (Hymenoptera, Formicidae). *Insectes Sociaux,* **28** (2), 182–90.

Wilson, E.O. and Farish, D.J. (1973) Predatory behaviour in the ant-like wasp *Methocha stygia* (Say) (Hymenoptera: Tiphiidae). *Animal Behaviour,* 21, 292–5.

Wilson, E.O. and Hölldobler, B. (1980) Sex differences in cooperative silk-spinning by weaver ant larvae. *Proceedings of The National Academy of Sciences, USA,* 77 (4), 2343–47.

Wilson, E.O. and Hunt, G.L. (1966) Habitat selection by the queens of two field-dwelling species of ant. *Ecology,* **47** (3), 485–7.

Wilson, E.O. and Regnier, F.E. (1971) The evolution of the alarm-defense system in the Formicine ants. *American Naturalist,* 105, 279–89.

Wilson, E.O., Carpenter, F.M. and Brown, W.L. (1967) The first mesozoic ants. *Science,* 157, 1038–40.

Winston, M.L. (1979) Intra-colony demography and reproductive rate of the africanized honeybee in South America. *Behavioral Ecology and Sociobiology,* 4, 279–92.

Winston, M.L. (1980) Swarming, afterswarming, and reproductive rate of unmanaged honeybee colonies (*Apis mellifera*). *Insectes Sociaux,* 27 (4), 391–8.

Winston, M.L. and Otis, G.W. (1978) Ages of bees in swarms and after swarms of the Africanized honeybee. *Journal of Apicultural Research,* 17 (3), 123–9.

Winston, M.L. and Taylor, O.R. (1980) Factors preceeding queen rearing in the Africanized honeybee (*Apis mellifera*) in South America. *Insectes Sociaux,* 27 (4), 289–304.

Winston, M.L., Otis, G.W. and Taylor, O.R. (1979) Absconding behaviour of the Africanized honeybee in South America. *Journal of Apicultural Research,* 18 (2), 85–94.

Winter, U. (1979) Untersuchungen zum raubzugverhalten der dulotischen ameise *Harpagoxenus sublaevis* (Nyl.) *Insectes Sociaux,* 26 (2), 121–32.

Winterbottom, S. (1981) The chemical basis for species and colony recognition in three species of myrmicine ants. (Hymenoptera: Formicidae). Doctoral thesis, University of Southampton.

Wirtz, P. (1973) *Mededelingen van de Landbouwhogeschool to Wageningen,* Agricultural University of Wageningen, Wageningen.

Wood, T.G. (1978) Food and feeding habits of termites, in *Production Ecology of ants and termites* (ed. M.V. Brian), Cambridge University Press, pp. 55–80.

Wood, T.G. (1981) Reproductive species of *Microterms* (Isoptera, Termitidae) in the Southern Guinea Savanna near Mokwa, Nigeria. Systematics Association Special Vol. *19, Biosystematics of Social Insects* (eds. P.E. Howse and J.L. Clement), Academic Press, London and New York, pp. 309–25.

Wood, T.G. and Sands, W. (1978) Termites in ecosystems, in *Production ecology of ants and termites* International Biological Programme Vol. 13, (ed. M.V. Brian), Cambridge University Press, Cambridge, pp. 245–92.

Wood, T.G., Johnson, R.A. and Ohiagu, C.E. (1977) Populations of termites (Isoptera) in natural and agricultural ecosystems in Southern Guinea Savanna near Mokwa, Nigeria. *Geo-Eco-Trop.,* 1 (2), 139–48.

Wood, T.G., Johnson, R.A. and Ohiagu, C.E. (1980) Termite damage and crop loss studies in Nigeria – a review of termite (Isoptera) damage to maize and estimation of damage, loss in yield and termite (*Microtermes*) abundance at Mokwa. *Tropical pest management,* 26 (3), 241–53.

Woodell, S.R.J. (1974) Anthill vegetation in a Norfolk salt marsh. *Oecologia,* 16 (3), 221–5.

Woyke, J. (1976) Brood-rearing efficiency and absconding in Indian honeybees. *Journal of Apicultural Research*, 15 (3/4), 133–43.

Woyke, J. (1980a) Effect of sex allele homo-heterozygosity on honeybee colony populations and on their honey production. I. Favourable development conditions and unrestricted queens. *Journal of Apicultural Research*, 19 (1), 51–63.

Woyke, J. (1980b) Genetic background of sexuality in the diploid drone honeybee. *Journal of Apicultural Research*, 19 (2), 89–95.

Wynne-Edwards, V.C. (1962) *Animal dispersion in relation to social behaviour* Oliver and Boyd, Edinburgh, London.

Yamauchi, K., Kinomura, K. and Miyake, S. (1981) Sociobiological studies of the polygynic ant *Lasius sakagamii* I General features of its polydomous system. *Insectes Sociaux*, 28 (3), 279–96.

Yasuno, M. (1964a) The study of the ant population in the grassland at Mt. Hakkôda. II The distribution pattern of ant nests at Kayano grassland. *Science Reports of the Tôhoku University, 4th Series, Biology*, 30 (1), 43–55.

Yasuno, M. (1964b) The study of the ant population in the grassland at Mt. Hakkôda, III The effect of the slave making ant *Polyergus samurai*, upon the nest distribution pattern of the slave ant, *Formica fusca japonica*. *Science Reports of the Tôhoku University, 4th Series, Biology*, 30 (2), 167–70.

Yasuno, M. (1965a) The study of ant population in the grassland at Mt. Hakkôda. IV The stability of the ant population. *Science Reports of the Tôhoku University, 4th Series, Biology*, 31 (2), 75–81.

Yasuno, M. (1965b) The study of ant population in the grassland at Mt. Hakkôda. V. The interspecific and intraspecific relation in the formation of the ant population, with special reference to the effect of the removal of *Formica truncorum yessensis*. *Science Reports of Tôkohu University, 4th Series, Biology*, 31 (3), 181–94.

Yasuno, M. (1965c) Territory of ants in the Kayano grassland at Mt. Hakkôda. *Science Reports of the Tôhoku University, 4th Series, Biology*, 31 (3), 195–206.

Yensen, N., Yensen, E. and Yensen, D. (1980) Intertidal ants from the Gulf of California, Mexico. *Annals of the Entomological Society of America*, 73, 266–9.

Yoshikawa, K. (1963) *Introductory studies on the life economy of polistine wasps*, Osaka City University.

Zakharov, A.A. (1980) Observer ants: storers of foraging area formation in *Formica rufa* L. (*Formicidae: Hymenoptera*). *Insectes Sociaux*, 27 (3), 203–11.

Zucchi, R. and Sakagami, S.F. (1972) Capacidade termo-reguladora em *Trigona spinipes* e em algumas outras espéces de abelhas sem ferrâo (Hymenoptera: Apidae: Meliponinae), in *Homenagem à W.E. Kerr*, Rio Clara, Brasil, pp. 301–9.

Author index

Dudzinski, M.L., see Robertson, P.L., 67
Duffield, R.M., see Olubajo, O., 138

Eastlake, O., 303
Edwards, J.P., 195, 216
Edwards, R., 105, 106, 125, 129, 130, 149
Ehrhardt, W., see Büschinger, A., 57, 228, 293
Eickwort, G.C., 39, 131
Eisenbraun, E.J., see McGurk, D.J., 228
Eisner, H., 152
Eldridge, D.W., see Brown, C.A., 133
Elmes, G.W., 174, 193, 195, 233, 234, 239,
    255, 256, 257, 258, 264, 265, 271,
    272, 273, 275, 288
    see also Brian, M.V., 26
Elton, E.T.G., 260
Emmert, W., 156, 157
Engel, H., 71, 139
Engels, E., 154
Engels, W., 154
Esch, H., 54, 55, 223
Esch, I., see Esch, H., 54, 55
Ettershank, G., 70, 140
Ettershank, J.A., see Ettershank, G., 70
Evans, H.E., 102, 103, 104, 123, 130, 220,
    306
Evans, S.L., see Wheeler, J.W., 139
Evershed, R.P., see Cammaerts, M.C., 54, 59,
    133, 250
Evesham, E.J.M., 157, 192, 193, 264

Fales, H.M., see Blum, M.S., 61
Farish, D.J., 16, 74
Fawcett, R.H., see McLellan, A.R., 177
Fell, D.R., 222
Ferguson, A.W., 56, 57, 188, 224, 249
Fisch, S.B., see Handel, S.N., 28, 29
Fisher, M., 171
Fisher, R.A., 237
Flanders, S.E., 198
Fletcher, D.J.C., 52, 54, 126, 191, 224, 229,
    259, 284
    see also Butler, C.G., 251
Fluker, S.S., 312
Forrest, J.M.S., see Hussain, A., 20
Fowler, H.G., 114
Free, J.B., 32, 33, 36, 37, 39, 40, 55, 56, 57,
    124, 132, 158, 163, 177, 188, 200, 249
    see also Ferguson, A.W., 224
Freeland, J., 150
Fresneau, D., 156
Friesen, L.J., 56
Frith, G., 306
Frogner, K.J., 184
Frost, J., see McGurk, D.J., 228
Fuchs, S., 132
Fukuda, H., 177
Fukushima, K., 280

Gallé, L., 290

Gamboa, G.J., 61, 114, 167, 198, 248, 265
    see also Terada, Y., 177
Garofalo, C.A., 45, 175, 178, 179
Gay, F.J., 119, 301
Gentry, J.B., 60
Gerig, L., see Rutz, W., 200
Ghiringelli, D., see Bernardi, C., 140
Gibo, D.L., 167
Gieselmann, M.J., see Martin, M.M., 31
Ginsberg, H.S., 39, 131
    see also Lofgren, C.S., 29, 76, 228, 270
Glancey, B.M., 154, 259
Glover, P.E., 86
Glunn, F.H., 29
Gobbel, M., 304
Göetsch, W., 152
Gontarski, H., 110, 163
Gorton, R.E., 306
Gösswald, K., 196
Gould, J.L., 56
Grassé, P.P., 88, 89, 90, 134
Gray, B., 150
Green, A., see Landolt, P.J., 188
Greenberg, L., 131, 249
Greenslade, P.J.M., 292, 293, 311
Grimal, A., 214
Gris, G.Ch., 261
Grob, K., see Huwyler, S., 63
Grogan, D.E., 149
Groth, I., see Bergström, G., 220, 281

Haas, A., 281
Haines, I.H., 79, 114
Haines, J.B., 79, 114
Halberstadt, K., 157
Hall, D.W., see Peacock, A.D., 203
Hamilton, W.D., 237, 244
Handel, S.N., 28, 29
Hangartner, W., 251
Hansell, M.H., 166
Harbo, J.R., 225
Hare, H., 237, 238, 240, 245
Harper, J.L., 262
Harris, W.V., 3, 85
Harrison, J.S., 60
Haskins, C.P., 75, 150, 168, 180, 227, 312
Haskins, E.F., 75, 150, 168, 180, 312
Hassell, M.P., 36, 42
Havarty, M.I., 143, 297
Hayashida, K., 78
Have, T.M. van der, see Boomsma, J.J., 193,
    238, 275
Hebling, N.J., 158
Heinrich, B., 37, 43, 124, 127
Hemmingsen, A.M., 97
Hepworth, D., 111
Herbers, J.M., 240
Herbert, E.W., 32, 157
Hermann, F.J., 30
Hermann, H.R., 3, 131, 132, 133, 135, 167

Lawson, K., 71
Ledoux, A., 98, 203
Lee, G.A. van der, see Boomsma, J.J., 193,
    238, 275
Lee, K.E., 12, 67, 79, 85, 116, 120, 297, 301
Leese, B.M., 30
LeMasne, G., 75, 163, 226
Lenoir, A., 43, 49, 155
Lensky, Y., 126, 188, 190
Lenz, M., 269
Lepage, M.G., 14, 52
Leprun, J.C., 90
Leroux, J.M., 71–72, 226, 315
Leston, D., 314
Letendre, M., 260
Leusink, A., 229
Leuthold, R.H., 54, 69, 90, 231
Lévieux, J., 16, 20, 26, 27, 76, 78, 114, 274,
    276, 308
Levin, B.R., 245
Lewis, T., 61
Leiberburg, I., 312
Light, S.F., 171
Lindauer, M., 51, 110, 225
Litte, M., 167
Locksley, H.D., 312
Lofgren, C.S., 29, 76, 228, 270
Löfquist, J., 140
Longhurst, C., 16, 52, 53, 303
Loon, A.J. van, 275
Louis, D., 276
Lövgren, B., 184
Lowenthal, 48
Lüscher, M., 171, 197, 207, 208, 210, 263
    see also Rutz, W., 200
Lynch, J.F., 292

Mabelis, A.A., 44, 62, 64, 230, 251, 260, 261
MacConnell, J.G., see Martin, M.M., 31
MacKay, W.P., 27
Mackensen, O., 33
McGinty, S., see Tomlinson, J., 56, 110
McGurk, D.J., 228
McLellan, A.R., 177
McMahan, E.A., 86
    see also Lenz, M., 269
Majer, J.D., 313–314, 316
Malotaux, M.E., see Honk, C.G.J., 199
Malyshev, S.I., 32, 128
Marikovsky, P.I., 140, 294
Marino, M.T., 242
Markin, G.P., 76, 84, 147, 148, 155, 170, 228
Markl, H., 52, 54, 55, 228
    see also Hölldobler, B., 54
Martin, B.B., see Glancey, B.M., 154
Martin, H., 110
Martin, M.M., 31
Maschwitz, U., 51, 52, 54, 62, 96, 130, 139,
    140, 143, 144, 149
    see also Hölldobler, B., 54

Masevicz, S., 239
Matsumoto, T., 116, 279, 307
Maurizio, 177
Maynard Smith, J., 40, 185, 203, 241,
    245
Messina, F.J., 20
Metcalf, R.A., 167, 265
Meudec, M., 118
Michener, C.D., 7, 32, 36, 75, 81, 98, 99, 100,
    102, 123, 124, 126, 153, 158, 162,
    240, 249, 280, 285
    see also Barrows, E.M., 248
Mintzer, A., 32, 62
Mirenda, J., 71
    see also Topoff, H., 71
Miyake, S., see Yamauchi, K., 259, 264
Möglich, M., 54, 57, 60, 119, 136, 305
    see also Hölldobler, B., 54
                Maschwitz, U., 51, 52
Moli, F. Le, 156, 296
Montagner, H., 108, 110, 199, 216
Moore, B.P., 142
Moraes, H.J. de, see Beig, D., 199
Morgan, E.D., 53, 54, 141
    see also Cammaerts, M.C., 53, 54, 59,
                133, 250
                Cammaert-Tricot, M.C., 53, 59,
                133
Morse, D.H., 38, 282
Morse, R.A., 108, 123–124, 224, 225
    see also Ambrose, J.T., 224
                Seeley, T.D., 224
Moser, J.C., see Robinson, S.W., 61
Moser, J.E., 135
Motro, M., 107
    see also Kugler, J., 107
Mountford, M.D., see Brian, M.V., 29, 275,
    288
Moure, J.S., see Sakagami, S.F., 176
Mudd, A., 31
    see also Peregrine, D.J., 152
Mudge, P.R., see Heinrich, B., 37
Muhlenberg, M., 144
Muir, D.A., 23

Nagin, R., 197
Naulleau, G., 110
Naumann, M.G., 198
Neih, M., see Regnier, F.E., 60
Nel, J.J.C., 158, 300
    see also Hewitt, P.H., 158
Nielsen, M.G., 178, 181, 234, 290
Niemela, P., 22
Nijhout, H.F., 211, 214
Nitschmann, J., 195
Noirot, C., 44, 88, 90, 142, 159, 208, 231,
    300
Noirot-Timothée, C., 159
Noonan, K.M., 240
Noy-Meir, I., see Ofer, J., 303

Subject index

Anal food, 159
Ants
  in deserts, 303–306
  in forests, 308–311
  in plantations, 311–316
Absconding
  in *Apis*, 126, 132, 285
  in *Formica fusca*, 119
Acacia, 26, 82, 130, 252–255, 269, 297
Acantholepis capensis, 315
Acantholepis custodiens, 117
Acanthomyops, 229
Acanthotermes acanthothorax, 144
Acer, 224
Aconitum columbianum, 40, 42, 281
Acromyrmex, 61, 77, 171, 310
  A. octinospinosus, 13, 171, 251
  A. versicolor, 61, 114
Adaiphrotermes, 80
Aenictus, 18, 133, 309
  A. gracilis, 18, 239, 309
  A. laviceps, 18, 309
Agapostemon, 131
Agrostis alba, 23
Agrostis tenuis, 23
Aleurodidae, 20, 26
Allodontotermes, 76
  A. giffardi, 231
Allognathotermes hypogeus, 231, 232
Amblyopone, 16
Amblyopone pallipes, 16, 74, 150, 157, 163
  168, 178
Amitermes, 70, 116, 123, 143, 231
  A. evuncifer, 208, 297, 302
  A. hastatus, 4
  A. meridionalis, 116
  A. vitiosus, 85
Amitermitidae, 307
Amitermitinae, 2, 4, 94, 143, 208
Ancistrotermes, 123, 308
  A. cavithorax, 79, 297, 300, 301, 302
  A. guineensis, 171
Andrena, 35
Andropogon, 14
Anergates atratulus, 227, 246
Anomma, 133, 191, 226
  A. kohli, 18, 71, 303
  A. nigricans, 71, 72, 134

  A. wilverthi, 71, 72
Anoplolepis custodiens, 114, 115, 313
Anoplolepis longipes, 79, 311, 312
Anthophoridae, 6, 7
Aphaenogaster fulva, 269
Aphaenogaster rudis, 29, 266, 268, 269, 291,
  292
Aphidae, 20
Aphis ulicis, 29, 275, 287
Apicotermes, 90
  A. desneuxi, 144, 180
  A. gurgulifex, 180, 298
Apidae, 6, 7, 36
Apinae, 6
Apini, 6, 7, 108 ff, 131, 153, 282
Apis, 102, 107, 108, 124, 131, 135, 154, 156,
  157, 158, 192, 200, 217, 220, 226,
  240, 251, 252, 283, 284, 306, 318
  A. cerana, 107, 125, 132, 188, 190, 283,
  284
  A. cerana indica, 126
  A. dorsata, 107, 123, 126, 132, 283, 284
  A. florea, 107, 124, 126, 132, 283, 284
  A. indica, 55
  A. mellifera, 6, 32, 54, 55, 56, 107, 109,
  125, 126, 132, 176, 178, 179, 188,
  190, 232, 237, 239, 249, 252, 262,
  263, 283, 284, 285, 306
  A. mellifera adansonii, 126, 132, 177, 178,
  223, 224, 284
  A. mellifera capensis, 200
  A. mellifera carnica, 55
  A. mellifera fasciata, 55
  A. mellifera ligustica, 55, 126, 285
  A. mellifera scutellata, 284
Apoidea, 6
Apomyrma, 16
Apotrigona nebulata, 123, 154
Aristida pungens, 28
Ascomycetes, 30
Aspergillus, 30
Aster, 37
Asteraceae, 23
Atopomyrmex, 26
Atta, 30, 50, 61, 77, 135, 139, 310
  A. cephalotes, 31, 45, 46, 47, 48, 61, 135,
  310
  A. mexicana, 31, 62

Tiphiidae, 16
Tiphioidea, 6
Tortrix viridana, 22
Trachymyrmex, 310
Trachymyrmex smithi, 30, 114, 305
Trifolium pratense, 37
Trifolium repens, 35, 38
Trigona, 51, 141, 142
   T. braunsi, 141
   T. fulviventris, 282
   T. fuscipennis, 282
   T. iridipennis, 283, 284
   T. nebulata, 123
   T. oyani, 99
   T. portoi, 99
   T. postica, 51
   T. recursa, 100
   T. spinifer, 123
   T. subnuda, 223, 263
   T. subterranea, 51, 141
   T. sylvestriana, 282
   T. testaceicornis, 282
Trinervitermes, 123, 142, 172, 276, 297, 302,
   303
   T. auritermes, 172
   T. bettonianus, 68, 231
   T. carbonarius, 172
   T. ebenerianus, 172
   T. geminatus, 14, 68, 115, 116, 144, 180,
      231, 277, 297, 298, 301, 302, 303
   T. occidentalis, 277
   T. oeconomus, 172, 277, 302
   T. suspensus, 172
   T. togoensis, 277, 301–302
   T. trinervius, 14, 277, 302
   T. trinervoides, 69, 300
Trophic eggs, 150–152, 154, 155, 156, 163,
   165, 200
Typhlopone, 71, 72

Ulex minor, 29, 275, 287
Ulmus, 35

Ventilation
  of nests, 119 ff.

Veromessor, 27, 70
Veromessor pergandei, 28, 60, 304, 305
Vespa, 51, 132, 176, 280
   V. analis, 280
   V. crabro, 280
   V. dybowskii, 280
   V. mandarinia, 280, 283
   V. orientalis, 106, 149, 188
   V. tropica, 280
   V. xanthoptera, 280
Vespidae, 6, 131
Vespinae, 7, 105, 107, 108, 125, 148, 149,
   167, 168, 199
Vespoidea, 6, 15, 32
Vespula, 18, 43, 51, 124, 129, 130, 132, 148,
   149, 168, 183, 189, 220, 234, 251
   V. germanica, 148, 184, 233, 252, 280
   V. norwegica, 280
   V. rufa, 105, 280
   V. sylvestris, 280
   V. vulgaris, 106, 149, 176, 183, 184, 200,
      280
Vicia cracca, 38
Vicia faba, 40
Viola, 29, 30
Vitamins
  essential, 10

Wasmannia auropunctata, 24, 25
Wasps
  in tropical rain forest, 306–308
  venom, 130
Wax, see Beeswax
Worker jelly, 217, 219

Xylocopini, 7

Yucca, 305

Zacryptoceros, 140, 214
Zacryptoceros varians, 138
Zootermopsis, 79, 206
   Z. angusticollis, 231
   Z. nevadensis, 67, 142, 158, 231